EQUATIONS OF STATE FOR SOLIDS AT HIGH PRESSURES AND TEMPERATURES

URAVNENIYA SOSTOYANIYA TVERDYKH TEL PRI VYSOKIKH DAVLENIYAKH
I TEMPERATURAKH

УРАВНЕНИЯ СОСТОЯНИЯ ТВЕРДЫХ ТЕЛ ПРИ ВЫСОКИХ ДАВЛЕНИЯХ
И ТЕМПЕРАТУРАХ

EQUATIONS OF STATE FOR SOLIDS AT HIGH PRESSURES AND TEMPERATURES

V. N. Zharkov and V. A. Kalinin

Laboratory of Theoretical Physics
O. Yu. Shmidt Institute of Physics of the Earth
Academy of Sciences of the USSR

Translated from Russian by

Albin Tybulewicz
Editor, *Soviet Physics-Semiconductors*

SPRINGER SCIENCE+BUSINESS MEDIA, LLC 1971

The original Russian text, published by Nauka Press in Moscow in 1968 for the O. Yu. Shmidt Institute of Physics of the Earth of the Academy of Sciences of the USSR, has been corrected by the authors for the present edition. The English translation is published under an agreement with Mezhdunarodnaya Kniga, the Soviet book export agency.

В. Н. Жарков, В. А. Калинин

**Уравнения состояния твердых тел
при высоких давлениях и температурах**

Library of Congress Catalog Card Number 77-120027

ISBN 978-1-4757-1519-4   ISBN 978-1-4757-1517-0 (eBook)
DOI 10.1007/978-1-4757-1517-0

PREFACE TO THE AMERICAN EDITION

We started our work on theoretical methods in the physics of high pressures (in connection with geophysical applications) in 1956, and we immediately encountered many problems. Naturally, we searched the published literature for solutions to these problems but whenever we failed to find a solution or when the solution did not satisfy us, we attempted to solve the problem ourselves. We realized that other investigators working in the physics of high pressures would probably encounter the same problems and doubts. Therefore, we decided to write this book in order to save our colleagues time and effort. Apart from the descriptions of experimental methods, the book deals only with those problems which we encountered in our own work.

All problems in high-pressure physics have, at present, only approximate solutions, which are very rough. Therefore, it is not surprising that different investigators approach the same problems in different ways. Our approach does not prejudge the issue and we are fully aware that there are other points of view. Our aim was always to solve a given problem on a physical basis. For example, the concept of the Grüneisen parameter needs further development but it is based on reliable physical ideas. On the other hand, Simon's equation for the melting curve has, in our opinion, no clear physical basis and is purely empirical. Equations of this type are useful in systematic presentation of the experimental material but they are unsuitable for any major extrapolation.

A couple of years have passed since the manuscript of this book was sent to press. Since then, additional experimental data have become available. A high proportion of these data can be analyzed by the methods described in the present book. However, their incorporation would not improve the book very greatly. One of the most pressing needs in high-pressure physics is to improve the accuracy of the measurements. When this barrier is surmounted, we shall then see which methods for the analysis of the experimental data should be refined and which should be replaced.

We are pleased to hear that our book will be published in the home country of P. W. Bridgman, and we are grateful to Plenum Publishing Corporation for arranging the translation.

Moscow, November 1969

V. N. Zharkov
V. A. Kalinin

PREFACE

The theoretical methods used in the high-pressure physics of solids are the subject of the present book. The development of these methods has been stimulated primarily by geophysics problems to which such methods have been applied in the first instance. Special attention is paid in the book to the equation of state, i.e., to the dependence of the pressure P on the volume V and the temperature T.

The equation of state is the basic relationship in high-pressure physics. It contains valuable information on the properties of a medium and it allows us to carry out transformations from one thermodynamic potential to another. Moreover, investigations of a certain property of a solid can be used to find the theoretical dependence of this property on the temperature and the lattice constant. The variables normally employed are P and T. The equation of state makes it possible to go over from the "theoretical" variables V and T to P and T.

The range of high pressures and temperatures considered in the present book is determined by the P—T conditions in the interior of the earth (P ≤ 3.5 · 10^6 bar, T ≤ 6000°K).* Unfortunately, such pressures and temperatures are outside the present capabilities of theoretical physics, i.e., practically none of the high-pressure and high-temperature problems can be solved in a systematic manner by the methods of modern theoretical physics. Consequently, we shall have to use semi-empirical methods. In other words, we shall first establish a functional dependence of a given quantity on the temperature and volume, and then we shall use the experimental data to determine the numerical parameters in this dependence. This approach is known as the method of potentials.

The method of potentials permits us to write the equation of state in the analytic form, which has definite advantages in the derivation of the equation and in its applications. This method is used to obtain equations of state for all those substances for which the necessary experimental data are available.

A detailed list of the topics dealt with in this book is given in the contents.

---

*1 bar = 10^6 dyn/cm^2 = 1.01972 kg_f/cm^2 = 0.986924 atm. The kilobar (1 kbar = 10^3 bar) and the megabar (1 Mbar = 10^6 bar) are usually employed in high-pressure physics.

CONTENTS

CHAPTER 1

EQUATION OF STATE AT T = 0.
METHOD OF POTENTIALS

1.1. Introduction

The solid state of matter differs from the gaseous and liquid states by a more compact and symmetrical distribution of atoms. The distance between the atoms in a solid is of the order of the atomic diameters and considerable forces act within a solid. At the very short interatomic distances found in solids, we must take into account not only the interaction of the nucleus and electrons within each atom but also the interaction between the nuclei and electrons of different atoms. A solid is a very complex quantum-mechanical system and, therefore, no systematic and consistent theoretical calculations can be carried out without some approximations. It is usual to consider models which differ from the real system by certain restrictions that are imposed on them. These restrictions can differ from problem to problem but they should apply, whenever possible, to the secondary processes which do not alter basically the main properties of the solid in question.

One such restriction is the adiabatic approximation, which allows us to consider separately the motion of electrons and nuclei in a solid [1]. This approximation is based on the observation that nuclei are much heavier than electrons and, consequently, the average velocity of nuclei is much less than the average velocity of electrons. Thus, a solid can be divided into two subsystems: electron and nuclear. When the electron subsystem is considered, it may be assumed that the nuclei are at rest although they need not occupy crystal lattice sites. If we take R to denote the vectors R_1, R_2, . . . , R_n, which describe the positions of n nuclei, we can use the solution of Schrödinger's equation, corresponding to the motion of N electrons in the field of n nuclei with the configuration R, to obtain a wave function which contains the vectors R as parameters. The energy eigenvalues also depend on the parameters R. In particular, this method can be used to determine the energy of the ground state of the electron subsystem, $E_0(R)$. When we consider the motion of the nuclei, we can assume that the state of the electron subsystem at any moment is described by the same wave function in which the positions of the nuclei at that moment are represented by R.

The adiabatic approximation is satisfied well by filled electron shells but it is inapplicable to conduction electrons in metals. The separation of neighboring energy levels in the conduction band of a metal is less than the excitation due to the motion of nuclei and this motion alters the wave functions of the conduction electrons.

We shall consider a crystalline solid at T = 0 and we shall ignore, for the time being, the motion of the nuclei due to zero-point vibrations of the crystal. We shall assume that the nuclei are at rest and are located at crystal lattice sites. Under these conditions, the equation of state for a solid can be obtained if the model of the solid satisfies only two basic conditions: the model must describe correctly the attractive forces which arise when nuclei move away from one another, and the repulsive forces when nuclei approach one another.

The binding energy of the atoms in a crystal is of the same order of magnitude as the binding energy of the valence electrons in an atom but it is less than the binding energy of the inner-shell electrons. Hence, we may conclude that the forces in a solid are primarily due to the interactions of the outer electrons and that the inner electron shells are practically unaffected by the various processes in a solid. This is also supported by the evidence that the x-ray spectra, corresponding to the transitions of electrons between deep energy levels, are practically identical for free atoms and for atoms in a solid. Therefore, instead of considering a system comprising all nuclei and electrons, we may consider, in approximation, two subsystems: one consisting of ions (formed from nuclei) and inner electrons, and the other consisting of outer electrons alone.

1.2. Forces in Solids

The forces in solids are basically electrostatic but their actual nature depends on the distribution of electrons between the various structure elements forming a crystal. By structure elements, we mean the particles into which atoms and molecules are converted when they form a solid.

These structure elements include positive ions, each consisting of a nucleus and its associated inner-shell electrons. These ions are formed, for example, from metal atoms which lose their valence electrons quite easily. The energy required for the detachment of an electron from an atom, and the consequent formation of a positive ion, is known as the ionization energy. Another example of a structure element is a negative ion, formed by the capture of excess electrons by an atom. Typical of the elements which form negative ions are halogens whose atoms capture excess electrons and assume the electron structure of inert-gas atoms. The energy evolved in the attachment of an electron to a neutral atom is known as the electron affinity. In inert-gas crystals, the structure elements are the atoms themselves since their electron structure is practically the same in the gaseous and crystalline states. Finally, the free-electron gas in the metal and the valence electrons participating in the formation of covalent bonds can also be regarded as structure elements.

The forces in a crystal can be regarded as acting between the structure elements. Let us now consider these forces in detail.

Coulomb Forces

The Coulomb forces can act only between charged structure elements. Such elements are positive ions, negative ions, and free electrons.

The Coulomb interaction of ions needs no further comment because the charge distribution in an ion is approximately spherical. We may assume that each ion has a radius outside which the electron density is negligibly small. Therefore, the interaction between ions is equivalent to the interaction between point charges. The Coulomb interaction between positive ions and a free-electron gas can be analyzed using the observation that the density of free (conduction) electrons is almost constant everywhere except in the immediate neighborhood of nuclei. The energy of this interaction is equal to the electrostatic energy of a system comprising positive point charges (which represent ions) and a uniformly distributed negative charge (representing the free-electron gas).

If r represents some characteristic dimension of the crystal lattice (for example, the distance between nearest neighbors), the dependence of the Coulomb energy on this dimension is of the very simple type

$$c / r, \qquad (1.1)$$

where c is a constant independent of r.

Quantum Repulsive Forces

When atoms or ions approach one another closely, their electron shells overlap and this gives rise to special quantum repulsive forces. These forces originate in two factors. First, they are due to the exchange effects between the electron shells of neighboring atoms. Secondly, they appear as a consequence of the Pauli principle.

Quantum-mechanical calculations of the energy of interaction of atoms and ions with filled electron shells [2-4] give an exponential dependence of the energy of repulsion on the distances between atoms:

$$a \exp(-r/\rho), \tag{1.2}$$

where a and ρ are ideally constants. In reality, a and ρ are slowly varying functions of r and this must be allowed for if r changes appreciably. The exponential nature of the repulsive forces is also supported by the experimental data on the scattering of atoms [5].

Different quantum repulsive forces act in a free-electron gas. These forces are governed by the Fermi energy. The dependence of the Fermi energy on the interatomic distance (in the case when the variation of r is accompanied by a change in the volume $V \propto r^3$) is of the form [see Eq. (C.33) in Appendix C]:

$$dr^{-2}, \tag{1.3}$$

where d is a constant.

Van der Waals Forces

The van der Waals forces act between atoms and ions with filled electron shells as well as between saturated molecules. They are due to the polarization of atoms, which results from the correlation of electron motion in the shells of neighboring atoms. Electrons of each of two interacting atoms are displaced relative to their nuclei. Thus, multipole (dipole, quadrupole, etc.) moments are induced in the atoms. The van der Waals forces are weak. They can be neglected in an analysis of the interaction between ions because they simply represent a small correction to the Coulomb interaction. However, the van der Waals forces dominate the interaction of neutral atoms of inert gases and of molecules with saturated bonds.

The dependence of the van der Waals energy on the interatomic distance is found from the multipole nature of the forces and, in the case of interaction of neutral atoms of molecules, it can be represented by a series of reciprocals of the even powers of r, beginning from the sixth:

$$-\frac{c}{r^6} - \frac{c_1}{r^8} - \frac{c_2}{r^{10}} + \dots \tag{1.4}$$

We must point out that this power law is valid in the asymptotic limit, i.e., for distances which are large compared with the atomic radii. However, since these forces rise rapidly with $1/r$, the series in Eq. (1.4) begins to diverge and the asymptotic law becomes invalid at short distances. Thus, the van der Waals forces cannot be described rigorously at short distances. However, it has been shown recently [6] that the van der Waals energy can be represented approximately in the form

$$-\xi(r) \frac{c}{r^6}, \tag{1.5}$$

where $\xi(r)$ is an approximation factor. For large values of r, this factor is equal to unity. When r decreases, the factor increases, which corresponds to the appearance of dipole—quadrupole (r^{-8}) and other terms of the series in Eq. (1.4). At low values of r, the factor $\xi(r)$ rapidly decreases to zero because of the compensating effect of higher terms.

Valence Binding Forces

Atoms of elements, which are one or two electrons short of the closed-shell configuration of inert gases, can easily capture additional electrons and form negative ions. Atoms, which have one, two, or even three electrons outside a closed shell, can lose these outer electrons quite easily and form positive ions. However, if an atom can form a closed-shell configuration by either capturing or losing the same number of electrons (this applies to elements in the middle groups of the periodic table), the formation of negative or positive ions is difficult. The difficulty is due to the fact that an atom cannot capture too many excess electrons because the excess negative charge can make the atom unstable by "loosening" its structure. On the other hand, an atom cannot lose many electrons because this would require too much energy. Such atoms are bound together by different means. They share their valence electrons and thus form a semblance of closed electron shells. Each shared electron belongs to two neighboring atoms. In other words, hybrid electron states are formed when atoms of this type approach one another, and the wave functions corresponding to these states are not small within the volume occupied by two neighboring atoms. According to the Pauli principle, each quantum state can be occupied by two electrons with antiparallel spins. A pair of such electrons binds two neighboring atoms together. Such bonds are known as covalent or homopolar. The average electron density in the covalent bonds can be quite considerable. A characteristic feature of the covalent bonds is their directional nature, i.e., atoms tend to join along definite directions.

A detailed theory of the covalent bonds is very complex. This is because atoms forming valence (covalent) crystals have many valence electrons and, therefore, calculations are much more complex than in the case of metals or ionic crystals.

1.3. Interaction Potentials in Various Types of Solid

Crystalline solids can be divided roughly into four types, in accordance with the nature of their interatomic forces. These four types are: ionic crystals, molecular crystals, valence (covalent) crystals, and metals.

Ionic and molecular crystals differ appreciably from valence and metal crystals. This is due to the differences in the structure of the outer electron shells. Basically, the structure elements in ionic and molecular crystals have completely filled electron shells, while the electron shells in valence and metal crystals are only partly filled. Physically, the difference between these two groups of crystals is explained as follows. When atoms or ions with filled shells approach each other closely, the wave functions of their electrons are only slightly affected, but in the case of atoms or ions with partly filled shells, the wave functions change appreciably. The outer-shell electrons in valence crystals and metals cannot be regarded as belonging to individual atoms. In valence crystals, these electrons are shared by neighboring atoms in the form of covalent bonds. The number of bonds which an atom can form is equal to the number of its valence electrons. Therefore, the number of nearest neighbors in valence crystals is equal to the number of valence bonds, while in ionic and molecular crystals the number of nearest neighbors is usually larger.

Ionic, molecular, and valence crystals have one common property: the occupied bands in their electron energy spectrum are separated by a forbidden band (energy gap) from the conduction band. Consequently, small perturbations due to the lattice vibrations of these crystals have little effect on the wave functions of the electrons and we can use the adiabatic approximation. In this sense, metals represent a special type of solid, because the adiabatic approximation cannot be applied to the conduction electrons.

We shall now consider in detail the interactions between structure elements in different types of crystal. Next, using the well-known qualitative dependences of the forces between structure elements on the distances between atoms, we shall derive the interaction potentials for each type of solid. Since rigorous calculations of these potentials are impossible, they will contain parameters deduced from empirical data.

A correct selection of the dependence of terms in an interaction potential on the interatomic distance makes it possible to reduce the number of parameters. However, such a reduction may yield a description which is insufficiently accurate for some experimental data. Therefore, the nature of a potential will depend on the purpose for which it is to be used. In this book we are interested in deriving equations of state, and we shall therefore be guided by the following two rules in selecting the various terms in the interaction potentials. First, these terms should represent the principal contributions to the cohesive energy; secondly, they should vary quite rapidly (compared with the rejected terms) when the interatomic distance is reduced. Suitable expressions for ionic crystals were derived by Born and Mayer [7]. Potentials for all four types of crystal were obtained by Davydov [8].

Ionic Crystals

Ionic crystals are those substances in which the main structure elements are positive and negative ions. They are usually inorganic compounds and most of them are good dielectrics. None of the pure elements forms ionic crystals, because ionic binding presupposes the transfer of electrons from one atom to another of a different kind, i.e., to an atom of a different element.

The ions are located at crystal lattice sites and the Coulomb attraction between ions of opposite sign is stronger than the Coulomb repulsion between ions of the same sign. The Coulomb interaction decreases relatively slowly with the distance. Consequently, the electrostatic energy of an ionic crystal must be calculated by summing the interactions of ion pairs throughout the lattice, but such summation can be carried out in an elementary manner. A general method for the calculation of the electrostatic energy of ionic crystals is presented in [1]. It gives an expression of the type

$$-\frac{\alpha q^2}{r}, \tag{1.6}$$

where q is the smallest ionic change; r is the shortest distance between ions of opposite sign; α is the Madelung constant, which depends only on the lattice structure.

Ions are drawn together not only by the Coulomb forces but also by the van der Waals forces. However, the contribution of the van der Waals forces to the cohesive energy is small [9] and it can be neglected.

The principal repulsive forces in ionic crystals are due to the overlap of the electron shells of the ions and, according to Eq. (1.2), they depend exponentially on the interatomic distance. Because the repulsive forces decrease rapidly with increasing interatomic distance, we need consider only the interaction of the nearest neighbors. However, ions of opposite sign can have different radii and we must then include the overlap not only of the nearest oppositely charged ions but also between the nearest identically charged ions which have larger radii. Therefore, the expression for the potential includes several exponential functions with different coefficients. However, the use of such potential shows that satisfactory results are still obtained if only one exponential function is used [10]. We thus obtain the following potential (potential energy) for ionic crystals:

$$E_\mathrm{p} = \frac{3A}{b\rho_0}\exp[b(1-x^{1/3})] - \frac{3K}{\rho_0}x^{-1/3},\tag{1.7}$$

where ρ_0 is the density under normal conditions (room temperature and atmospheric pressure); $x_0 = V/V_0$ is the dimensionless volume; $V_0 = 1/\rho_0$ is the volume of unit mass of matter under normal conditions; $V \propto r^3$; A, b, and K are constants which still have to be determined. The van der Waals energy and the energies of other less-important interactions are included in the two terms of Eq. (1.7), but since these interactions are small, the parameters A, b, and K should not differ very greatly from their theoretical values for the Coulomb interaction.

Using the thermodynamic equation

$$P = -\left(\frac{\partial F}{\partial V}\right)_T = -\rho_0\left(\frac{\partial F}{\partial x}\right)_T,\tag{1.8}$$

which relates the pressure P to the derivative of the free energy F (with respect to the volume), we can obtain an expression for the potential pressure, P_p, due to the potential energy of an ionic crystal given by Eq. (1.7):

$$P_\mathrm{p} = Ax^{-2/3}\exp[b(1-x^{1/3})] - Kx^{-4/3}.\tag{1.9}$$

Equation (1.9) is derived on the assumption that, at T = 0, the free energy and the potential energy are equal. By the addition, to Eq. (1.9), of a term representing the zero-point vibrations (see Sec. 1.4), we obtain the isotherm of an ionic crystal at T = 0 (it is often called the cold-compression isotherm).

An important property of a solid is its bulk modulus:

$$K_T = -V\left(\frac{\partial P}{\partial V}\right)_T = -x\left(\frac{\partial P}{\partial x}\right)_T,\tag{1.10}$$

whose potential component is

$$K_\mathrm{p} = \frac{Ax^{-2/3}}{3}(bx^{1/3}+2)\exp[b(1-x^{1/3})] - \frac{4}{3}Kx^{-4/3}.\tag{1.11}$$

The bulk modulus at T = 0 is obtained by adding — to Eq. (1.11) — a term corresponding to the zero-point vibrations. We must point out here a factor which always arises in the method of potentials. The potential of repulsive forces varies with distance much more rapidly than the potential of attractive forces. This follows from physical considerations and from the nature of the potential given by Eq. (1.7). Therefore, the relative importance of the first term in Eq. (1.7) increases after each differentiation of this equation with respect to x. It is known that the cohesive energy of ionic crystals is governed primarily by the electrostatic interaction of ions [11], i.e., by the second term in Eq. (1.7). However, in the expression for the pressure, given by Eq. (1.9), both terms are comparable, and in the definition of the bulk modulus (1.11) the first term predominates.

If the cohesive energy is taken as the criterion of the fitness of the potential, a serious error can be made in the determination of the potential of the repulsive forces. In the equation of state at high pressures, the dominant term is the first one in Eq. (1.9) and, therefore, the fitness of the equation of state is governed primarily by the correctness of the first term.

Molecular Crystals

Molecular crystals are solids formed from atoms with filled electron shells (inert-gas configuration) and from molecules with saturated bonds. These crystals have low binding energies and low melting points. They usually have dielectric properties. The electron states in these crystals are practically identical with those in free atoms of the same substance. The energy gap between the valence and conduction bands is of the same order as the separation between the corresponding atomic levels.

The attractive forces in molecular crystals are the van der Waals forces. The dependence of these forces on the interatomic distance is given by Eq. (1.4) or, more accurately, by Eq. (1.5). In the case of small variations of the lattice constants, we need retain only the term proportional to r^{-6}. Since the van der Waals forces decrease rapidly with distance, it is sufficient to consider the interaction of the nearest neighbors. As in ionic crystals, the repulsive forces are due to the overlap of electron shells.

Thus, the potential energy of a molecular crystal is of the form:

$$E_p = \frac{3A}{b\rho_0} \exp[b(1 - x^{1/6})] - \frac{K}{2\rho_0} x^{-2}. \tag{1.12}$$

Here, x is the dimensionless volume; A, b, and K are parameters which have to be determined. The nature of the argument of the exponential function can be found from Eq. (1.2) by suitable computations.

By analogy with Eqs. (1.9) and (1.11), we can obtain expressions for the potential pressure

$$P_p = Ax^{-2/3}\exp[b(1 - x^{1/6})] - Kx^{-3} \tag{1.13}$$

and for the potential component of the bulk modulus

$$K_p = \frac{Ax^{-1/3}}{3}(bx^{1/6} + 2)\exp[b(1 - x^{1/6})] - 3Kx^{-3} \tag{1.14}$$

of a molecular crystal.

Metals

A characteristic property of metals is the presence of a partly filled valence band. The Fermi level passes inside the valence band and electrons can be excited from the filled levels near the Fermi level to unoccupied levels in the valence band. The energy required for such a transition is negligibly small and this allows us to consider the valence electrons in a metal as completely free.

The principal structure elements of metals are positive ions and a free-electron gas (the conduction electrons). The attractive forces are due to the Coulomb interaction between free electrons and positive ions and to the energy exchange between free electrons. These two terms can be amalgamated to form a single term of the c/r type [9]. The Coulomb repulsion between the positive ions is proportional to 1/r, and it reduces the coefficient c in the attractive-force term. Moreover, the repulsion in metals is due to the overlap of the electron shells of the positive ions, and it is of the form $a\exp(-r/\rho)$, similar to the corresponding interaction in ionic and molecular crystals. The Fermi kinetic energy of the free electrons also gives rise to repulsion, and its dependence on the interatomic distance is given by Eq. (1.3).

The resultant potential for metals is of the form

$$E_p = a\exp[b(1 - x^{1/6})] + dx^{-2/3} - cx^{-1/3}, \tag{1.15}$$

where a, b, c, and d are constants. In metals with a slight overlap of the electron shells of the positive ions, for example in alkali metals, the repulsive force is mainly due to the Fermi kinetic energy of the conduction electrons. In this case, the first term of Eq. (1.15) is small and can be neglected. Its influence is manifested only at high pressures. In other cases, when the overlap forces are dominant, we must reject the second term in Eq. (1.15) so that the potential energy of a metal becomes

$$E_p = \frac{3A}{b\rho_0} \exp[b(1 - x^{1/3})] - \frac{3K}{\rho_0} x^{-1/3}, \qquad (1.16)$$

which is identical with the form of the potential for ionic crystals. Consequently, the potential pressure and the potential component of the bulk modulus are again given by Eqs. (1.9) and (1.11). Thus, the equation of state for the majority of metals at T = 0 is formally identical with the equation of state for ionic crystals. This does not apply to alkali metals, but we shall not consider them in detail.

The expression (1.16) has been used frequently to determine the equations of state for ionic crystals and metals [8, 10, 12-18]. A group led by L. V. Al'tshuler has also recently [19-21] used this potential, preferring it to the nonpotential method.

Valence Crystals

Let us examine the considerations [8] on which the potential of valence (covalent) crystals is based. The attractive forces in such crystals are due to covalent bonds. These forces depend exponentially on the interatomic distance and the corresponding term in the potential is of the form $c \exp(-r/\rho)$. The Coulomb repulsion between ions (i.e., between atoms which have lost their outer electrons) is screened by an exponential distribution of the negative space charge of the valence electrons. This interaction should be represented by the term $(a/r) \exp(-r/\rho)$. The coefficients ρ in both terms should be identical because the exponential factors are due to the same electrons. In addition to these interactions, we must include also the repulsion of the filled electron shells of ions, described by the expression $a_1 \exp(-r/\rho_1)$. However, in the case of small variation of r, this term is unimportant since a comparison of the covalent radii [22] and of the radii of ions calculated by the Hartree — Fock method [23] shows that the overlap of the electron shells of ions is slight. It follows that the expression for the potential of valence crystals can be written in the form *

$$E_p = (a x^{-1/3} - c) \exp[b(1 - x^{1/3})], \qquad (1.17)$$

where a, b, and c are parameters which need to be determined.

1.4. Contribution of Zero-Point Vibrations
to the Equation of State

According to the concepts of classical mechanics, all motion in a crystal should stop at T = 0. In particular, atoms must occupy crystal lattice sites. However, in quantum theory the concept of complete rest has no meaning. If the atoms could become immobile, we would know exactly their coordinates and velocities, but this contradicts the principle of indeterminacy. Consequently, atoms must move even at absolute zero and, since they occupy equilibrium positions, their motion must be vibrational. This motion is known as the zero-point vibrations.

Let us consider the contribution of the zero-point vibrations to the energy of a crystal. A crystal, consisting of N atoms, is mechanically equivalent to 3N harmonic oscillators. If the mass of each oscillator is m and its restoring force coefficient is K, it follows that its frequency is $\omega_0 = (K/m)^{\frac{1}{2}}$. We are interested in an oscillator at T = 0, when its momentum p and dis-

---

*The potential of Eq. (1.17) has not yet been used in calculations and it is too early to judge its fitness.

placement l have the lowest possible values. These quantities are related by the principle of indeterminacy $pl \approx \hbar/2$. The energy of the oscillator $p^2/2m + Kl^2/2$, found using the principle of indeterminacy, has its lowest value at $p = (K\hbar^2 m/4)^{1/4}$, given by

$$\frac{\hbar}{2}\left(\frac{K}{m}\right)^{1/2} = \frac{\hbar\omega_0}{2}.$$

If all 3N frequencies in a crystal are identical (the Einstein approximation), the energy of the zero-point vibrations is $3N\hbar\omega_0/2$. Actually, a solid is characterized not by a single vibration frequency but by a whole spectrum and the energy of the zero-point vibrations is given by the sum

$$E_{zv} = \sum_{i=1}^{3N} \frac{\hbar\omega_i}{2}.$$

For the majority of solids, the value of E_{zv} is small compared with the cohesive energy of a crystal. However, in molecular crystals the zero-point vibration energy is important and in the case of helium the value of E_{zv} is so high that at atmospheric pressure helium remains liquid even at T = 0.

We shall assume that the vibration spectrum of a solid is described by the Debye theory (see Sec. 2.5). Then, the energy of the zero-point vibrations per unit mass of a crystal is given by the relationship

$$E_{zv} = \frac{9}{8}\frac{R}{\mu}\theta, \tag{1.18}$$

where R is the universal gas constant; μ is the molecular weight; θ is the Debye temperature, which depends on the volume.

The contribution of the zero-point vibrations to the equation of state is given by the term

$$P_{zv} = \frac{9}{8}\frac{R}{\mu}\theta\rho_0\frac{\gamma}{x}, \tag{1.19}$$

where ρ_0 is the density at atmospheric pressure; x is the dimensionless volume; $\gamma = -\partial\ln\theta/\partial\ln x$ is the Grüneisen parameter (see Chapter 2). The pressure corresponding to the zero-point vibrations of a solid, P_{zv}, is obtained from Eq. (1.18) using Eq. (1.8).

A correction to the bulk modulus due to the zero-point vibrations is obtained from Eqs. (1.19) and (1.10):

$$K_{zv} = \frac{9}{8}\frac{R}{\mu}\theta\rho_0\frac{\gamma}{x}\left(\gamma + 1 - \frac{\partial\ln\gamma}{\partial\ln x}\right). \tag{1.20}$$

Lattice vibrations will be considered in greater detail in the next chapter. Here, we shall simply mention that the pressure due to zero-point vibrations is several kilobars at atmospheric pressure and that it increases with decreasing volume.

1.5. Murnaghan—Birch Equation of State

In contrast to the atomic approach used in the preceding sections, Birch [24] used the phenomenological theory of finite deformations put forward by Murnaghan [25].

We shall use η_k to denote the coordinates of points in a body before its deformation (Lagrange variables). The same points after deformation have the coordinates ξ_k (Euler variables). If $dl^2 = d\eta_k d\eta_k$ is the square of an element of length before deformation (as usual, summation is carried out over the repeated indices), the square of the same element after deformation is $dL^2 = d\xi_k d\xi_k$. We shall describe the deformation (strain) using the Euler variables:

$$dL^2 - dl^2 = 2\varepsilon_{ij}\, d\xi_i\, d\xi_j, \tag{1.21}$$

where

$$\varepsilon_{ij} = \frac{1}{2}\left(\delta_{ij} - \frac{\partial \eta_k}{\partial \xi_i}\frac{\partial \eta_k}{\partial \xi_j}\right) \tag{1.22}$$

is the Almansi strain tensor [26].

We shall consider uniform strain

$$\xi_k = \eta_k(1 + \alpha). \tag{1.23}$$

Compression always corresponds to $\alpha < 0$. In this case, the tensor of Eq. (1.22) becomes

$$\varepsilon_{ij} = \varepsilon\delta_{ij},$$

where

$$\varepsilon = \frac{1}{2}\left(1 - \frac{1}{(1+\alpha)^2}\right). \tag{1.24}$$

The relative change in volume due to uniform strain is given by the relationship

$$\frac{\partial V_\eta}{\partial V_\xi} = \frac{\rho_\xi}{\rho_\eta} = \frac{\partial(\eta_1, \eta_2, \eta_3)}{\partial(\xi_1, \xi_2, \xi_3)} = (1+\alpha)^{-3}, \tag{1.25}$$

where ρ_η is the density before deformation and ρ_ξ is the density after deformation. It follows from Eqs. (1.24) and (1.25) that

$$\frac{\rho_\xi}{\rho_\eta} = (1 - 2\varepsilon)^{3/2}. \tag{1.26}$$

This relationship is exact and is not restricted by any assumptions about the smallness of the strains.

We shall expand the free energy F as a power series of ε:

$$F = a_0 + a_1\varepsilon + a_2\varepsilon^2 + a_3\varepsilon^3 + \ldots, \tag{1.27}$$

where the coefficients a_l depend only on temperature. The convergence of the series in Eq. (1.27) is self-evident only at low values of ε. At high values of ε, we cannot restrict ourselves to terms up to the third power, as has been done by Birch. In the case of such high values of the parameter, we must use a large number of terms, which is not a practical proposition.

It follows from Eq. (1.27) that the pressure is

$$P = -\left(\frac{\partial F}{\partial V}\right)_T = -(a_1 + 2a_2\varepsilon + 3a_3\varepsilon^2 + \ldots)\frac{d\varepsilon}{dV}.$$

We shall use Eq. (1.26), which yields

$$\frac{d\varepsilon}{dV} = \frac{\rho_\eta}{3}(1 - 2\varepsilon)^{5/2};$$

we then find that the pressure is given by

$$P = -\frac{\rho_\eta}{3}(1 - 2\varepsilon)^{5/2}(a_1 + 2a_2\varepsilon + 3a_3\varepsilon^2 + \ldots).$$

The isothermal bulk modulus is

$$K_r = -V\left(\frac{\partial P}{\partial V}\right)_T = -V\left(\frac{\partial P}{\partial \varepsilon}\right)_T \frac{d\varepsilon}{dV} = \frac{\rho\eta^2}{9\rho_\xi}(1-2\varepsilon)[(2a_2-5a_1)+2(3a_3-7a_2)\varepsilon+\ldots].$$

In the absence of deformation ($\varepsilon = 0$), the bulk modulus is

$$K_{T0} = \frac{\rho\eta}{9}(2a_2-5a_1).$$

Expressing ε of Eq. (1.26) in terms of ρ_ξ/ρ_η, we obtain

$$\varepsilon = \frac{1}{2}\left(1-\left(\frac{\rho_\xi}{\rho_\eta}\right)^{2/3}\right),$$

and then the expression for the pressure becomes

$$P = -\frac{3K_{T0}}{2a_2-5a_1}\left[a_1+a_2\left(1-\left(\frac{\rho_\xi}{\rho_\eta}\right)^{2/3}\right)+\frac{3}{4}a_3\left(1-\left(\frac{\rho_\xi}{\rho_\eta}\right)^{2/3}\right)^2+\ldots\right]\left(\frac{\rho_\xi}{\rho_\eta}\right)^{5/3}.$$

If we assume that the initial undeformed state η is the state at $P = 0$, we can show that $a_1 = 0$; then, the final expression for the pressure becomes

$$P = \frac{3}{2}K_{T0}(x^{-7/3}-x^{-5/3})\left[1-\frac{3}{4}\frac{a_3}{a_2}(x^{-2/3}-1)+\ldots\right], \tag{1.28}$$

where, as before, we have used $x = \rho_0/\rho = \rho_\eta/\rho_\xi$.

The expression (1.28) is the Birch equation. The parameters K_{T0} and a_3/a_2 are found experimentally.

If the deformation is described by means of the Lagrange variables, we obtain — by analogy with Eq. (1.21) — the following expression:

$$dL^2 - dl^2 = 2u_{ij}\,d\eta_i\,d\eta_j,$$

where

$$u_{ij} = \frac{1}{2}\left(\frac{\partial \xi_k}{\partial \eta_i}\frac{\partial \xi_k}{\partial \eta_j}-\delta_{ij}\right) \tag{1.29}$$

is the Green's strain tensor [26]. In the case of uniform strain, given by Eq. (1.23), we have the tensor

$$u_{ij} = u\delta_{ij},$$

where

$$u = \frac{1}{2}[(1+a)^2-1].$$

Expanding again the free energy F as a series in powers of u,

$$F = b_0 + b_1 u + b_2 u^2 + b_3 u^3 + \ldots,$$

we obtain the following equation for the pressure:

$$P = \frac{3}{2}K_{T0}(x^{-1/3}-x^{1/3})\left[1-\frac{3}{4}\frac{b_3}{b_2}(1-x^{2/3})+\ldots\right]. \tag{1.30}$$

Comparing Eqs. (1.28) and (1.30), we find that the basically different dependences of the pressure on the relative volume are obtained solely because of the different definition of the strain tensor. Formally, we cannot give preference to Eq. (1.28) over Eq. (1.30), or conversely. However, Eq. (1.28) has the advantage that it describes the experimental data more accurately at values of x close to unity.

1.6. Other Types of Potential

Mie was the first to derive the equation of state for a solid using classical statistical mechanics [27]. He used the van der Waals method for the derivation of the equation of state for a liquid, and considered monatomic isotropic solids subjected to hydrostatic pressures. Mie employed the Clausius virial theorem

$$3PV = 2\bar{E}_k + \overline{\sum rf(r)}. \qquad (1.31)$$

Here, V is the volume of a crystal; E_k is the kinetic energy of the lattice vibrations; r is the distance between two atoms; $f(r)$ is the force acting between them. Summation is carried out over all pairs of particles and each pair is counted only once. The interatomic forces in Eq. (1.31) are assumed to be of the pair and central type. The bar over the second term denotes statistical-mechanics averaging.

Selection of the function $f(r)$ or of the interaction potential $\varphi(r)$ of two particles in a lattice must be based on some special assumptions. Mie represented the interaction force $f(r)$ as consisting of two components: an attractive force, representing the cohesion of particles in a crystal, and a repulsive force, preventing the atoms from "falling on top of one another." It is obvious that the repulsive forces should decrease more rapidly with the distance than the attractive forces. Hence, the potential can be written in the form

$$\varphi(r) = \frac{a}{r^n} - \frac{b}{r^m} \qquad (n > m). \qquad (1.32)$$

This expression is the simplest possible law for the description of the interaction between atoms. At great distances, the attractive forces, described by the term with the lower power of r, are dominant, whereas at short distances the repulsive forces, described by the higher power of r, are important. *

The potential energy of a gram-atom of a crystal at equilibrium can be represented in the form

$$\Phi_0 = \Phi_0^{(1)} + \Phi_0^{(2)} = \frac{A}{V^{n/3}} - \frac{B}{V^{m/3}}, \qquad (1.33)$$

where the constants A and B, depending on the nature of interatomic forces and on the lattice structure, are given by the relationships

$$A = a(\beta N_0)^{n/3} \sum \left(\frac{\delta}{r_e}\right)^n, \quad B = b(\beta N_0)^{m/3} \sum \left(\frac{\delta}{r_e}\right)^m,$$

where δ is the lattice constant; $V = \beta N_0 \delta^3$ is the volume of 1 g-atom; β is the structure constant; N_0 is Avogadro's number; r_e is the equilibrium distance between atoms of a given pair. Summation is carried out over all atom pairs and each pair is counted only once.

---

*Mie assumed that m = 3, because he regarded it as essential that the attractive forces be attributed to the van der Waals interaction. This reasoning is incorrect, since the van der Waals forces are proportional to r^{-6}.

Since the particles in the crystal lattice vibrate about their equilibrium positions, the potential energy of a crystal Φ can be expanded as a series of powers of the displacements of these particles. In this expansion, the first-order terms are equal to zero. Neglecting terms of the third and higher orders of smallness, we find that the second-order terms represent the potential energy of the lattice vibrations. According to the virial theorem, the average value of these terms is equal to the average value of the kinetic energy of the lattice vibrations $\overline{E}_k$, so that

$$\overline{\Phi} = \overline{\Phi}^{(1)} + \overline{\Phi}^{(2)} = \Phi_0^{(1)} + \Phi_0^{(2)} + \overline{E}_k$$

This energy is distributed in some way between the potential energies of the attractive and repulsive forces. Let us assume that

$$\overline{\Phi^{(1)}} = \Phi_0^{(1)} + \eta \overline{E}_k; \qquad \overline{\Phi^{(2)}} = \Phi_0^{(2)} + (1 - \eta) \overline{E}_k,$$

where η is a proper fraction.

The force $f(r)$, acting between two particles, is equal to the derivative of the potential $\varphi(r)$ with its sign reversed. It follows that

$$\overline{\sum rf(r)} = n\overline{\sum \frac{a}{r^n}} - m\overline{\sum \frac{b}{r^m}} = n\overline{\Phi^{(1)}} + m\overline{\Phi^{(2)}} = n\Phi_0^{(1)} + m\Phi_0^{(2)} + [n\eta + m(1-\eta)]\overline{E}_k.$$

Substituting the above expression into Eq. (1.31), we obtain the relationship

$$PV = G(V) + 2\gamma \overline{E}_k, \tag{1.34}$$

where

$$G(V) = \frac{n}{3} \frac{A}{V^{n/3}} - \frac{m}{3} \frac{B}{V^{m/3}} = -V \frac{\partial \Phi_0(V)}{\partial V},$$

$$\gamma = \frac{1}{3}\left[1 + \frac{n}{3}\eta + \frac{m}{2}(1-\eta)\right].$$

The Mie equation of state is obtained from Eq. (1.34) by substituting in it m = 3, $\eta = 1$, and using the classical value $\frac{3}{2}RT$ for the kinetic energy $\overline{E}_k$.

Grüneisen [28] gave a somewhat different derivation of the Mie equation of state. Moreover, he assumed that the power exponent m for the attractive forces is arbitrary and he left open the question of the temperature dependence of the vibration energy $2\overline{E}_k$. Grüneisen also checked experimentally some consequences of the theory. In particular, he established that at low temperatures the ratio of the thermal expansion coefficient of a metal to its specific heat is practically independent of temperature. This law can be derived as follows.

When a solid is heated at constant volume, all the thermal energy supplied to it is transformed into the lattice vibration energy, so that

$$2\overline{E}_k = \int_0^T c_V \, dT,$$

where c_V is the specific heat at constant volume. It follows from this relationship and from Eq. (1.34) that

$$\left(\frac{\partial P}{\partial T}\right)_V = \frac{\gamma}{V} \frac{\partial (2\overline{E}_k)}{\partial T} = \frac{\gamma}{V} c_V.$$

On the other hand,

$$\left(\frac{\partial P}{\partial T}\right)_V = -\frac{\left(\frac{\partial V}{\partial T}\right)_P}{\left(\frac{\partial V}{\partial P}\right)_T} = -\left(\frac{\partial V}{\partial T}\right)_P \left(\frac{\partial P}{\partial V}\right)_T;$$

which yields the following expression for the Grüneisen parameter:

$$\gamma = \frac{V \varkappa K_T}{c_V}, \tag{1.35}$$

where $\varkappa = \frac{1}{V}\left(\frac{\partial V}{\partial T}\right)_P$ is the thermal expansion coefficient; $K_T = -V\left(\frac{\partial P}{\partial V}\right)_T$ is the isothermal bulk modulus.

If we assume that γ is independent of temperature, we find that at low temperatures V and K_T reach some finite limits, while $\varkappa$ and c_V tend to zero. Consequently, at low temperatures, V and K_T depend weakly on the temperature and the Grüneisen law is obtained as the first approximation.

Mie and Grüneisen derived qualitative properties of crystals from formal assumptions about the nature of the forces acting between particles. Later investigators attempted to justify physically the theoretical dependence of the attractive forces on the interatomic distance. The repulsive forces were still expressed formally by power laws. This continued until the exponential nature of the attractive forces was discovered.

However, the power-law form of the potential is simpler, more convenient in calculations, and does not differ basically from the exponential expression. Moreover, in the case of small variations of the interatomic distances, the true potential can be approximated satisfactorily by a power law. That is why the power-law form of the potential (rather than the exponential form) is still being used.

The first investigations were carried out on ionic crystals. Attempts were made to determine primarily the parameters of the repulsive-force potential. Thus, Born [29] used the potential

$$\Phi(V) = \frac{A}{V^{n/3}} - \frac{B}{V^{1/3}}, \tag{1.36}$$

in which the second term was used to describe the Coulomb interaction of ions. Considering a crystal at P = 0, T = 0, and neglecting the energy of the zero-point vibrations of the lattice, Born found that

$$P = -\frac{\partial \Phi}{\partial V} = \frac{n}{3}\frac{A}{V_0^{n/3+1}} - \frac{1}{3}\frac{B}{V_0^{1/3}} = 0$$

and

$$K = -V\frac{\partial^2 \Phi}{\partial V^2} = \frac{n}{3}\left(\frac{n}{3}+1\right)\frac{A}{V^{n/3+1}} - \frac{4}{9}\frac{B}{V^{1/3}},$$

where K is the bulk modulus, found experimentally. These two equations yield

$$n = 1 + \frac{9K}{B}V_0^{1/3} \qquad A = \frac{B}{n}V_0^{(n-1)/3},$$

where the constant B is known and is defined in terms of the Madelung constant; V_0 is the equilibrium volume of the crystal. The determination of the energy of repulsion thus reduces to the measurement of the bulk modulus and of the density $\rho_0 = \mu / V_0$, where μ is the molecular weight. The expressions obtained for n and A are valid at absolute zero when the energy of the zero-point vibrations is neglected. In practice, one uses the room-temperature values of ρ_0 and K. The values of n calculated in this way for alkali-halide compounds are close to 9. Slater [30] refined somewhat the values of n. He did this by measuring K at two different temperatures and extrapolating linearly to absolute zero. He still found that the values of n were of the same order as those calculated by simpler methods.

Lennard-Jones [31] determined the law governing the repulsive forces of inert-gas atoms by considering their gas — kinetic behavior. He used the fact that the experimental value of the second virial coefficient could be expressed relatively simply in terms of the interatomic interaction potential $\varphi(r)$ [32]. Lennard-Jones used the following potential:

$$\varphi(r) = \frac{a}{r^n} - \frac{c}{r^\delta}. \qquad (1.37)$$

The second term in this potential represents the van der Waals forces.

The work of Lennard-Jones showed that the measured value of the second virial coefficient could be made to agree with the calculated value by varying the exponent n of the repulsive forces within wide limits: from 9 to 14, in fact.

Born and Mayer [7] tried to update the old theory of ionic crystals using the results of quantum mechanics. They employed the following exponential law for the repulsive forces:

$$b_{ij} \exp(-r/\rho), \qquad (1.38)$$

where b_{ij} and ρ are constants. Moreover, they found that one could use $\rho = 0.333 \cdot 10^{-8}$ cm for alkali-halide compounds when the constant b_{ij} was calculated from

$$b_{ij} = bc_{ij} \exp[(r_i + r_j)/\rho],$$

where b is a new constant; r_i and r_j are the ionic radii; the coefficient c_{ij} takes into account the dependence of the repulsive forces on ion charges found by Pauling [33]; this coefficient is given by

$$c_{ij} = 1 + \frac{z_i}{n_i} + \frac{z_j}{n_j},$$

where z_i and z_j are the valences of two interacting ions; n_i and n_j are the numbers of outer electrons in ions i and j. The valence n of electronegative ions should be negative. Therefore, the potential of the repulsive forces in an ionic crystal, consisting of two types of ion, is

$$\Phi^{(r)} = Nb\left\{ Mc_{12} \exp[(r_1 + r_2 - r)/\rho] + \frac{1}{2} M'[c_{11} \exp(2r_1/\rho) + c_{22} \exp(2r_2/\rho)] \exp(-\delta r/\rho) \right\}, \qquad (1.39)$$

where N is the number of molecules in the crystal; M is the number of nearest neighbors of opposite sign; M' is the number of nearest neighbors of the same sign; δ is the ratio of the distance between ions of the same sign to the distance between ions of opposite sign.

Massey and Mohr [34] analyzed the available values of the viscosity of gaseous helium and showed that the interaction energy can be approximated satisfactorily using the following formula for the potential:

$$\varphi(r) = a \exp(-br) - cr^{-6}. \qquad (1.40)$$

The same potential was used by Kalinin [35] to describe the data on the compressibility of solid argon.

Bardeen [36] obtained the equation of state for alkali metals using the following potential:

$$\varphi(r) = \frac{a}{r^3} + \frac{b}{r^2} - \frac{c}{r}. \tag{1.41}$$

Alkali metals have one S valence electron per atom and their singly charged positive ions have the electron structure of inert gases. At room temperature and at atmospheric pressure, they have the bcc crystal structure. The distance between the nearest neighbors in these metals is approximately four times greater than the ionic radius, so that a large part of the volume of a crystal is occupied by valence electrons, which may be assumed to be approximately free. The Fermi kinetic energy of these electrons is proportional to r^{-2}, which corresponds to the second term in the potential given by Eq. (1.41). The third term in Eq. (1.41) is the Coulomb interaction between ions and the free-electron gas. The additional kinetic energy of the conduction electrons, due to the presence of positive ions, is proportional to r^{-3}. This energy is represented [9] by the first term in Eq. (1.41).

Bardeen determined the coefficients a, b, and c from the experimental values of the cohesive energy, the equilibrium volume of a crystal, and the compressibility. The pressure dependences obtained for the volume were compared with the measurements made by Bridgman up to $4 \cdot 10^3$ bar, and extrapolated to absolute zero. It was found that the discrepancy between the theoretical and experimental curves grew with increasing pressure. Moreover, the discrepancy grew with increasing mass of the ions forming the lattice. It was obvious from these results that the repulsive forces between ions, which depended exponentially on the lattice constant and which were ignored in Eq. (1.41), became important in the case of the compression of alkali metals by 20-30%. Bardeen found that the coefficients b of potassium and rubidium were negative, which had no physical meaning.

A detailed review of early work on the theory of the crystalline state is given in [37, 9].

Recent investigations show that there are two main ways in which the method of potentials can be used to determine the equations of state for solids. One of them is the extension and development of the classical approach to the problem, founded on the work of Mie, Grüneisen, Born, and others. The interatomic potential has two or three arbitrary parameters, which are determined from the experimental data. Each term in this potential has a definite physical meaning. The expressions for the potential are frequently of the two-term type, corresponding to the principal interactions of repulsion and attraction. Sometimes corrections are added to these principal terms. However, corrections are used very rarely because the two-term potential is usually an approximate function of the interatomic distance and there is no point in making small corrections.

Let us consider some applications of this approach. Bradburn [38] assumed that the interatomic interaction potential is of the type given by Eq. (1.32) and obtained the following high-temperature equation of state for monatomic cubic crystals:

$$P(x) = \rho_{00} \left[N_1 u F(x) + \frac{RT}{\mu} G(x) \right], \tag{1.42}$$

where $x = \rho / \rho_{00}$ is the dimensionless volume;

$$F(x) = a(x^{-n/3} - x^{-m/3}); \qquad G(x) = c + b \frac{x^{(m-n)/3}}{1 + I(x^{(m-n)/3} - 1)} ;$$

$$a = \frac{mn}{6(n-m)} \left(\frac{S_m^0}{S_n^0} \right)^{m/(n-m)} S_m^0 ;$$

$$b = \frac{1}{2} I(n-m); \qquad c = 1 + \frac{m}{2};$$

$$I = \left[1 - \frac{m-1}{n-1} \cdot \frac{S_n^0 S_{m+2}^0}{S_m^0 S_{n+2}^0} \right]^{-1};$$

R is the universal gas constant; μ is the molecular weight; N_1 is the number of atoms per unit mass; ρ_{00} is the density at T = 0, P = 0; u is the dissociation energy of an isolated pair of particles; S_p^0 are the lattice sums calculated and tabulated by Misra [39] for all integers p from 4 to 15 for the three cubic Bravais lattices.

Fürth [40] used the equation of state (1.42) and developed a method for calculating the power exponents m and n from the experimental values of the sublimation energy, the compressibility, and the thermal expansion coefficient, and from the dependences of these quantities on the pressure and temperature. He applied this method to a large number of elements. His values of m and n were very similar for elements belonging to the same group in the periodic table.

Pack, Evans, and James [41] took into account the quantum-mechanical requirement of an exponential increase in the pressure with decreasing interatomic distance, and they used the following equation of state for metals:

$$P_p(x) = A x^{2/3} \left\{ \exp[b(1 - x^{1/3})] - 1 \right\}. \tag{1.43}$$

The corresponding potential

$$\Phi(x) = \frac{3A}{\rho_0} \left\{ \frac{1}{b} \left(x^{1/3} + \frac{4}{b} x + \frac{12}{b^2} x^{2/3} + \frac{24}{b^3} x^{1/3} + \frac{24}{b^4} \right) \exp[b(1 - x^{1/3})] + \frac{1}{5} x^{5/3} \right\}$$

was obtained by integration of Eq. (1.43). They found that the function $\Phi(x)$ increased without limit when x was increased.

Prieto [42] obtained the equation of state for silver using the potential

$$\Phi(x) = A \exp[b(1 - x^{1/3})] - K x^{-1/3} + B x^{-1} + C x^{-2/3}, \tag{1.44}$$

which differed from the Bardeen potential (1.41) by an additional term representing the exponential repulsion due to the overlap of the electron shells of the ions.

Another use of the method of potentials in the determination of the equations of state of solids is based on the approach in which the terms in the potential are not interpreted from the atomic point of view.

Some workers [43, 44] have used the empirical relationship

$$K_T = K_0 + rP, \tag{1.45}$$

which is valid at low pressures. Integrating this equation, they obtained

$$P = \frac{K_0}{r} \left[\left(\frac{P_0 r}{K_0} + 1 \right) x^{-r} - 1 \right], \tag{1.46}$$

assuming that V = V_0 at P = P_0. The constants K_0 and r were found from the experimental data. This equation was applied to various types of solid.

Kormer et al. [45-48] used a polynomial expression for the potential

$$\Phi(x) = \frac{3}{\rho_{00}} \sum_{n=1} \frac{a_n}{n} y^{-n/3}, \tag{1.47}$$

where ρ_{00} is the density at P = 0, T = 0; $y = \rho_{00}/\rho$. The coefficients a_n were determined from the experimental data and the obtained equation of state had to agree with the results yielded by a statistical model of a solid at high pressures. In practice, calculations were carried out using polynomials with six or seven terms and it was found that the values of the coefficients in a six-term polynomial differed considerably from the corresponding values in a seven-term polynomial. No assumptions were made in Eq. (1.47) about the nature of the interatomic interactions in solids and, therefore, this equation could be applied to metals and ionic crystals.

Literature Cited

1.   M. Born and K. Huang, Dynamical Theory of Crystal Lattices, Oxford University Press (1954).
2.   J. C. Slater, Phys. Rev., 32:339 (1928).
3.   V. A. Kalinin, Izv. Akad. Nauk SSSR, Ser. Geofiz., No. 2, p. 333 (1960).
4.   P. E. Phillipson, Phys. Rev., 125:1981 (1962).
5.   H. S. W. Massey and E. H. S. Burhop, Electronic and Ionic Impact Phenomena, Oxford University Press (1952).
6.   V. P. Trubitsyn, Fiz. Tverd. Tela, 7:3443 (1965).
7.   M. Born and J. Mayer, Z. Physik, 75:1 (1932).
8.   B. I. Davydov, Izv. Akad. Nauk SSSR, Ser. Geofiz., No. 12, p. 1411 (1956).
9.   F. Seitz, Modern Theory of Solids, McGraw-Hill, New York (1940).
10. V. A. Kalinin, Trudy Inst. Fiziki Zemli Akad. Nauk SSSR, No. 11(178), p. 67 (1960).
11. C. Kittel, Introduction to Solid State Physics, 3rd ed., Wiley, New York (1966).
12. V. N. Zharkov, Trudy Inst. Fiziki Zemli Akad. Nauk SSSR, No. 11(178), p. 14 (1960).
13. V. N. Zharkov and V. A. Kalinin, Dokl. Akad. Nauk SSSR, 135:811 (1960).
14. V. N. Zharkov and V. A. Kalinin, Dokl. Akad. Nauk SSSR, 145:551 (1962).
15. V. N. Zharkov and V. A. Kalinin, Izv. Akad. Nauk SSSR, Ser. Geofiz., No. 3, p. 298 (1962).
16. V. N. Zharkov, Trudy Inst. Fiziki Zemli Akad. Nauk SSSR, No. 20(187), p. 3 (1962).
17. V. A. Kalinin, "Equations of state for solids at high pressures and their applications to some geophysics problems," Dissertation for Candidate's Degree [in Russian], Moscow (1963).
18. V. N. Zharkov, "Investigations in geophysics," Doctoral Dissertation [in Russian], Moscow (1964).
19. L. V. Al'tshuler, L. V. Kuleshova, and M. N. Pavlovskii, Zh. Éksp. Teor. Fiz., 39:16 (1960).
20. L. V. Al'tshuler, M. N. Pavlovskii, L. V. Kuleshova, and G. V. Simakov, Fiz. Tverd. Tela, 5:279 (1963).
21. L. V. Al'tshuler, A. A. Bakanova, and R. F. Trunin, Zh. Éksp. Teor. Fiz., 42:91 (1962).
22. L. Pauling, The Nature of the Chemical Bond and the Structure of Molecules and Crystals, 3rd ed., Cornell University Press, New York (1960).
23. D. R. Hartree, The Calculation of Atomic Structures, Wiley, New York (1957).
24. F. Birch, J. Geophys. Res., 57:227 (1952).
25. F. D. Murnaghan, Finite Deformation of an Elastic Solid, Wiley, New York (1951).
26. W. Prager, Introduction to Mechanics of Continua, Ginn and Co., Boston (1961).
27. G. Mie, Ann. Physik, 11:657 (1903).
28. E. Grüneisen, Ann. Physik, 26:393 (1908).
29. M. Born, Atomtheorie des festen Zustandes, Teubner, Leipzig (1923).
30. J. C. Slater, Phys. Rev., 23:488 (1924).
31. J. E. Lennard-Jones, Proc. Roy. Soc. (London), A106:441, 463, 709 (1924).
32. L. D. Landau and E. M. Lifshitz, Statistical Physics, 2nd ed., Pergamon Press, Oxford (1968).
33. L. Pauling, Z. Krist., 67:377 (1928).

34. H. S. W. Massey and C. B. O. Mohr, Proc. Roy. Soc. (London), A144:188 (1934).
35. V. A. Kalinin, Zh. Éksp. Teor. Fiz., 34:229 (1958).
36. J. Bardeen, J. Chem. Phys., 6:367 (1938).
37. M. Born and M. Goeppert-Mayer, in: Handbuch der Physik, Vol. 24, Part 2, Ch. IV, Springer-Verlag, Berlin (1933), p. 623.
38. M. Bradburn, Proc. Cambridge Phil. Soc., 39:113 (1943).
39. R. D. Misra, Proc. Cambridge Phil. Soc., 36:173 (1940).
40. R. Fürth, Proc. Roy. Soc. (London), A183:87 (1944).
41. D. C. Pack, W. M. Evans, and H. J. James, Proc. Phys. Soc. (London), 60:1 (1948).
42. F. E. Prieto, Phys. Rev., 129:37 (1963).
43. M. Kornfel'd, Usp. Fiz. Nauk, 54:315 (1954).
44. Yu. N. Ryabinin, Zh. Tekh. Fiz., 30:739 (1960).
45. S. B. Kormer and V. D. Urlin, Dokl. Akad. Nauk SSSR, 131:542 (1960).
46. S. B. Kormer, V. D. Urlin, and L. T. Popova, Fiz. Tverd. Tela, 3:2131 (1961).
47. S. B. Kormer, A. I. Funtikov, V. D. Urlin, and A. N. Kolesnikova, Zh. Éksp. Teor. Fiz., 42:686 (1962).
48. S. B. Kormer, M. V. Sinitsyn, A. I. Funtikov, V. D. Urlin, and A. V. Blinov, Zh. Éksp. Teor. Fiz., 47:1202 (1964).

CHAPTER 2

EQUATION OF STATE AT $T \neq 0$.
GRÜNEISEN PARAMETER

2.1. Introduction

At a temperature T other than that of absolute zero, every atom in a solid has additional kinetic energy. This energy represents the vibrational motion of atoms about their equilibrium positions at lattice sites. This energy, and consequently the amplitude of the vibrations, increases with increasing temperature. A substance is solid if its temperature is sufficiently low. The value of kT should be small compared with the binding energy of atoms in a crystal. Consequently, the amplitudes of thermal vibrations in a solid are always small compared with the interatomic distances (lattice constants).

If the displacements of atoms from their equilibrium positions are sufficiently small, the potential energy of a crystal can be expanded as a power series in terms of these displacements. There should be no linear terms in this series since there is no net force acting on an atom in its equilibrium position. Dropping terms higher than those of the second order (because of the smallness of the displacements), we obtain a system with a potential energy proportional to the squares of the displacements. In such a system, atoms perform small vibrations about their equilibrium positions.

The spectrum of a solid depends on its volume. This is because any change in volume alters the force constants, i.e., the coefficients which stand in front of the squares of displacements in the expansion of the potential energy in terms of the displacements. Moreover, the contraction or compression of a solid at a constant temperature may increase the importance of the anharmonicity (see Chapter 3). However, we shall assume here that a change in the volume alters only the spectrum of the vibrations and we shall postulate that the vibrations themselves remain harmonic. This is known as the quasiharmonic approximation.

2.2. Free Energy of a Crystal Lattice
in the Quasiharmonic Approximation

Let us assume that a crystal consists of N unit cells, each of which contains ν atoms. The total number of degrees of freedom of such a crystal is $3N\nu$, three of which represent the translational motion and three the rotational motion of the crystal as a whole. The number of vibrational degrees of freedom is $3N\nu - 6$, but since $3N\nu \gg 6$, it is usually assumed (for the sake of simplicity) that the number of vibrational degrees of freedom is $3N\nu$.

The equations of motion for the atoms in such a system can be transformed to normal coordinates. In these coordinates, the vibrations of the atoms in a crystal are equivalent, from the mechanical point of view, to the vibrations of a system of $3N\nu$ independent harmonic oscillators.

The free energy of a harmonic oscillator F_α (α is the number of the oscillator) is obtained using the expression

$$F_\alpha = -kT \ln \sum_n \exp(-E_n/kT), \tag{2.1}$$

which follows from the Gibbs distribution [1], and the expression for the energy of a quantum oscillator

$$E_n = \hbar\omega(n + {}^1/_2) + \varphi_p, \tag{2.2}$$

where φ_p is the potential energy of the oscillator at its equilibrium position. Substituting Eq. (2.2) into Eq. (2.1), and using the formula for the sum of an infinite decreasing geometric series, we obtain

$$F_\alpha = \varphi_{p\alpha} + \frac{1}{2}\,\hbar\omega_\alpha + kT \ln[1 - \exp(-\hbar\omega_\alpha/kT)], \tag{2.3}$$

where k is the Boltzmann constant and $\hbar$ is the Planck constant.

The lattice component of the free energy of a crystal, i.e., the free energy of a system of $3N\nu$ harmonic oscillators, can be written in the form

$$F = E_p + \frac{1}{2}\sum_\alpha \hbar\omega_\alpha + kT \sum_\alpha \ln[1 - \exp(-\hbar\omega_\alpha/kT)]. \tag{2.4}$$

Here, the summation is carried out over all $3N\nu$ normal vibrations. The term $E_p = \Sigma\varphi_{p\alpha}$ is the potential energy of a crystal, which depends only on its volume (see Sec. 1.3). The frequencies of the oscillators, ω_α, also depend on the volume.

All other thermodynamic quantities can be obtained, in principle, from the free energy of Eq. (2.4). The sums in Eq. (2.4) cannot be calculated for the general case, since their values depend considerably on the actual distribution of frequencies in the vibration spectrum of a given solid. However, the summation can be carried out in the low- and high-temperature limits. This applies only to the second sum in Eq. (2.4), which depends on T. The value of the first sum, corresponding to the energy of zero-point vibrations of a crystal, is independent of T. The first sum can be neglected (because of its smallness) or included in the term E_p, because ω_α, like E_p, depends only on the volume of a solid.

2.3. Low Temperatures

At low values of kT, the only important terms in the second sum of Eq. (2.4) are those with low frequencies: $\hbar\omega \lessgtr kT$. These low-frequency vibrations are elastic waves whose wavelengths are much longer than the interatomic distances. Consequently, a solid can be regarded as a continuous medium.

Three volume waves can be propagated in an isotropic continuous medium: one longitudinal wave (velocity c_l) and two transverse waves of different polarizations but with the same velocity c_t. The frequency of these waves ω is related to the modulus of the wave vector **k** by the linear equation $\omega = ck$, in which the coefficient of proportionality c is the velocity c_l or c_t.

Let us consider a macroscopic solid in the form of a cube whose edge is L. The problem of free vibrations of such a solid can be solved on the basis of the theory of elasticity. The displacement of a point in a medium in a standing wave along the x axis of a cube is given by the relationship

$$u_x = a \sin k_x x \sin \omega t,$$

where k_x is the x-component of the wave vector. Similar expressions are obtained for the displacements along the other two axes. Since the faces of a cube are not free, the components of the wave vector can assume only the discrete values $k_x = 2\pi n/L$, where n is an integer (if a sample is sufficiently large, the frequency distribution of normal vibrations ceases to depend on the shape of the sample and on specific boundary conditions [2, 3]). However, since L is large, the values of k are effectively a continuous set. We shall use the linear relationship between ω and k and replace summation over α in the second sum of Eq. (2.4) with integration with respect to ω.

In order to carry out this integration, we must know the spectral density $Z(\omega)$, which can be obtained from the following considerations. Each allowed vibration corresponds to a volume $(2\pi/L)^3$ in the wave-vector space. The number of vibrations of a given type with wave vectors within the range from k to k + dk is equal to the volume of a spherical layer in the k space, $4\pi k^2 dk$, divided by the volume $(2\pi/L)^3$ corresponding to one vibration, i.e.,

$$V \frac{4\pi k^2}{(2\pi)^3} dk,$$

where $V = L^3$ is the macroscopic volume of the cube. We can now calculate quite simply the number of longitudinal vibrations in the frequency range from ω to $\omega + d\omega$:

$$Z_l(\omega) d\omega = \frac{4\pi V \omega^2}{(2\pi c_l)^3} d\omega. \tag{2.5}$$

Since each value of ω corresponds to two transverse vibrations of different polarization, we obtain the following expression for the number of transverse vibrations in the interval from ω to $\omega + d\omega$:

$$Z_t(\omega) d\omega = 2 \frac{4\pi V \omega^2}{(2\pi c_t)^3} d\omega. \tag{2.6}$$

Thus, the spectral density for an isotropic elastic body is given by the relationship

$$Z(\omega) = \frac{V \omega^2}{2\pi^2} \left(\frac{1}{c_l^3} + \frac{2}{c_t^3} \right). \tag{2.7}$$

We shall introduce the average velocity of sound $\bar{c}$, defined by the formula

$$\frac{3}{\bar{c}^3} = \frac{1}{c_l^3} + \frac{2}{c_t^3}. \tag{2.8}$$

We can now write Eq. (2.7) in the form

$$Z(\omega) = \frac{3V \omega^2}{2\pi^2 \bar{c}^3}. \tag{2.9}$$

The expression for the low-temperature value of the free energy of a crystal [Eq. (2.4)], whose spectral density is given by Eq. (2.9), is of the form

$$F = E_p + \frac{1}{2} \sum_\alpha \hbar \omega_\alpha + kT \frac{3V}{2\pi^2 \bar{c}^3} \int_0^\infty \ln[1 - \exp(-\hbar\omega/kT)] \omega^2 d\omega. \tag{2.10}$$

Here, integration is extended to infinity since at low values of kT the integral converges rapidly and large values of ω do not affect significantly the value of the integral.

We shall use E_{zv} to denote the energy of the zero-point vibrations, and we shall introduce a new integration variable $y = \hbar\omega/kT$. Integrating Eq. (2.10) by parts, we obtain

$$F = E_p + E_{zv} - \frac{V(kT)^4}{2\pi^2 \bar{c}^3 \hbar^3} \int_0^\infty \frac{y^3 \, dy}{e^y - 1}.$$

Substituting the value of the integral, which is $\pi^4/15$, we obtain the following expression for the free energy:

$$F = E_p + E_{zv} - \frac{\pi^2 V(kT)^4}{30(\bar{c}\hbar)^3}. \tag{2.11}$$

This temperature dependence of F corresponds to the T^3 law of the specific heat. The temperature dependence given by Eq. (2.11) is in good agreement with experiment.

The equation of state for a crystal at low temperatures follows directly from Eq. (2.11). Using the thermodynamic relationship $P = -(\partial F/\partial V)_T$, we find that

$$P = P_p + P_{zv} + \gamma \frac{\pi^2(kT)^4}{10(\bar{c}\hbar)^3}, \tag{2.12}$$

where P_p and P_{zv} correspond to the terms E_p and E_{zv} in Eq. (2.11). The last term in Eq. (2.12) describes the thermal motion. It is due to the thermal vibrations of the lattice and is proportional to the energy density of these vibrations. The coefficient of proportionality γ is known as the Grüneisen parameter and, in the present case, is defined by

$$\gamma = -\frac{\partial \ln(\bar{c}/V^{1/3})}{\partial \ln V}. \tag{2.13}$$

The equation of state thus obtained applies also to anisotropic bodies. However, in this case, the value of $\bar{c}$ is no longer given by the simple expression (2.8) but must be found numerically.

Equation (2.12) has exactly the same structure as Eq. (1.34). All equations of state similar to Eq. (1.34) will be called the Mie — Grüneisen equations.

2.4 High Temperatures

In the limiting case of high temperatures, when $\hbar\omega_\alpha < kT$, the last term in Eq. (2.4) can be expanded as a series in powers of a small parameter $\hbar\omega_\alpha/kT$:

$$kT \sum_\alpha \ln(1 - e^{-\hbar\omega_\alpha/kT}) \approx kT \sum_\alpha \ln \frac{\hbar\omega_\alpha}{kT} - \frac{1}{2} \sum_\alpha \hbar\omega_\alpha. \tag{2.14}$$

The second term on the right-hand side of Eq. (2.14) is exactly equal to the energy of the zero-point lattice vibrations, but its sign is opposite. Thus, the high-temperature value of the free energy of the crystal lattice is given by

$$F = E_p + kT \sum_\alpha \ln \frac{\hbar\omega_\alpha}{kT};$$

introducing an average frequency $\bar{\omega}$ in accordance with the formula

$$\ln \bar{\omega} = \frac{1}{3N\nu} \sum_\alpha \ln \omega_\alpha, \tag{2.15}$$

we finally obtain

$$F = E_p - 3N\nu \, kT \ln kT + 3N\nu \, kT \ln \hbar\bar{\omega}. \tag{2.16}$$

Like the average velocity of sound $\bar{c}$ in Eq. (2.11), the average frequency $\bar{\omega}$ depends on the volume V of a given solid.

The expression obtained for the free energy corresponds to a constant value of the specific heat $c_V = 3N\nu k$, which agrees with the experimental law of Dulong and Petit. This law predicts that the specific heat of monatomic solids ($\nu = 1$) should be equal to 3Nk and it describes well the room-temperature specific heat of many elements.

The high-temperature equation of state of a crystal lattice follows directly from Eq. (2.16):

$$P = P_\mathrm{p} + \gamma \frac{3N\nu kT}{V}. \tag{2.17}$$

It is evident from Eq. (2.17) that the thermal pressure in the high-temperature limit is (as in the low-temperature limit) proportional to the thermal energy density, i.e., once again we have a Mie — Grüneisen equation of state. The Grüneisen parameter of Eq. (2.17) is defined by the relationship

$$\gamma = -\frac{\partial \ln \bar{\omega}}{\partial \ln V}. \tag{2.18}$$

When the temperature is increased, the amplitudes of the lattice vibrations also increase. At sufficiently high temperatures, the anharmonicity of the lattice vibrations becomes appreciable and this gives rise to an additional term in Eq. (2.17) (this will be considered in Chapter 3).

2.5. Debye and Einstein Approximations

The normal frequencies (eigenfrequencies) of a macroscopic solid consisting of a large number of atoms have a practically continuous distribution. Therefore, summation in Eq. (2.4) can be replaced by integration with respect to frequencies ω. Such a substitution introduces explicitly the spectral density $Z(\omega)$ under the integral sign:

$$F = E_\mathrm{p} + \int_0^\infty \left\{ \frac{1}{2}\hbar\omega + kT \ln[1 - \exp(-\hbar\omega/kT)] \right\} Z(\omega)\, d\omega, \tag{2.19}$$

and this density affects considerably those properties of a crystal which are due to the lattice vibrations. Since the number of normal vibrations of a crystal is finite and equal to $3N\nu$, the function $Z(\omega)$ must satisfy the normalization condition

$$\int_0^\infty Z(\omega)\, d\omega = 3N\nu. \tag{2.20}$$

The actual form of the function $Z(\omega)$ is quite difficult to determine and such a determination has not yet been carried out in the general form. Therefore, model representations, which make it possible to calculate approximately $Z(\omega)$ are used very frequently.

The most widely used is the Debye theory [4], which describes quite well the thermal properties of solids, particularly their specific heat, at all temperatures. It is known (see Sec. 2.3) that the spectrum of low frequencies can be obtained from the elastic constants. According to Debye, $Z(\omega) \propto \omega^2$ at low frequencies and, approximately, at high frequencies. In order to satisfy the normalization condition of Eq. (2.20), the elastic wave spectrum must be limited by some maximum frequency ω_D. Consequently, we obtain the following expression for the spectral density:

$$Z(\omega) = \begin{cases} 9N\nu\dfrac{\omega^2}{\omega_D{}^3} & \text{for } 0 \leqslant \omega \leqslant \omega_D, \\ 0 & \text{for } \omega > \omega_D. \end{cases} \tag{2.21}$$

We shall introduce a characteristic temperature θ_D, known as the Debye temperature and related to ω_D by the expression

$$k\theta_D = \hbar\omega_D. \tag{2.22}$$

Substituting Eq. (2.21) in Eq. (2.19) for the free energy of a crystal, we obtain

$$F = E_p + N\nu kT\left\{\frac{9}{8}\frac{\theta_D}{T} + 3\ln[1 - \exp(-\theta_D/T)] - D\left(\frac{\theta_D}{T}\right)\right\}, \tag{2.23}$$

where

$$D\left(\frac{\theta_D}{T}\right) = \frac{3T^3}{\theta_D{}^3}\int_0^{\theta_D/T}\frac{z^3 dz}{e^z - 1} \tag{2.24}$$

is the Debye function. A fuller treatment is given in Appendix B.

It is evident that the Debye theory should describe correctly the thermal properties of crystals at low temperatures. This follows from the observation that only the low-frequency part of the spectrum is important at low temperatures and the Debye spectrum is an extrapolation of the low-frequency elastic-wave spectrum throughout the full range of frequencies.

At high temperatures, all normal vibrations of a crystal are excited and the total number of vibrations, rather than the specific nature of the spectrum density, is the important parameter. Because of the normalization condition (2.20), the Debye theory gives the correct number of vibrations and, therefore, it also describes correctly the high-temperature thermal properties of crystals.

Thus, Eq. (2.23) is an interpolation formula which describes correctly the properties of crystals in the low- and high-temperature limits and which is a good approximation in the intermediate range of temperatures. The equation of state, corresponding to the free energy of Eq. (2.23), is of the form

$$P = P_p + \frac{\gamma}{V}N\nu k\left[\frac{9}{8}\theta_D + 3TD\left(\frac{\theta_D}{T}\right)\right], \tag{2.25}$$

where the Grüneisen parameter γ is defined by the relationship:

$$\gamma = -\frac{\partial\ln\theta_D}{\partial\ln V}. \tag{2.26}$$

At high temperatures, Eq. (2.25) becomes identical with Eq. (2.17) if $\bar{\omega}$ in Eq. (2.18) is replaced with ω_D.

The Debye temperature θ_D is usually found from the experimental values of the low-temperature specific heat (in this case, the Debye temperature is used as a fitting parameter) or from the velocities of elastic waves in a solid

$$\theta_D = \bar{c}\frac{\hbar}{k}\left(\frac{6\pi^2 N\nu}{V}\right)^{1/3}, \tag{2.27}$$

where $\bar{c}$ is the average velocity of sound given by Eq. (2.8). This expression for θ_D is obtained from Eqs. (2.9) and (2.21). The specific-heat and velocity-of-sound methods give similar values of θ_D [3, 5].

Einstein [6] was the first to apply the quantum theory to a solid (this was a few years before Debye's theory). Einstein assumed that all the frequencies of the atoms are identical and equal to ω_E. This assumption implies that the atoms are vibrating independently of one another. The spectral density in the Einstein model is

$$Z_E(\omega) = 3N\nu\delta(\omega - \omega_E). \tag{2.28}$$

This spectral density also satisfies the normalization condition of Eq. (2.20) and, therefore, the Einstein model should describe correctly the thermal properties of solids at high temperatures.

Substituting Eq. (2.28) into Eq. (2.19), we obtain the following expression for the free energy of a crystal in the Einstein model:

$$F = E_p + 3N\nu\left\{\frac{1}{2}\hbar\omega_E + kT\ln[1 - \exp(-\hbar\omega_E/kT)]\right\}. \tag{2.29}$$

The equation of state follows from Eq. (2.29):

$$P = P_p + \frac{\gamma}{V}3N\nu\left[\frac{1}{2}\hbar\omega_E + \frac{\hbar\omega_E}{\exp(\hbar\omega_E/kT) - 1}\right], \tag{2.30}$$

where the Grüneisen parameter is

$$\gamma = -\frac{\partial\ln\omega_E}{\partial\ln V}. \tag{2.31}$$

The Einstein model is unrealistic in the case of monatomic solids. This is because each atom is held at its equilibrium position by its interactions with other atoms in the lattice and, therefore, it cannot vibrate independently of other atoms. However, the Einstein model can be used to describe approximately the optical-frequency part of the vibration spectrum of a multiatomic solid.

2.6. Phonons

We shall now describe qualitatively the lattice vibrations of a crystal. Instead of considering the vibrations of single atoms about their equilibrium positions, we can discuss their collective motion. This collective motion is in the form of waves, particularly standing waves, which are frequently called normal vibrations of the lattice.

Let us assume that a unit cell of a crystal consists of atoms which, in general, may differ from one another in their physical properties and in their positions in the lattice. If a crystal consists of N unit cells, we can speak of ν different groups of atoms, each of which consists of N identical particles. The vibrations of all the atoms in any one such group are identical but the vibrations of atoms belonging to different groups are, in general, different. In view of this, we can speak of ν different types or modes of lattice vibration. Since the displacement of an atom is specified by three components, it is obvious that each group of identical atoms in a crystal can perform vibrations of three types, which differ from one another by the directions of the atomic displacements. Thus, 3ν different types of vibration can be excited in a crystal.

When the problem of the crystal lattice vibtations is solved rigorously it is found, in agreement with our qualitative conclusion, that any wave vector **k** corresponds, in general, to 3ν different values of the frequency ω. In other words, we can say that the frequency $\omega = \omega(\mathbf{k})$ is a multivalued function of the wave vector and that it has 3ν branches.

These branches include those which correspond to conventional low-frequency elastic waves in a crystal. In real crystals there are only three such types of wave, which differ from one another by their dependence of ω on $\mathbf{k}$; for all three types the frequency ω is a homogeneous first-degree function of the components of the vector $\mathbf{k}$, which vanishes at $\mathbf{k} = 0$. In the isotropic case, these three waves correspond to one longitudinal wave and two transverse waves with different polarizations. These three types of wave are known as the acoustical waves and the corresponding three branches of $\omega = \omega(\mathbf{k})$ are known as the acoustical modes. A characteristic property of these acoustical waves is the vibration of the lattice as a continuous medium. In the limiting case of infinitely long wavelengths, these vibrations reduce to parallel motion of the lattice as a whole. It is evident that only the acoustical vibrations can be excited in a monatomic crystal ($\nu = 1$).

The other $3(\nu - 1)$ types of wave have the following property: their frequency does not vanish as $\mathbf{k} \to 0$ but tends to a fixed limit. These waves (lattice vibrations) are known as the optical waves and the corresponding branches of the function $\omega = \omega(\mathbf{k})$ are known as the optical modes. The optical modes describe the vibrations of atoms relative to one another in the same unit cell of a crystal. If atoms within a cell interact with one another more strongly than atoms in neighboring cells, the optical vibration frequencies are governed primarily by internal forces. In this case, the influence of the rest of the lattice is slight and the optical modes extend over a limited range of frequencies which is quite different from the range of frequencies of the acoustical modes.

We shall now consider lattice vibrations from the quantum-theoretical point of view.

Each normal vibration of a crystal can be regarded as a standing wave and it can be expanded into two traveling waves moving in opposite directions. By analogy with photons (electromagnetic field quanta), we can introduce the concept of lattice vibration quanta known as phonons. Phonons are quasiparticles of definite energy and direction of motion. The energy of a phonon is related to its frequency by the expression

$$\varepsilon = \hbar\omega, \tag{2.32}$$

and the wave vector $\mathbf{k}$ defines a quantity known as the crystal momentum of a phonon $\mathbf{p}$:

$$\mathbf{p} = \hbar\mathbf{k}. \tag{2.33}$$

Crystal momentum has much in common with ordinary momentum but there are several important differences between these two concepts. The velocity of a phonon is equal to the group velocity of the corresponding classical waves $\mathbf{v} = \partial\omega/\partial\mathbf{k}$. Using Eqs. (2.32) and (2.33), we shall rewrite the velocity formula in the form

$$\mathbf{v} = \partial\varepsilon/\partial\mathbf{p}. \tag{2.34}$$

This expression for the phonon velocity is identical with the usual relationship between the energy, momentum, and velocity of a free particle.

The relationship between the frequency and the wave vector $\omega = \omega(\mathbf{k})$ applies also to the dependence of the phonon energy on the phonon crystal momentum. In particular, the function $\varepsilon = \varepsilon(\mathbf{p})$ has, in general, 3ν different branches, i.e., phonons of 3ν different types can exist in a crystal.

The phonon concept makes it possible to reduce the lattice vibration problem to the problem of a phonon gas filling the whole volume of the lattice. The behavior of such a gas can be described quite well by the same mathematical expressions which are used in the theory of black-body radiation.

However, we must remember that the phonon concept is used in the harmonic approximation, in which the potential energy of the lattice is assumed to be a quadratic function of the displacements of atoms from their equilibrium positions. In this approximation, the phonons move freely and do not interact with each other.

The inclusion of anharmonic terms in the expansion of the potential energy gives rise to various types of phonon collision. These collisions establish the thermal equilibrium motion of the lattice.

A full treatment of phonons and phonon spectra is given in Ziman's book [7].

2.7. Approximate Methods for the Determination of the Grüneisen Parameter

We have shown earlier that, in the high- and low-temperature limits, the quasiharmonic approximation yields the Mie — Grüneisen equation of state. The interpolation theory of Debye extends this equation to intermediate temperatures.

The right-hand part of the Mie — Grüneisen equation consists of two terms:

$$P = P_{\mathrm{p}}(V) + \frac{\gamma}{V} E_{\mathrm{k}}.$$

The first of these terms represents the potential pressure P_{p} and depends only on the volume. The second term is due to the lattice vibrations and is proportional to the energy E_{k} of these vibrations.

At low temperatures, the contribution of the second term to the total pressure is small and the equation of state is governed primarily by the potential term $P_{\mathrm{p}}(V)$. The importance of the thermal motion (lattice vibrations) increases with increasing temperature. Eventually, it becomes comparable with or greater than the potential pressure. Under these conditions, the volume dependence of the thermal pressure becomes important.

At high temperatures, the thermal energy of a crystal can be easily calculated, but the volume dependence of the Grüneisen parameter γ has hardly been investigated. We shall now consider this dependence.

It is shown in Chapter 1 that the Mie — Grüneisen equation of state yields directly Eq. (1.35):

$$\gamma_0 = \frac{V \varkappa K_T}{c_V},$$

which defines the Grüneisen parameter in terms of measurable thermodynamic properties. We shall call this quantity the thermodynamic value of the Grüneisen parameter.

The thermal expansion coefficient $\varkappa$ and the specific heat c_V have not yet been determined experimentally at high pressures. Consequently, the above expression cannot be used to determine the volume dependence of γ. We can only use this equation to determine the thermodynamic Grüneisen parameter γ_0 at normal (atmospheric) pressure.

In principle, if we know the interatomic interaction potential, we can calculate the frequency spectrum of a crystal and its volume dependence, and then we can find the dependence $\gamma(V)$. However, this approach is so mathematically complex that it has not yet been used. In view of this, we must employ various model representations which give approximate dependences $\gamma(V)$.

Slater Formula

Slater [8] and Landau and Stanyukovich [9] used the elastic medium model and, assuming that Poisson's ratio σ was independent of volume, they obtained the following expression for the Grüneisen parameter:

$$\gamma = -\frac{V}{2}\frac{\partial^2 P_p/\partial V^2}{\partial P_p/\partial V} - \frac{2}{3}. \tag{2.35}$$

This result can be obtained easily from the Debye theory described in Sec. 2.5. We shall use well-known expressions for the velocities of longitudinal waves

$$c_l = \sqrt{\frac{3K_S}{\rho}\frac{1-\sigma_S}{1+\sigma_S}} \tag{2.36}$$

and of transverse waves

$$c_t = \sqrt{\frac{3K_S}{2\rho}\frac{1-2\sigma_S}{1+\sigma_S}}. \tag{2.37}$$

The adiabatic bulk modulus K_S and the adiabatic Poisson's ratio σ_S are related with the corresponding isothermal quantities by the simple expressions [10]:

$$\frac{1}{K_S} = \frac{1}{K_T} - \frac{VT\varkappa^2}{c_P} \qquad \sigma_S = \sigma + (1+\sigma)(1-2\sigma)\frac{K_T VT\varkappa^2}{3c_P}. \tag{2.38}$$

In order to clarify the dependence $\theta_D(V)$, we must take c_l and c_t at T = 0. Then, the adiabatic moduli are equal to the isothermal values and θ_D is found from Eq. (2.27):

$$\theta_D = aV^{1/6}K^{1/2}\zeta(\sigma),$$

where

$$a = 3^{5/6}\frac{\hbar}{k}(6\pi^2 N\nu)^{1/3}, \tag{2.39}$$

$$\zeta(\sigma) = \left(\frac{1-\sigma}{1+\sigma}\right)^{1/2}\left[1 + 2^{5/2}\left(\frac{1-\sigma}{1-2\sigma}\right)^{3/2}\right]^{-1/6}.$$

Since, in the case considered, $K_T = -V(\partial P_p/\partial V)_T$, we find that substitution of Eq. (2.39) in Eq. (2.26) yields

$$\gamma = -\frac{V}{2}\frac{\partial^2 P_p/\partial V^2}{\partial P_p/\partial V} - \frac{2}{3} + \delta, \tag{2.40}$$

where

$$\delta = -\frac{\partial \ln \zeta}{\partial \ln V} = \frac{V}{1-\sigma^2}\left[1 + \frac{2^{1/2}(1-\sigma^2)}{(1-2\sigma)^2}\frac{\left(\frac{1-\sigma}{1-2\sigma}\right)^{1/2}}{1+2^{5/2}\left(\frac{1-\sigma}{1-2\sigma}\right)^{3/2}}\right]\frac{\partial\sigma}{\partial V}. \tag{2.41}$$

If we assume that σ is constant [8, 9], we obtain the Slater formula (2.35) (the influence of the volume dependence of Poisson's ratio on the Grüneisen parameter is considered in [11]).

Slater [8] and Gilvarry [12] used the first and second derivatives at zero pressure (obtained from Bridgman's data on the compressibility) to calculate γ by means of Eq. (2.35) and they compared the results with the thermodynamic values of the Grüneisen parameter. They obtained good agreement for the majority of metals.

Dugdale — MacDonald Formula

Dugdale and MacDonald [13] suggested that Eq. (2.35) should be modified, and they obtained the following expression for the Grüneisen parameter:

$$\gamma = -\frac{V}{2}\frac{\partial^2(P_p V^{2/3})/\partial V^2}{\partial(P_p V^{2/3})/\partial V} - \frac{1}{3}. \tag{2.42}$$

However, their analysis is based on some erroneous assumptions. Later, Rice, McQueen, and Walsh [14] gave a different derivation of Eq. (2.42). They obtained it for a cubic lattice, assuming that all the force constants have the same volume dependences. The values of γ obtained from Eq. (2.42) at zero pressure are in good agreement with the thermodynamic values γ_0.

Grüneisen Parameter in the Free-Volume Approximation

Vashchenko and Zubarev [15] considered the vibrations of atoms in the spherically symmetrical field of their neighbors (free-volume theory). They obtained the following expression for the Grüneisen parameter:

$$\gamma = -\frac{V}{2}\frac{\partial^2(P_p V^{4/3})/\partial V^2}{\partial(P_p V^{4/3})/\partial V}. \tag{2.43}$$

Generalized Formula for γ

All three formulas (2.35), (2.42), and (2.43) can be combined in one expression:

$$\gamma = -\frac{V}{2}\frac{\partial^2(P_p V^{2m/3})/\partial V^2}{\partial(P_p V^{2m/3})/\partial V} + \frac{1}{3}(m-2), \tag{2.44}$$

which reduces to Eq. (2.35) when m = 0, to Eq. (2.42) when m = 1, and to Eq. (2.43) when m = 2. At zero pressure, these formulas give different values of the Grüneisen parameter, which are related by the following expression:

$$\gamma_{m=0} = \gamma_{m=1} + \frac{1}{3} = \gamma_{m=2} + \frac{2}{3}.$$

The values of γ yielded by Eq. (2.44) under normal (atmospheric) conditions do not always agree well with the thermodynamic Grüneisen parameter of Eq. (1.35). Since γ given by Eq. (2.44) is obtained from model considerations for m = 0, 1, or 2, and the value of γ_0 is an intrinsic property of a substance, it is logical to equate γ of Eq. (2.44) with γ_0. We shall show in Chapter 5 that this procedure is permissible when an additional normalizing constant δ is introduced into Eq. (2.44). Then, the Grüneisen parameter becomes:

$$\gamma = -\frac{V}{2}\frac{\partial^2(P_p x^{2m/3})/\partial V^2}{\partial(P_p x^{2m/3})/\partial V} + \frac{1}{3}(m-2) + \delta. \tag{2.45}$$

We shall use this expression later in the book.

Grüneisen Parameter Deduced from the Mean-Square Frequency $\overline{\omega^2}$

Zharkov [16] calculated the Grüneisen parameter γ from the mean-square frequency $\overline{\omega^2}$ of the crystal lattice vibrations:

$$\gamma = -\frac{1}{2}\frac{\partial \ln \overline{\omega^2}}{\partial \ln V}. \tag{2.46}$$

This calculation was carried out for ionic and molecular crystals. It was assumed that the interaction between atoms or ions was of the pair and central type. The following formula was obtained for ionic crystals:

$$\gamma = \frac{xy\left(1+\dfrac{2}{xy}-\dfrac{2}{x^2y^2}\right)A_{+-}e^{-xy}+\dfrac{\bar{z}}{z}x_1y\left(1+\dfrac{2}{x_1y}-\dfrac{2}{x_1^2y^2}\right)A_M e^{-x_1 y}}{6\left\{\left(1-\dfrac{2}{xy}\right)A_{+-}e^{-xy}+\dfrac{\bar{z}}{z}\left(1-\dfrac{2}{x_1y}\right)A_M e^{-x_1 y}\right\}}. \tag{2.47}$$

This approach is described in more detail in Chapter 6 [the symbols used in Eq. (2.47) are explained in that chapter]. Here, we shall draw the reader's attention to the following point.

The Grüneisen parameter of Eq. (2.35), obtained from the Debye theory (curve 2 in Fig. 6.24), is governed primarily by the low-frequency part of the lattice vibration spectrum. On the other hand, γ calculated from $\overline{\omega^2}$, using Eq. (2.46), depends considerably on the high-frequency part of the spectrum (curves 4-6 in Fig. 6.24). The true Grüneisen parameter, which reflects correctly the whole lattice vibration spectrum, should lie between curve 2 and curves 4-6 in Fig. 6.24. It is evident from Fig. 6.24 that curve 3, representing the parameter γ calculated using Eq. (2.42), lies in this region. It does not follow that Eq. (2.42) gives the true volume dependence of the Grüneisen parameter γ but, because of its intermediate position, it is probably the best approximation of the dependence $\gamma(x)$.

Literature Cited

1. L. D. Landau and E. M. Lifshitz, Statistical Physics, 2nd ed., Pergamon Press, Oxford (1968).
2. R. E. Peierls, Quantum Theory of Solids, Clarendon Press, Oxford (1955).
3. M. Born and K. Huang, Dynamical Theory of Crystal Lattices, Clarendon Press, Oxford (1954).
4. P. Debye, Ann. Physik, 39:789 (1912).
5. C. Kittel, Introduction to Solid State Physics, 3rd ed., Wiley, New York (1966).
6. A. Einstein, Ann. Physik, 22:180 (1907).
7. J. M. Ziman, Electrons and Phonons, Clarendon Press, Oxford (1960).
8. J. C. Slater, Introduction to Chemical Physics, McGraw-Hill, New York (1939).
9. L. D. Landau and K. P. Stanyukovich, Dokl. Akad. Nauk SSSR, 46:399 (1945).
10. L. D. Landau and E. M. Lifshitz, Theory of Elasticity, 2nd ed., Pergamon Press, Oxford (1969).
11. V. N. Zharkov, Izv. Akad. Nauk SSSR, Ser. Geofiz., No. 10, p. 1417 (1960).
12. J. J. Gilvarry, Phys. Rev., 102:331 (1956).
13. J. S. Dugdale and D. K. C. MacDonald, Phys. Rev., 89:832 (1953).
14. M. H. Rice, R. G. McQueen, and J. M. Walsh, Solid State Phys., 6:1 (1958).
15. V. Ya. Vashchenko and V. N. Zubarev, Fiz. Tverd. Tela, 5:886 (1963).
16. V. N. Zharkov, Dokl. Akad. Nauk SSSR, 154:302 (1964).

CHAPTER 3

HIGH-TEMPERATURE CORRECTIONS
TO EQUATION OF STATE

3.1. Introduction

The two-term equation of state, considered in the preceding chapter, is found to become unsatisfactory as temperature is increased. The equation must be supplemented by additional terms, which will be considered in the present chapter.

We shall write the free energy of a crystal in the form

$$F(x, T) = E_p(x) + \theta f(\theta/T) + f_1(x, T), \tag{3.1}$$

where θ is the characteristic temperature; $E_p(x)$ is the potential component of the free energy, which depends only on the volume. The second term is due to phonons and it has been considered in detail in the preceding chapter. The third term is usually small compared with the second over a wide range of temperatures and pressures. This term represents the following phenomena: 1) anharmonicity, i.e., inclusion of third- and fourth-order terms in the expansion of the potential energy in terms of the displacements of atoms from their equilibrium positions; 2) the thermal excitation of the conduction electrons in metals; 3) the formation of point defects (Frenkel, Schottky, and other defects); 4) the thermal ionization and formation of excitons in dielectrics.

The free energy $F(x, T)$ is represented in the form of Eq. (3.1) because over a wide range of pressures and temperatures we have

$$E_p(x) \gg \begin{cases} \theta f(\theta/T), \\ f_1(x, T), \end{cases} \tag{3.2}$$

i.e., $E_p(x)$ is the main term in the expression for the free energy. The temperature-dependent term in the free energy can be divided into two components because there is a range of temperatures in which the third term in Eq. (3.1) is small compared with the second. Using $\overline{T}$ to denote the temperature above which the inclusion of $f_1(x, T)$ is necessary, we can write

$$f_1(x, T) \ll \theta f(\theta/T) \quad \text{at} \quad T < \overline{T}. \tag{3.3}$$

A study of $f_1(x, T)$ is not only of interest in itself but is also necessary in the determination of the range of volumes and temperatures in the (x, T) plane in which the two-term expression for the free energy can be used.

The present chapter is concerned not only with the high-temperature corrections but also with the melting curves.

3.2. Anharmonicity [1-5]

The anharmonicity of the vibrations of atoms and ions in solids begins to affect the thermodynamic properties only at fairly high temperatures. In the majority of solids, this range of temperatures lies above the Debye temperature and represents the classical limit. The temperature dependence of the free energy can be found using a simple model. This model is closely related to the Einstein model of a solid discussed in Chapter 2.

Let us consider a particle in a one-dimensional potential well. The energy of this particle can be written in the form

$$H = \frac{1}{2m} p^2 + V(z),$$

where m and p are, respectively, the mass and momentum of the particle; z is the coordinate; $V(z)$ is the potential of the field in which motion takes place. The first derivative of the potential $V(z)$ vanishes at the equilibrium position $z = z_0$, which corresponds to the energy minimum of the system considered. The generality of the treatment is not affected by assuming that $z_0 = 0$. We shall measure the energy from the constant value $V(0)$, and we shall expand the potential in terms of the coordinate up to the fourth power of z. Then, the energy of an anharmonic oscillator becomes

$$H = H_0 + H_1,$$
$$H_0 = \frac{1}{2m} p^2 + \frac{1}{2} az^2, \qquad H_1 = bz^3 + cz^4,$$

where the following notation is used:

$$a = V^{II} = m\omega^2, \quad b = \frac{1}{6} V^{III}, \quad c = \frac{1}{24} V^{IV},$$

H_0 is the energy of a harmonic oscillator, and H_1 is the perturbation of the energy due to the anharmonicity of the vibrations. The analogy with the Einstein model of a solid can be seen from the following considerations. In an Einstein solid, all 3N frequencies of the vibrations are assumed to be equal (N is the number of particles in a solid). Therefore, the thermodynamic functions can be obtained by making a calculation at one frequency and multiplying the result by the number of degrees of freedom 3N. Our calculation of the anharmonicity yields its value for one degree of freedom. Multiplying the results obtained by the number of degrees of freedom (3N), we obtain the anharmonicity of an Einstein solid.

The sum-over-states (partition function)

$$Z = (2\pi mkT)^{-1/2} \int\int dp\, dz \exp[-\beta(H_0 + H_1)], \quad \beta = 1/kT$$

can be expanded in terms of H_1

$$Z = (2\pi mkT)^{1/2} \int_{-\infty}^{\infty} dz \exp\left(-\frac{1}{2}\beta az^2\right)\left[1 - \beta H_1 + \frac{1}{2}(\beta H_1)^2 - \ldots\right]$$

and calculated easily using the value of the integral [6]

$$I_n = \int_0^{\infty} y^{2n} \exp(-qy^2)\, dy = \frac{(2n-1)!!}{2(2q)^n}\sqrt{\frac{\pi}{q}} \quad (q > 0).$$

This procedure yields

$$Z = (2\pi mkT)^{1/2} \sqrt{\frac{2\pi kT}{a}} \left[1 + kT\left(\frac{3c}{a^2} - \frac{15}{2}\frac{b^2}{a^3} \right) \right].$$

The free energy is given by

$$F = -\frac{1}{\beta}\ln Z = -kT\left[\ln\frac{2\pi kT}{\sqrt{a}}\sqrt{m} + kT\left(\frac{3c}{a^2} - \frac{15}{2}\frac{b^2}{a^3} \right) \right].$$

Consequently, the contribution of the anharmonicity to the free energy of a solid (per each degree of freedom) is of the form

$$F_a = -\frac{3c}{a^2}(kT)^2 + \frac{15}{2}\frac{b^2}{a^3}(kT)^2. \tag{3.4}$$

We can see that when the anharmonic contribution to the free energy is calculated to within T^2, the third-order terms in the expansion of the potential energy in terms of the displacements from the equilibrium position contribute to the second approximation in the perturbation theory, and the fourth-order terms in this expansion contribute to the first approximation.

In the method of potentials, the anharmonic contribution to the free energy of a crystal can be written in the form

$$F_a = -\frac{A_a(x)}{\rho_0}T^2, \tag{3.5}$$

where $A_a(x)$ is some function of volume, which has to be determined. The sign of the function $A_a(x)$ can be positive or negative, and cannot be determined theoretically.

The nature of the function $A_a(x)$ for molecular and ionic crystals, in which the interaction between particles in the crystal lattice is known to be practically of the pair type, is considered in [4, 5]. It is found that, as in Eq. (3.4), the terms in the expression for the free energy F_a are proportional to T^2 and are of the form

$$F_a = F_3 + F_4, \qquad F_3 = -\frac{9}{4}s^3 N_C(kT)^2 Q_3, \ F_4 = \frac{9}{8}s^2 N_C(kT)^2 Q_4, \tag{3.6}$$

where N_C is the number of unit cells in a crystal; s is the number of atoms in each unit cell; Q_3 is a quantity representing third-order terms in the expansion of the total potential energy of a crystal Φ; Q_4 represents the fourth-order terms in this expansion. The main results of these calculations can be summarized by the formula

$$Q_3,\ Q_4 \propto \exp(\varkappa x^{1/4}), \tag{3.7}$$

where $\varkappa$ is some average value of the argument of the repulsive force potential.

Equation (3.7) can be obtained qualitatively as follows: [see Eq. (3.4)]

$$Q_3 \propto \frac{|\Phi^{III}|^2}{(\overline{\omega^2})^3}, \qquad Q_4 \propto \frac{\Phi^{IV}}{(\overline{\omega^2})^2}, \tag{3.8}$$

where Φ^{III}, Φ^{IV} are the third and fourth derivatives of the potential energy of a crystal; $\overline{\omega^2}$ is the mean-square frequency. The frequency $\overline{\omega^2}$ is proportional to the second derivative of the potential, Φ^{III} is proportional to the third derivative, and Φ^{IV} is proportional to the fourth derivative. The exponential term becomes prominent in the second derivative and its importance increases in higher derivatives. Therefore, Eq. (3.7) follows immediately from Eq. (3.8). This means that the anharmonicity of crystals with an exponential repulsive potential decreases rapidly with decreasing volume.

We shall now determine the "anharmonic analog" of the Grüneisen parameter using the standard formula

$$\gamma_a = \frac{x P_a}{\rho_0 E_a}. \tag{3.9}$$

The "anharmonic" component of pressure P_a and energy E_a can be easily calculated using F_a given by Eq. (3.5):

$$P_a = -\left(\frac{\partial F_a}{\partial V}\right)_T = T^2 \left(\frac{\partial A_a}{\partial x}\right)_T = \gamma_a x^{-1} A_a T^2, \tag{3.10}$$

$$E_a = -T^2 \left(\frac{\partial}{\partial T} \frac{F_a}{T}\right)_V = \frac{A_a}{\rho_0} T^2. \tag{3.11}$$

Substituting Eqs. (3.10) and (3.11) into Eq. (3.9), we obtain an expression for the anharmonic analog of the Grüneisen parameter in terms of the function of volume $A_a(x)$:

$$\gamma_a = \frac{\partial \ln A_a}{\partial \ln x} = \left(\frac{\partial \ln F_a}{\partial \ln x}\right)_T. \tag{3.12}$$

Equations (3.5)-(3.8) yield the following estimate of γ_a for crystals with exponential repulsive forces:

$$\gamma_a \sim \varkappa/3. \tag{3.13}$$

Since $\varkappa \sim 6$-12, it follows that $\gamma_a \sim 2$-4, i.e., the anharmonic Grüneisen parameter is large compared with the parameter for a perfect lattice.

3.3. Thermally Excited Conduction Electrons
in Metals

The quantum theory of metals was developed soon after the discovery of quantum mechanics. This theory made it possible to obtain correct temperature dependences of the thermodynamic quantities and of some transport parameters of metals [7, 8]. The success was achieved by the application of the Fermi — Dirac statistics to the conduction electrons in metals. It was initially assumed that these conduction electrons can be regarded as an ideal degenerate Fermi gas. The thermodynamics of an ideal degenerate Fermi gas is given in many handbooks and it is summarized for convenience in Appendix C. Until quite recently, it was not clear why the behavior of strongly interacting electrons in metals can be described very satisfactorily by formulas applicable to a gas. The situation became clearer after development of a theory of the Fermi liquid by Landau [2]. This theory justifies the formulas of the gas approximation, but it modifies somewhat the principal concepts and alters the terminology.

Electrons in a metal do not form a Fermi gas but a Fermi liquid of quasiparticles which are acted upon by the self-consistent field of other electrons and by the crystal lattice field. The number of such particles is exactly equal to the number of electrons in the conduction band and the charge of each particle is equal to the electronic charge. Therefore, for the sake of convenience, these quasiparticles can still be called electrons but the term now has a somewhat different meaning. In the early stages of the development of the quantum electron theory of metals, it was assumed that, at absolute zero, electrons are condensed in a Fermi sphere on whose surface the energy is equal to the chemical potential (see Appendix C). Later theoretical and experimental investigations of the electron theory of metals [9] established that the condensation of electrons at T = 0 into a Fermi sphere is an exceptionally rare event. The Fermi surfaces in which electrons are condensed at T = 0 are considerably more complex than a sphere. The various types of Fermi surface encountered in metals are reviewed in [9].

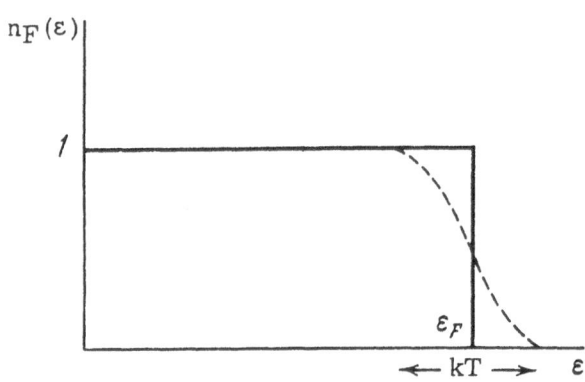

Fig. 3.1. Fermi — Dirac distribution function for a metal.

However, the thermodynamics of the conduction electrons is independent of the complex nature of Fermi surfaces in metals and it is not affected by the fact that electrons in metals do not form a Fermi gas but a Fermi liquid. This is because the Fermi — Dirac distribution [Eq. (C.41)] describes particles in a Fermi gas as well as in a Fermi liquid. At sufficiently low temperatures, the Fermi — Dirac distribution is a step function, and this has analytic consequences in the rule for the calculation of Fermi integrals, given by Eq. (C.47). Consequently, the energy of thermally excited conduction electrons is proportional to the square of the temperature, and this governs the temperature dependence of all other thermodynamic quantities. However, we must point out that this is valid only at sufficiently low temperatures, much lower than the degeneracy temperature T_0, which is $\sim(1\text{-}5) \cdot 10^4$ °K for the majority of metals.

Let us consider the situation using the diagram shown in Fig. 3.1.

According to the Pauli principle, all states right up to the Fermi energy ε_F are occupied at T = 0. The distribution of the states is in the form of an abrupt step and it is independent of any model assumption. At T ≠ 0 but sufficiently low to satisfy T ≪ ε_F/k, the edge of the step-like distribution broadens by an amount kT. Consequently, the number of uncondensed particles in the high-energy tail of the distribution is of the order of kT. The energy of these excited particles, measured from the Fermi energy ε_F, is also of the order kT. Thus, the order of magnitude of the total energy of thermally excited particles in the high-energy tail is $\sim (kT)^2$.

After these preliminary explanations, we shall now derive the relevant formulas.

The only quantity which we must know in the thermodynamics of the conduction electrons is the density of electron states per unit energy interval, $\nu(\varepsilon)$. According to a general quantum-mechanical formula, the number of electron states in an S-th band in an element of volume of the momentum space is

$$dn_{\mathbf{p}}^{(S)} = \frac{2V d\mathbf{p}}{(2\pi\hbar)^3}, \qquad d\mathbf{p} = dp_x dp_y dp_z. \tag{3.14}$$

The state of an electron is governed by its position on the constant-energy surface $\varepsilon_S(\mathbf{p}) = \varepsilon$. An element of volume $d\mathbf{p}$ is equal to $dSd\varepsilon/v_S$, where dS is an element of area on the constant-energy surface $\varepsilon_S(\mathbf{p}) = \varepsilon$, and $v_S = |\nabla_{\mathbf{p}}\varepsilon_S(\mathbf{p})|$ is the modulus of the electron velocity.

The dependence of the energy of an electron on its momentum $\varepsilon = \varepsilon_S(\mathbf{p})$ for each of the S overlapping electron bands in a metal is, in general, specific to each metal and fairly complex [9]. This dependence is known as the dispersion law. Integrating Eq. (3.14) over the constant-energy surface, we find the number of electron states $dn_\varepsilon^{(S)}$ per energy interval $d\varepsilon$:

$$dn_{\varepsilon}^{(S)} = \nu_S(\varepsilon) d\varepsilon, \tag{3.15}$$

where

$$\nu_S(\varepsilon) = \frac{2V}{(2\pi\hbar)^3} \oint \frac{dS}{v_S} \tag{3.16}$$

is the density of electron states in the S-th band per unit energy interval. To clarify the meaning of the function ν, we shall give its value for a gas of free electrons [Eq. (C.43)]:

$$\nu(\varepsilon) = \frac{\sqrt{2}}{\pi^2} \frac{V}{\hbar^3} m^{3/2} \varepsilon^{1/2}, \qquad \varepsilon(p) = \frac{p^2}{2m}. \tag{3.17}$$

If the constant-energy surfaces form a triaxial ellipsoid with its center at the point $\varepsilon = 0$, we find that

$$\varepsilon = \frac{p_1^2}{2m_1} + \frac{p_2^2}{2m_2} + \frac{p_3^2}{2m_3}, \tag{3.18}$$

where m_1, m_2, m_3 are the effective masses; p_1, p_2, p_3 are the projections of the momentum along the principal axes of the ellipsoid. Then, the density of states is given by the simple formula

$$\nu(\varepsilon) = \frac{\sqrt{2}}{\pi^2} \frac{V}{\hbar^3} \sqrt{m_1 m_2 m_3} \, \varepsilon^{1/2}. \tag{3.19}$$

A more detailed classification of the densities of states $\nu(\varepsilon)$ for different dispersion laws is given in [9]. Further calculations of the thermodynamic quantities, to within T^2, are similar to those given in Appendix C for an ideal Fermi gas.

The distribution function of quasiparticles in a metal, which we shall call electrons, is the Fermi—Dirac distribution function (often called the Fermi function):

$$n_F(\varepsilon) = \frac{1}{\exp[(\varepsilon - \mu)/kT] + 1}, \tag{3.20}$$

where T is the temperature and μ is the chemical potential, defined by the normalization condition

$$N = \sum_s \int \frac{\nu_s(\varepsilon)\, d\varepsilon}{\exp[(\varepsilon - \mu)/kT] + 1}, \tag{3.21}$$

where N is the total number of the conduction electrons. Integration can be extended over all values of the energy since $\nu_S(\varepsilon) \neq 0$ only within the S-th overlapping energy band. Using the definition

$$\nu(\varepsilon) = \sum_s \nu_s(\varepsilon), \tag{3.22}$$

we can omit the summation symbol in Eq. (3.21), so that

$$N = \int \nu(\varepsilon) n_F(\varepsilon)\, d\varepsilon. \tag{3.21a}$$

We shall calculate first the chemical potential μ to within T^2. At T = 0, the Fermi—Dirac distribution degenerates into the step shown in Fig. 3.1, and μ becomes equal to the Fermi energy ε_F (see Appendix C):

$$N = \int_0^{\varepsilon_F} \nu(\varepsilon)\, d\varepsilon \quad \text{at} \quad T = 0. \tag{3.23}$$

At $T \neq 0$, the chemical potential μ differs from ε_F by a small quantity of the order of T^2:

$$\mu = \varepsilon_F + \Delta\mu, \qquad \Delta\mu \sim T^2.$$

The value of $\Delta\mu$ can be easily determined using the general formula for the calculation of the Fermi integrals [Eq. (C.47)] and applying it to Eq. (3.21a):

$$N = \int_0^{\varepsilon_F} \nu(\varepsilon)\,d\varepsilon + \int_{\varepsilon_F}^{\varepsilon_F+\Delta\mu} \nu(\varepsilon)\,d\varepsilon + \frac{\pi^2}{6}(kT)^2 \left(\frac{d\nu(\varepsilon)}{d\varepsilon}\right)_{\varepsilon_F}.$$

According to Eq. (3.23), we can divide both sides of the above equation by N. Replacing the variables in the second integral by $\varepsilon' = \varepsilon - \varepsilon_F$, we obtain

$$\int_0^{\Delta\mu} \nu(\varepsilon'+\varepsilon_F)\,d\varepsilon' = -\frac{\pi^2}{6}(kT)^2 \left(\frac{d\nu(\varepsilon)}{d\varepsilon}\right)_{\varepsilon_F}.$$

Carrying out calculations to within T^2 and using the expression

$$\nu(\varepsilon_F + \varepsilon') = \nu(\varepsilon_F) + O(T^2),$$

we find that $\nu(\varepsilon_F + \varepsilon')$ in the integrand can be replaced with $\nu(\varepsilon_F)$ and taken outside the integral. Then,

$$\left. \begin{aligned} \Delta\mu &= -\frac{\pi^2}{6}(kT)^2 \frac{1}{\nu(\varepsilon_F)} \left(\frac{d\nu(\varepsilon)}{d\varepsilon}\right)_{\varepsilon_F} \\[2mm] \mu &= \varepsilon_F - \frac{\pi^2}{6}(kT)^2 \frac{1}{\nu(\varepsilon_F)} \left(\frac{d\nu(\varepsilon)}{d\varepsilon}\right)_{\varepsilon_F} \end{aligned} \right\} \tag{3.24}$$

We shall now calculate the energy

$$E = \int \varepsilon\nu(\varepsilon)\,n_F(\varepsilon)\,d\varepsilon. \tag{3.25}$$

We shall apply Eq. (C.47) to Eq. (3.25):

$$E = \int_0^{\mu} \varepsilon\nu(\varepsilon)\,d\varepsilon + \frac{\pi^2}{6}(kT)^2 \left(\frac{d[\varepsilon\nu(\varepsilon)]}{d\varepsilon}\right)_{\varepsilon_F}.$$

Dividing the integral, as before, into regions from 0 to ε_F and from ε_F to $\varepsilon_F + \Delta\mu$, we obtain (to within T^2)

$$E = \int_0^{\varepsilon_F} \varepsilon\nu(\varepsilon)\,d\varepsilon + \varepsilon_F\nu(\varepsilon_F)\Delta\mu + \frac{\pi^2}{6}(kT)^2 \left(\frac{d[\varepsilon\nu(\varepsilon)]}{d\varepsilon}\right)_{\varepsilon_F}.$$

Replacing $\Delta\mu$ with its value given by Eq. (3.24), we obtain the final expression

$$\left. \begin{aligned} E &= E_0 + E_{Te} = E_0 + \frac{\pi^2}{6}\nu(\varepsilon_F)(kT)^2, \\[2mm] E_0 &= \int_0^{\varepsilon_F} \varepsilon\nu(\varepsilon)\,d\varepsilon. \end{aligned} \right\} \tag{3.26}$$

The specific heat at constant volume c_V is

$$c_{ve} = \left(\frac{\partial E}{\partial T}\right)_V = \frac{\pi^2 k^2}{3}\nu(\varepsilon_F)\,T. \tag{3.27}$$

According to the Nernst theorem, entropy vanishes at $T = 0$ and, therefore, using Eq. (A.3), we find that entropy is given by

$$S_e = \int_0^T \frac{c_{ve}}{T} dT = c_{ve} = \frac{\pi^2 k^2}{3} \nu(\varepsilon_F) T. \tag{3.28}$$

Using Eqs. (A.8), (3.26), and (3.28), we can determine the free energy

$$F = E_0 - \frac{\pi^2}{6} \nu(\varepsilon_F)(kT)^2. \tag{3.29}$$

The pressure due to the conduction electrons consists of the zero-point pressure P_0 and the thermal pressure of the excited conduction electrons

$$P = -\left(\frac{\partial F}{\partial V}\right)_T = P_0 + P_{Te} = -\frac{\partial E_0}{\partial V} + \frac{\pi^2}{6}\left(\frac{\partial \nu(\varepsilon_F)}{\partial V}\right)_T (kT)^2. \tag{3.30}$$

We shall define the "electron analog" of the Grüneisen parameter using the formula

$$\gamma_e = \frac{x}{\rho_0} \frac{P_{Te}}{E_{Te}}. \tag{3.31}$$

Substituting E_{Te} from Eq. (3.26) and P_{Te} from Eq. (3.30), we obtain

$$\gamma_e = \left(\frac{\partial \ln \nu(\varepsilon_F)}{\partial \ln x}\right)_T. \tag{3.32}$$

We shall now calculate γ_e for the simplest dispersion laws given by Eqs. (3.17) and (3.19). First, we shall determine the volume dependences of the corresponding Fermi energies ε_F using Eq. (3.23). The dispersion law (3.17) gives the expression

$$\varepsilon_F = (3\pi^2)^{2/3} \frac{\hbar^2}{2m} \left(\frac{N}{V}\right)^{2/3} \tag{3.33}$$

and the law (3.19) yields correspondingly

$$\varepsilon_F = (3\pi^2)^{2/3} \frac{\hbar^2}{2\sqrt[3]{m_1 m_2 m_3}} \left(\frac{N}{V}\right)^{2/3}. \tag{3.34}$$

In both cases, the value of γ_e is the same:

$$\gamma_e = 2/3. \tag{3.35}$$

This result is a consequence of the fact that, in the dispersion laws (3.17) and (3.19), $\nu(\varepsilon) \propto \varepsilon^{\frac{1}{2}}$. The electron Grüneisen parameter given above is equal to the value γ_e in Eq. (C.59) for an ideal electron Fermi gas. The value of γ_e for other dispersion laws may not be equal to $2/3$ and therefore there may be deviations from the general relationship $PV = 2E/3$ of Eq. (C.56), which applies to an ideal electron Fermi gas.

We shall now calculate the derivative $(\partial S_e / \partial V)_T$, which — according to Eqs. (3.28) and (3.32) — is given by

$$\left(\frac{\partial S_e}{\partial V}\right)_T = \rho_0 \left(\frac{\partial S_e}{\partial x}\right)_T = \frac{c_{ve}\rho_0\gamma_e}{x}. \tag{3.36}$$

Next, using Eq. (A.18), we obtain

$$\left(\frac{\partial \check{S}}{\partial V}\right)_T = \left(\frac{\partial P}{\partial T}\right)_x = \frac{\varkappa}{\beta_T}. \tag{3.37}$$

Dividing entropy into the lattice and electron components, $S = S_l + S_e$, repeating this procedure for the thermal expansion coefficient $\varkappa = \varkappa_l + \varkappa_e$, and using Eqs. (3.36) and (3.37), we obtain the following expression for the electron Grüneisen coefficient:

$$\gamma_e = \frac{\varkappa_e x}{\rho_0 c_{ve} \beta_T}, \tag{3.38}$$

where β_T is the total isothermal compressibility of a crystal. It is known (see Chapter 2) that at low temperatures $T \ll \theta$ the lattice components of the specific heat and of the thermal expansion coefficient are proportional to the cube of the temperature. According to Eqs. (3.27), (3.32), and (3.38), the electron components of the same quantities are proportional to the first power of T. Therefore, at sufficiently low temperatures, the electron components of the specific heat and of the thermal expansion coefficient become dominant. This makes it possible to determine experimentally the low-temperature values of the density of states $\nu(\varepsilon)_F$ and the electron analog of the Grüneisen parameter. The results of such determinations are given in Table 5.1, which shows that the value of γ_e of many metals differs considerably from $2/3$.

The influence of the thermally excited conduction electrons on the thermodynamic parameters of metals is manifested not only at low temperatures. It is appreciable also at fairly high temperatures. This is because the high-temperature values of the thermal phonon energy and of the thermal phonon pressure increase proportionally to T but the corresponding electron quantities are proportional to T^2. Therefore, allowance for the components due to thermally excited electrons becomes important at temperatures of the order of several thousand degrees. This is particularly important in the determination of the equation of state from dynamic experimental data obtained using strong shock waves. This point will be discussed more fully in Chapter 5. We shall only mention here that, in principle, the electron parameters can be determined also from the high-temperature data. The first such determination is reported in [12].

In the phenomenological approach, used in the present book, the basic equation (3.29) is written in the form

$$F_e = -\frac{A_e(x)}{\rho_0} T^2, \tag{3.39}$$

where

$$A_e = \frac{\pi^2 k^2}{6} \rho_0 \nu(\varepsilon_F). \tag{3.40}$$

The function $A_e(x)$ in Eq. (3.39) is assumed to be some unknown function of volume, which has to be found. The electron Grüneisen parameter γ_e [Eqs. (3.31) and (3.32)], expressed in terms of $A_e(x)$, is of the form

$$\gamma_e = \frac{d \ln A_e(x)}{d \ln x}. \tag{3.41}$$

We shall approximate A_e using the following two trial functions [12, 13]:

$$A_{e1}(x) = A_{e01} x^{n_1 + 2/3} \tag{3.42}$$

or

$$A_{e2}(x) = A_{e02} x^{2/3} \exp\left[B_1(x^{1/3} - 1)\right], \tag{3.43}$$

in which the two numerical constants (A_{e01}, n_1) or (A_{e02}, B_1) are found from low-temperature and high-temperature experimental data, respectively. The factor $x^{2/3}$ in Eq. (3.43) is shown separate for the sake of clarity. When $n_1 = 0$ or $B_1 = 0$, Eqs. (3.42) and (3.43) give expressions of the type (3.17) and (3.19), yielding $\gamma_e = \frac{2}{3}$. In general, Eqs. (3.42) and (3.43) give the following formulas for the electron analog of the Grüneisen parameter:

$$\gamma_{e1} = n_1 + \frac{2}{3},$$
(3.41a)

$$\gamma_{e2} = \frac{2}{3} + \frac{B_1}{3} x^{1/3}.$$
(3.41b)

The thermal pressure of the conduction electrons is

$$P_{Te} = -\left(\frac{\partial F_e}{\partial V}\right)_T = A_e \gamma_e x^{-1} T^2,$$
(3.44)

and the contribution of the conduction electrons to the isothermal bulk modulus is denoted by K_{Te}:

$$K_{Te} = -x\left(\frac{\partial P_{Te}}{\partial x}\right)_T = -P_{Te}\left\{\gamma_e - 1 + \left(\frac{\partial \ln \gamma_e}{\partial \ln x}\right)_T\right\} = -P_{Te}\begin{cases} (n_1 - \frac{1}{3}), \\ -\frac{1}{3} + \frac{B_1}{3} x^{1/3} + \frac{1}{3\left(1 + \frac{2}{B_1 x^{1/3}}\right)}. \end{cases}$$
(3.45)

The derivative $(\partial P_{Te}/\partial T)_x$ and the electron component of the thermal expansion coefficient $\alpha_e = (\partial P_{Te}/\partial T)_x (1/K_T)$ are given by

$$\left(\frac{\partial P_{Te}}{\partial T}\right)_x = \frac{2P_{Te}}{T}, \quad \alpha_e = \frac{2P_{Te}}{TK_T} = \bar{\alpha}_e T.$$
(3.46)

Here, K_T is the total isothermal bulk modulus and $\bar{\alpha}_e$ is a quantity which is approximately independent of temperature.

It is interesting to note that when $B_1 > 1$ (or $n > \frac{1}{3}$) the presence of the thermally excited conduction electrons reduces the value of the bulk modulus K_T. Physically, this corresponds to an increase in the degeneracy temperature $T_0 \sim \varepsilon_F/k$ and a corresponding decrease in the relative temperature T/T_0 of the electron liquid. We recall that all the thermodynamic quantities are proportional to $(T/T_0)^n$, where $n = 1$-2.

3.4. Phenomenological Theory of the Equation of State Accurate to Within T^2

We shall now describe a method for determining the anharmonicity parameter A_a and the electron parameter A_e as well as their derivatives with respect to the relative volume $x = V/V_0 = \rho_0/\rho$. These quantities can be determined from low-temperature ($T \ll \theta/20$) or high-temperature ($T > \theta$) experimental data obtained at atmospheric pressure.

The simultaneous determination of these parameters is based on the observation that the contributions of the anharmonicity and of the thermally excited conduction electrons to the high-temperature free energy are proportional to T^2. Consequently, these effects cannot be separated at $T > \theta$. However, they can be separated using experimental data for the specific heat and the thermal expansion coefficient near $T \sim 0°K$, because at such temperatures the influence of the anharmonicity can be neglected.

It is known that the Debye approximation is fairly reliable in the classical-temperature limit defined by $T > \theta$. Bearing in mind that we shall use the high-temperature data (specific heat at constant pressure c_P, thermal expansion coefficient α, and adiabatic bulk modulus K_S), we shall write the expressions for the free energy F and for the internal energy E per unit mass in the following form:

$$F = E_p(x) + \frac{9}{8}\frac{R\theta}{M} + \frac{RT}{M}[3\ln(1 - e^{-\theta/T}) - D(\theta/T)] - \frac{A_a + A_e}{\rho_0}T^2; \tag{3.47}$$

$$E = E_p + \frac{9}{8}\frac{R\theta}{M} + \frac{3RT}{M}D(\theta/T) + \frac{(A_a + A_e)}{\rho_0}T^2, \tag{3.48}$$

or, expanding in terms of $(\theta/T) < 1$, we obtain

$$F = E_p - \frac{RT}{M}\left[1 - 3\ln\frac{\theta}{T} - \frac{3}{40}\frac{\theta^2}{T^2}\right] - \frac{(A_a + A_e)}{\rho_0}T^2, \tag{3.47a}$$

$$E = E_p + \frac{3RT}{M}\left[1 + \frac{1}{20}\frac{\theta^2}{T^2}\right] + \frac{(A_a + A_e)}{\rho_0}T^2, \tag{3.48a}$$

$$c_V = \left(\frac{\partial E}{\partial T}\right)_x = c_{V0}\left\{1 - \frac{1}{20}\frac{\theta^2}{T^2} + \frac{2(A_a + A_e)}{\rho_0 c_{V0}}T\right\}, \quad c_{V0} = \frac{3R}{M}, \tag{3.49}$$

where M is the molecular weight.

The pressure is given by

$$P = -\rho_0\left(\frac{\partial F}{\partial x}\right)_T = P_p(x) + \frac{3RT\gamma\rho_0}{xM}\left[1 + \frac{1}{20}\frac{\theta^2}{T^2}\right] + (A_a' + A_e')T^2, \tag{3.50}$$

where

$$\gamma = -\frac{d\ln\theta}{d\ln x} \tag{3.51}$$

and the lattice Grüneisen parameter and a prime denotes a derivative with respect to x. Introducing γ_a of Eq. (3.12) and γ_e of Eq. (3.41) into Eq. (3.50), we obtain

$$P = P_p(x) + \frac{3RT\gamma\rho_0}{xM}\left[1 + \frac{1}{20}\frac{\theta^2}{T^2}\right] + (A_a\gamma_a + A_e\gamma_e)x^{-1}T^2. \tag{3.50a}$$

Using x-ray diffraction and other data on the temperature dependence of the lattice constant at atmospheric pressure (P = 0) and assuming that the functions $P_p(x)$, $\gamma(x)$, $\theta(x)$ are known, we obtain from Eq. (3.50a)

$$\frac{A_a}{x}\gamma_a = -\frac{A_e}{x}\gamma_e - \frac{P_p(x) + \frac{3R\rho_0\gamma}{xM}T\left(1 + \frac{1}{20}\frac{\theta^2}{T^2}\right)}{T^2} \tag{3.52}$$

where A_e and γ_e are assumed to be known from measurements of the specific heat and thermal expansion near $T \sim 0°K$.

We must make the following comments about Eq. (3.52). When

$$A_e\gamma_e \gg A_a\gamma_a, \tag{3.53}$$

Eq. (3.52) can be used to determine $A_e\gamma_e$ from the high-temperature data. This makes it possible, in principle, to determine the "stability" of the quantity $A_e\gamma_e$ in a wide range of temperatures.

Next, following the treatment given in Sec. 3.2, the quantity γ_a can be found reliably using Eq. (3.13):

$$\gamma_a \sim \varkappa/3,$$

where $\varkappa$ is the argument in the exponential law of repulsion

$$\exp(-\varkappa x^{1/3}). \tag{3.54}$$

Consequently, Eqs. (3.13) or (3.52) can be used to determine the value of A_a in those cases when the independent determination of this quantity from the specific heat c_p is impossible because of the insufficient accuracy of the specific heat data (this will be discussed later).

We shall now assume the quantity $(A_a\gamma_a + A_e\gamma_e)$ to be known or we shall replace it using Eq. (3.52). We must remember that in the latter case the formulas obtained are valid only at atmospheric pressure $(P = 0)$.

The isothermal bulk modulus K_T is

$$K_T = -x(\partial P/\partial x)_T = K_p + K_{TT}, \qquad K_p = -x\frac{\partial P}{\partial x}_{\text{P.}}; \tag{3.55}$$

$$K_{TT} = \frac{3R\gamma\rho_0 T}{Mx}\left\{\left(1-\frac{d\ln\gamma}{d\ln x}\right)\left(1+\frac{1}{20}\frac{\theta^2}{T^2}\right)+\frac{1}{10}\gamma\frac{\theta^2}{T^2}\right\} - x(A_a'' + A_e'')T^2.$$

At present, there are no sufficiently reliable experimental data which could be used to determine K_{TT} and other quantities which occur in the expression enclosed by braces in Eq. (3.55). Had such data been available we could have determined the second derivatives $(A_a'' + A_e'')$ in accordance with Eq. (3.55). We shall now determine $(\partial P/\partial T)_x$:

$$\left(\frac{\partial P}{\partial T}\right)_x = \frac{3R\gamma\rho_0}{xM}\left[1-\frac{1}{20}\frac{\theta^2}{T^2}\right]+2x^{-1}(A_a\gamma_a+A_e\gamma_e)T = -\frac{2}{T}P_p(x)-\frac{3R\gamma\rho_0}{xM}\left[1+\frac{3}{20}\frac{\theta^2}{T^2}\right]. \tag{3.56}$$

It follows from Eq. (3.56) that the derivative $(\partial P/\partial T)_x$ can be determined from known quantities.

Using the thermodynamic relationship

$$K_T = \frac{(\partial P/\partial T)_x}{\alpha} \tag{3.57}$$

we can express K_T in terms of those quantities which can be found from experimental data. We shall define the adiabatic bulk modulus K_S using the thermodynamic relationship

$$K_S = K_T + \frac{xT}{\rho_0 c_V}\left(\frac{\partial P}{\partial T}\right)_x^2. \tag{3.58}$$

Substituting the relevant values, we obtain

$$K_S = K_p + \frac{c_{V0}\rho_0\gamma T}{x}\left\{\left(1-\frac{d\ln\gamma}{d\ln x}\right)\left(1+\frac{1}{20}\frac{\theta^2}{T^2}\right)+\frac{1}{10}\gamma\frac{\theta^2}{T^2}\right\} -$$

$$-x(A_a''+A_e'')T^2+\frac{xT}{\rho_0 c_{V0}\left\{1-\frac{1}{20}\frac{\theta^2}{T^2}+\frac{2(A_a+A_e)T}{\rho_0 c_{V0}}\right\}}\begin{cases}\left[\frac{\gamma\rho_0 c_{V0}}{x}\left(1-\frac{1}{20}\frac{\theta^2}{T^2}\right)+2x^{-1}(A_a\gamma_a+A_e\gamma_e)T\right]^2\\ \text{or}\\ \left[\frac{2}{T}P_p(x)+\frac{c_{V0}\gamma\rho_0}{x}\left(1+\frac{3}{20}\frac{\theta^2}{T^2}\right)\right]^2.\end{cases} \tag{3.59}$$

Finally, we shall calculate c_P using the thermodynamic relationship

$$c_P = c_V + \frac{Tx}{\rho_0}\frac{(\partial P/\partial T)_x^2}{K_T}. \tag{3.60}$$

Eliminating K_T by means of Eq. (3.57), we obtain

$$c_P = c_V + \frac{Tx}{\rho_0}a\left(\frac{\partial P}{\partial T}\right)_x = c_{V0}\left\{1 - [1 + \gamma aT]\frac{1}{20}\frac{\theta^2}{T^2} + \gamma aT\left[1 + \frac{2(A_a + A_e)}{\rho_0\gamma ac_{V0}}\left(1 + \frac{A_a\gamma_a + A_e\gamma_e}{A_a + A_e}aT\right)\right]\right\}, \quad (3.61)$$

or

$$c_P \approx c_{V0}\left\{1 - \frac{1}{20}\frac{\theta^2}{T^2} + \gamma aT\left[1 + \frac{2(A_a + A_e)}{_0\rho_0\gamma ac_{V0}}\right]\right\}. \quad (3.61a)$$

Equation (3.61a) shows clearly that at $(\theta/T) < 1$ the specific heat at constant pressure c_P is of the form

$$c_P = a + bT - cT^{-2},$$

where

$$a = c_{V0} = \frac{3R}{M}, \qquad b = c_{V0}\gamma a\left[1 + \frac{2(A_a + A_e)}{\rho_0\gamma ac_{V0}}\right], \qquad c = c_{V0}\frac{\theta^2}{20}. \quad (3.62)$$

In the foregoing derivation, we have assumed that

$$\gamma aT \ll 1, \qquad aT\frac{A_a\gamma_a + A_e\gamma_e}{A_a + A_e} \ll 1. \quad (3.63)$$

When A_a, A_e, γ_a, γ_e have been determined, they can be approximated by a simple power or exponential function:

$$A_a \propto A_{a0}x^n \qquad \text{or} \qquad A_a \propto A_{a0}\exp[B(1 - x^{1/3})] \quad (3.64)$$

and, correspondingly,

$$A_e \propto A_{e0}x^{n_1} \qquad \text{or} \qquad A_e \propto A_{e0}\exp[B_1(1 - x^{1/3})]. \quad (3.65)$$

The use of an exponential law for A_a is justified in Chapter 6. We shall now consider the renormalization of the Grüneisen parameter which is closely associated with the phenomenological approach used so far. We shall show later that the determination of the equation of state from dynamic experimental data can be divided into two stages: a) the potential pressure (cold-compression isotherm) $P_p(x)$ is found from the low-temperature part of the Hugoniot curve at $x > 0.8-0.75$, using the Debye approximation $A_a = A_e = 0$; b) the effect of nonzero values of A_a and A_e on the equation of state determined in the first stage is investigated in the high-temperature region $x < 0.75$.

When the Grüneisen parameter is normalized using the thermodynamic value

$$\gamma = \frac{K_S a}{c_P\rho} = \frac{K_T a}{c_V\rho}; \qquad \gamma_0 = \left(\frac{K_T a}{c_V\rho}\right)_{x=1}, \quad (3.66)$$

we must subtract the contribution of the anharmonicity ($A_a \neq 0$) and of the thermally excited conduction electrons ($A_e \neq 0$) to the various quantities involved.

Using the thermodynamic relationships given by Eq. (3.57), we shall write γ in the form

$$\gamma = \frac{(\partial P/\partial T)_x}{c_V\rho}. \quad (3.67)$$

We shall divide $(\partial P/\partial T)_x$ and c_V into the Debye component and the components due to the anharmonicity and the thermally excited conduction electrons:

$$\left(\frac{\partial P}{\partial T}\right)_x = c_{v0}\rho\gamma_1\left\{4D(\theta/T) - 3\frac{\theta}{T}\frac{1}{e^{\theta/T}-1}\right\}+$$

$$+ 2x^{-1}(A_a\gamma_a + A_e\gamma_e)T = \left(\frac{\partial P}{\partial T}\right)_{xd} + \left(\frac{\partial P}{\partial T}\right)_{xa} + \left(\frac{\partial P}{\partial T}\right)_{xe} = \left(\frac{\partial P}{\partial T}\right)_{xd}(1+\Delta_1+\Delta_1');$$

$$c_v = c_{v0}\left\{4D - 3\frac{\theta}{T}\frac{1}{e^{\theta/T}-1}\right\} + \frac{2(A_a+A_e)}{\rho_0}T = c_{vd} + c_{va} + c_{ve} = c_{v0}(1+\Delta_2+\Delta_2'). \quad (3.68)$$

We shall denote the required value of the Grüneisen parameter by γ_1:

$$\gamma_1 = \frac{(\partial P/\partial T)_{xd}}{c_{vd}\rho}. \qquad (3.69)$$

Then, γ_1 can be defined in terms of the thermodynamic value of Eq. (3.66) and in terms of the parameters A_a, A_e, γ_a, γ_e in the following way:

$$\gamma_1 = \gamma_0\frac{1 + \dfrac{2(A_a + A_e)T}{c_{v0}\rho_0\left[4D - \dfrac{3\theta}{T}(e^{\theta/T}-1)^{-1}\right]}}{1 + \dfrac{2(A_a\gamma_a + A_e\gamma_e)}{c_{v0}\rho_0\gamma_0\left[4D - 3\dfrac{\theta}{T}(e^{\theta/T}-1)^{-1}\right]}}. \qquad (3.70)$$

When $T > \theta$, Eq. (3.70) simplifies to:

$$\gamma_1 = \gamma_0\frac{1 + \dfrac{2(A_a + A_e)T}{c_{v0}\rho_0\left[1 - \dfrac{1}{20}\dfrac{\theta^2}{T^2}\right]}}{1 + \dfrac{2(A_a\gamma_a + A_e\gamma_e)T}{c_{v0}\rho_0\gamma_0\left[1 - \dfrac{1}{20}\dfrac{\theta^2}{T^2}\right]}}. \qquad (3.71)$$

In the majority of cases, the corrections to unity on the right-hand side of Eq. (3.70) or Eq. (3.71) are so small that the expression for the Grüneisen parameter can be expanded in terms of these corrections:

$$\gamma_1 = \gamma_0(1-\Delta),$$

$$\Delta = \begin{cases} \dfrac{2T\left[A_a\left(\dfrac{\gamma_a}{\gamma_0}-1\right) + A_e\left(\dfrac{\gamma_e}{\gamma_0}-1\right)\right]}{c_{v0}\rho_0\left[4D - 3\dfrac{\theta}{T}(e^{\theta/T}-1)^{-1}\right]}, & T < \theta; \\[4mm] \dfrac{2T\left[A_a\left(\dfrac{\gamma_a}{\gamma_0}-1\right) + A_e\left(\dfrac{\gamma_e}{\gamma_0}-1\right)\right]}{c_{v0}\rho_0\left(1 - \dfrac{1}{20}\dfrac{\theta^2}{T^2}\right)}, & T > \theta. \end{cases} \qquad (3.72)$$

3.5. Thermal Defects

At sufficiently high temperatures, the thermodynamic functions of solids may be affected by the contribution of the thermal defects, as well as by the anharmonicity and the thermally

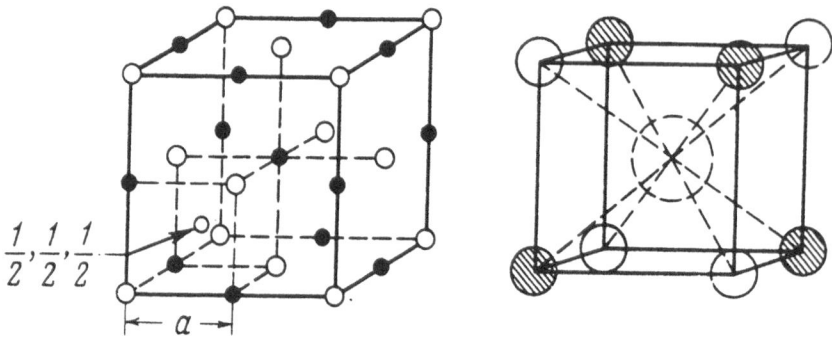

Fig. 3.2. Ion at an interstice in the NaCl lattice.

excited conduction electrons. This follows from many experimental observations, including de-fect-induced anomalies in the specific heat and thermal expansion coefficient near the melting point.

The concept of thermal defects in solids was introduced by Frenkel [14]. Frenkel reasoned as follows. The amplitudes of the thermal vibrations of lattice atoms increase as the temperature rises. A solid always has a few atoms (or ions) with very large vibration ampli-tudes. Under the influence of thermal fluctuations, these atoms (ions) can be ejected from their normal positions (lattice sites) into interstices. Subsequently, thermal fluctuations may shift such interstitial atoms sufficiently far from their original sites so that their interaction with the vacancies left behind can be neglected (the energy of such interactions should be much less than kT). Consequently, a Frenkel defect consists of a pair of noninteracting particles. One of them is an interstitial atom and the other is a vacancy which behaves as a quasiparticle. Figure 3.2 shows the position of an interstitial atom in the lattice of NaCl. Naturally, a re-verse process in which a Frenkel defect is annihilated can also occur in a solid: in this case, an interstitial atom reverts to a regular lattice site. Under thermal equilibrium conditions, the processes of defect formation and recombination (annihilation) occur all the time, but the concentration of defects remains constant.

The presence of thermal defects in solids is a direct consequence of the fundamental as-sumptions of the Boltzmann statistics. Let us assume that e denotes the energy of formation of a thermal defect. Numerically, this energy is equal to the change in the energy of a crystal due to the formation of one defect. It follows that the number of defects in a crystal at a given fixed temperature should be proportional to $\exp(-e/kT)$. Under thermal equilibrium condi-tions, the value of e is a function of volume and temperature, $e = e(V, T)$, or of pressure and temperature, $e = e(P, T)$, etc.

Experiments and theoretical calculations show that the value of e is such that, in the majority of solids, the concentration of thermal defects c_d reaches values $\sim 10^{-4}$-10^{-3} (0.01-0.1%) at temperatures close to the melting point T_{mp}. Thermal defects in solids play an im-portant role only at high temperatures $(T > \frac{2}{3}T_{mp})$. This is because, at lower temperatures, the concentration of thermal defects is less than the concentration of accidentally present im-purities (which is usually $\sim 10^{-5}$-10^{-6}).

Another important type of defect is known as the Schottky defect (Fig. 3.3). In this case, an atom or an ion which leaves its regular site is not located at an interstice but on the sur-face of a crystal or a crack (it is now known that dislocations can also act as sources and sinks of Schottky defects, i.e., of vacancies). Consequently, a Schottky defect in a monatomic solid is simply a vacancy. In ionic crystals, for example in NaCl, a Schottky defect consists of an anion vacancy (in the Cl^- sublattice) and a cation vacancy (in the Na^{2+} sublattice); this is due to the requirements of electrical neutrality.

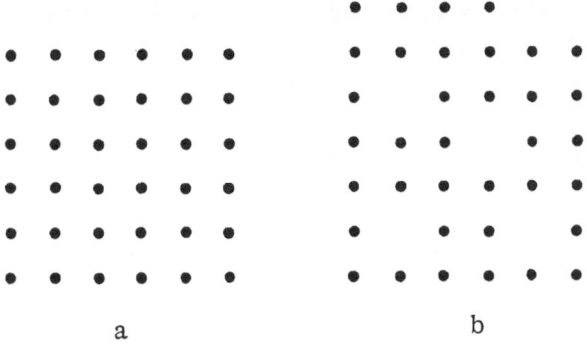

a   b

Fig. 3.3. Schottky defects. a) Ideal lattice; b) lattice with
Schottky defects.

Thermodynamic Formulas

We shall derive first the formula which gives the concentration of thermal defects under normal conditions (P ~ 0) [15-18]. For the sake of simplicity, we shall consider the formation of a vacancy in a monatomic crystal in which all N atoms are located at lattice sites at T = 0. We shall use the following symbols: n, the number of thermal defects; e, the energy of formation of a thermal defect; v_d, the change in volume of a crystal due to the formation of one thermal defect; $\Delta V = nv_d$; $h = e + Pv_d$, the enthalpy of formation of a thermal defect; $\Delta H = nh$; s_1, the configurational entropy (or mixing entropy) of one defect; $\Delta S_1 = ns_1$; s_2, the thermal entropy of the formation of one defect; $\Delta S_2 = ns_2$; $s = s_1 + s_2$, the increase in the entropy of a crystal due to the formation of one defect; $\Delta S = ns$; $f = e - Ts$, the free energy of formation of one defect; $\Delta F = nf$; $g = h - Ts$, the thermodynamic potential of formation of one defect; $\Delta G = ng$.

The increase in the configurational entropy due to the formation of n vacancies is

$$\Delta S_1 = k \ln \frac{(N+n)!}{N!n!},$$

where N and n are large numbers; therefore, using Stirling's formula

$$\ln N! \approx N \ln N,$$

we can obtain a simpler expression

$$\Delta S_1 = kn(1 + c_d - \ln c_d) \approx kn(1 - \ln c_d), \quad c_d = n/N,$$
$$s_1 = k(1 - \ln c_d). \tag{3.73}$$

When a defect (a vacancy) is formed, the frequency of the lattice vibrations changes as well and this gives rise to a change in the entropy of a crystal (ΔS_2). At P = 0, the state of thermodynamic equilibrium is defined by a minimum of the free energy F. The change in the free energy of a crystal due to the formation of defects is

$$\Delta F = \Delta E - T\Delta S = n\{e - T[k(1 - \ln c_d) + s_2]\}.$$

The concentration of thermal defects is found from the condition of thermal equilibrium $(\partial \Delta F / \partial n)_{V,T} = 0$:

$$e - T[k(1 - \ln c_d) + s_2] + kT = 0,$$

or

$$c_d = \exp[(s_2/k) - (e/kT)]. \tag{3.74}$$

In the more general case, when $P \neq 0$, the state of equilibrium is given by the minimum of the thermodynamic potential G. Defining ΔG as

$$\Delta G = \Delta H - T\Delta S = n\{h - T[k(1 - \ln c_d) + s_2]\},$$

we can find c_d from the condition $(\partial \Delta G / \partial n)_{P,T} = 0$

$$h - T[s_2 - k \ln c_d] = 0,$$

or

$$c_d = \exp[(s_2/k) - (h/kT)]. \tag{3.75}$$

Equation (3.75) for c_d is quite general. When $P = 0$, it transforms into Eq. (3.74), because $h = e + Pv_d$.

Using Eqs. (3.75) and (3.73), we obtain the following expressions for s_1 and s:

$$s_1 = k\left(1 - \frac{s_2}{k} + \frac{h}{kT}\right), \qquad s = s_1 + s_2 = k\left(1 + \frac{h}{kT}\right). \tag{3.76}$$

The expression for ΔG assumes the following very simple form:

$$\Delta G = nh - nTs = -nkT. \tag{3.77}$$

We shall now use the relationships

$$\Delta S = n(s_1 + s_2) = -\left(\frac{\partial \Delta G}{\partial T}\right)_P, \tag{3.78}$$

$$\Delta V = nv_d = \left(\frac{\partial \Delta G}{\partial P}\right)_T, \tag{3.79}$$

which are direct consequences of the general thermodynamic formulas (A.13) and our definitions. Simple calculations based on Eqs. (3.75)-(3.77) show that Eqs. (3.78) and (3.79) yield the conditions

$$\left(\frac{\partial e}{\partial T}\right)_P = T\left(\frac{\partial s_2}{\partial T}\right)_P - P\left(\frac{\partial v_d}{\partial T}\right)_P, \tag{3.80a}$$

$$\left(\frac{\partial e}{\partial P}\right)_T = T\left(\frac{\partial s_2}{\partial P}\right)_T - P\left(\frac{\partial v_d}{\partial P}\right)_T. \tag{3.80b}$$

Multiplying Eq. (3.80a) by dT and Eq. (3.80b) by dP and adding, we obtain the following fundamental relationship for the thermodynamics of defects:

$$de = Tds_2 - Pdv_d. \tag{3.80}$$

We shall now use Eq. (3.80) to find the differential of the enthalpy $h = e + Pv_d$:

$$dh = Tds_2 + v_d dP; \tag{3.81}$$

$$\left(\frac{\partial h}{\partial T}\right)_P = T\left(\frac{\partial s_2}{\partial T}\right)_P; \tag{3.81a}$$

$$\left(\frac{\partial h}{\partial P}\right)_T = T\left(\frac{\partial s_2}{\partial P}\right)_T + v_{\mathrm{d}}. \tag{3.81b}$$

Using Eq. (3.76) and the definition of the enthalpy of a defect h, we can transform the free energy ΔF to a simpler form

$$\Delta F = \Delta E - T\Delta S = -n(kT + Pv_{\mathrm{d}}),$$
$$f = -(kT + Pv_{\mathrm{d}}). \tag{3.82}$$

We shall now use the formulas

$$\Delta S = n(s_1 + s_2) = -\left(\frac{\partial \Delta F}{\partial T}\right)_V,$$
$$\Delta P = -\left(\frac{\partial \Delta F}{\partial V}\right)_T, \tag{3.83}$$

which, like Eqs. (3.78) and (3.79), are direct consequences of the general thermodynamic formulas (A.10) and our definitions. The differentiation of ΔF with respect to T gives

$$\left(\frac{\partial \Delta F}{\partial T}\right)_V = -n\left\{\left[\frac{h}{T} - v_{\mathrm{d}}\left(\frac{\partial P}{\partial T}\right)_V\right]\left(1 + \frac{Pv_{\mathrm{d}}}{kT}\right) + k + \frac{\partial}{\partial T}(Pv_{\mathrm{d}})v\right\}. \tag{3.84}$$

Substituting this expression in Eq. (3.83) and cancelling, we obtain

$$\left(\frac{\partial \ln v_{\mathrm{d}}}{\partial \ln T}\right)_V = -\frac{1}{kT}\left[h - v_{\mathrm{d}}T\left(\frac{\partial P}{\partial T}\right)_V\right]. \tag{3.85}$$

The importance of Eq. (3.85) is due to the following circumstances. The most important parameter in the theory of thermal defects v_{d} cannot be calculated easily by theoretical methods. However, it can be determined from experimental data. Equation (3.85) can be used to determine the temperature dependence $v_{\mathrm{d}} = v_{\mathrm{d}}(V, T)$, which is not associated with the thermal expansion of the crystal. Differentiating ΔF with respect to V at constant temperature, we can find the contribution of defects ΔP to the equation of state of a crystal:

$$\Delta P = \frac{nv_{\mathrm{d}}}{V}P\left\{\left(\frac{Pv_{\mathrm{d}}}{kT}\right)\left(\frac{K_T}{P}\right) + \left(\frac{\partial \ln v_{\mathrm{d}}}{\partial \ln V}\right)_T\right\}, \tag{3.86}$$

where K_T is the isothermal bulk modulus. The quantity ΔP is the thermal pressure of the defects. Under normal conditions, this pressure is equal to zero but it increases rapidly with increasing pressure P.

We shall now obtain some order-of-magnitude estimates of the thermal pressure of defects. We shall estimate first the value of the terms enclosed by braces in Eq. (3.86). We shall assume that

$$v_{\mathrm{d}} \sim 10^{-23}, \quad K_T \sim 3P, \quad \left(\frac{\partial \ln v_{\mathrm{d}}}{\partial \ln V}\right)_T \sim 1,$$
$$P \sim l_P \cdot 10^{11}, \quad T \sim l_T \cdot 10^3. \tag{3.87}$$

Here, l_P denotes the pressure expressed in hundreds of kilobars or hundreds of kiloatmospheres, and l_T is the temperature in thousands of degrees Kelvin. Substitution of Eq. (3.87) into Eq. (3.86) gives

$$\frac{\Delta P}{P} \sim c_{\mathrm{d}}\frac{v_{\mathrm{d}}}{v_{\mathrm{a}}}\left\{20\frac{l_P}{l_T} + 1\right\}, \quad v_{\mathrm{a}} = \frac{V}{N}. \tag{3.88}$$

At a pressure of 1 Mbar, $l_p = 10$; near the melting point $l_T \sim 3\text{-}5$, $v_d/v_a \sim 1$. Assuming that the concentration of defects remains constant along the melting curve (see Sec. 3.5) and is $c_d \sim 10^{-4}\text{-}10^{-3}$, we obtain

$$\frac{\Delta P}{P} \sim (6 - 4) \cdot 10 c_d \sim (6 - 4) \cdot 10^{-3} - 10^{-2}.$$

Thus, the contribution of the thermal pressure of the defects to the total pressure does not exceed several percent even at high pressures near the melting point. It follows that, in this respect, thermal defects of the lattice make a smaller contribution to the equation of state than the contributions of the anharmonicity and the thermally excited conduction electrons.

We shall now determine the Grüneisen parameter of the thermal defects using the standard relationship

$$\gamma = \frac{V \Delta P}{\Delta E}. \tag{3.89}$$

Substituting ΔP from Eq. (3.86) and $\Delta E = ne$ into Eq. (3.89), we obtain

$$\gamma_d = \frac{v_d P}{e}\left[\left(\frac{K_T}{P}\right)\left(\frac{P v_d}{kT}\right) + \left(\frac{\partial \ln v_d}{\partial \ln V}\right)_T\right]. \tag{3.90}$$

At some pressures and temperatures, the Grüneisen parameters γ_d can be a very complex function of volume.

We shall now consider a formula, similar to Eqs. (3.78), (3.79), and (3.84), associated with the derivatives of the enthalpy ΔH:

$$\Delta V = \left(\frac{\partial \Delta H}{\partial P}\right)_s, \tag{3.91}$$

$$\Delta c_P = \left(\frac{\partial \Delta H}{\partial T}\right)_P. \tag{3.92}$$

Equation (3.91) gives rise to an additional thermodynamic relationship for the defects

$$T\left(\frac{\partial s_2}{\partial P}\right)_s = \frac{h}{kT}\left[v_d - \frac{h}{T}\left(\frac{\partial T}{\partial P}\right)_s\right]. \tag{3.93}$$

Evaluation of Eq. (3.92) gives

$$\Delta c_P = c_d Nk\left[\left(\frac{h}{kT}\right)^2 + \frac{1}{k}\left(\frac{\partial h}{\partial T}\right)_P + \frac{h}{k^2}\left(\frac{\partial s_2}{\partial T}\right)_P - \frac{1}{T}\left(\frac{\partial h}{\partial T}\right)_P\right],$$

or [see Eq. (3.81a)]

$$\Delta c_P = c_d Nk\left[\left(\frac{h}{kT}\right)^2 + \frac{1}{k}\left(\frac{\partial h}{\partial T}\right)_P\right]. \tag{3.94}$$

Near the melting point, we have $(h^2/kT) \sim 10^2$, which gives rise to the specific-heat anomaly due to thermal defects. Finally, the contribution of the defects to the thermal expansion coefficient is

$$\Delta \alpha = \frac{1}{V}\left(\frac{\partial \Delta V}{\partial T}\right)_P = \frac{c_d}{T}\frac{v_d}{v_a}\left[\frac{h}{kT} + \left(\frac{\partial \ln v}{\partial \ln T}\right)_P\right], \quad v_a = \frac{V}{N}. \tag{3.95}$$

The second terms on the right-hand side of Eqs. (3.94) and (3.95) are small and, therefore, they can be omitted in approximate estimates of Δc_P and $\Delta \alpha$.

Approximate Relationships

We shall now consider the phenomenological methods, which can be used, together with experimental data, to estimate the pressure dependence of c_d. First, we shall consider the temperature dependence of h. We shall assume that h depends weakly on T. Then,

$$h(P, T) = h_0(P_0) + \left(\frac{\partial h}{\partial T}\right)_{P_0} T; \qquad h_0(P) \equiv h(P, T = 0), \tag{3.96}$$

where h_0 is the enthalpy, calculated for the cold-compression isotherm (T = 0). The enthalpy h_0 is a function of pressure $P_0 = P_0(V)$ or volume V corresponding to the cold-compression isotherm. Substitution of h from Eq. (3.96) into Eq. (3.75) gives the factor

$$B = \exp\left\{-\frac{1}{k}(\partial h/\partial T)_{P_0}\right\},$$

which is independent of temperature. In general, s_2 is also a function of P and T. Assuming that s_2 depends weakly on T and retaining only the terms linear in T, we find that

$$s_2(P, T) = s_{20}(P_0) + \left(\frac{\partial s_2}{\partial T}\right)_{P_0} T; \qquad s_{20}(P) \equiv s_2(P, T = 0), \tag{3.97}$$

where $s_{20}(P) \equiv s_2(P_0)$ is the value of s_2 on the cold-compression isotherm. The substitution of Eq. (3.97) in Eq. (3.75) gives the factor

$$B' = \exp\left\{\frac{T}{k}\left(\frac{\partial s_2}{\partial T}\right)_{P_0}\right\}.$$

According to the general thermodynamic relationship (3.81a), the product of B and B' is identically equal to unity: $BB' \equiv 1$. Consequently, Eq. (3.75) simplifies to

$$c_d = \exp[(s_{20}/k) - (h_0/kT)], \tag{3.75a}$$

where s_{20} and h_0 are functions only of volume and are calculated for the cold-compression isotherm.

We shall now consider in detail the physical meaning of s_2. For the sake of simplicity, we shall assume that we are dealing with the classical limit $T > \theta$. Then, the thermal (vibrational) contribution to the entropy can be represented in the form

$$S = 3kN\left\{1 + \frac{1}{3N}\sum_{\alpha=1}^{3N}\ln\frac{kT}{h\nu_\alpha}\right\}, \tag{3.98}$$

where ν_α are the frequencies of normal vibrations of a solid. We shall use ν_α' to denote the frequencies of normal vibrations in a solid containing one defect and the corresponding entropy will be denoted by S'. Then,

$$s_2 = S' - S = k\sum_{\alpha=1}^{3N}\ln\frac{\nu_\alpha}{\nu_\alpha'}. \tag{3.99}$$

Equation (3.99) shows clearly the physical meaning of s_2. To make this clearer, we shall carry out a further simplification. We shall assume that all frequencies are equal, i.e., we shall use the Einstein model of a solid. Then,

$$S = 3kN\left\{1 + \ln\frac{kT}{h\nu}\right\}, \qquad \nu_\alpha = \nu.$$

Finally, we shall assume that the formation of a defect affects significantly only those frequencies which are associated with the vibrations of atoms nearest to that defect and directed along the lines joining these atoms with the defect. Then, denoting the number of nearest neighbors of a defect by q, we obtain the required formula for s_2:

$$s_2 = kq \ln(\nu/\nu'), \qquad (3.100)$$

where ν' represents the values of frequencies ν'_α of q atoms which have suffered appreciable changes due to the formation of a defect. All other values of ν'_α are assumed to be equal to the unperturbed frequencies ν_α. Physical considerations show that the following conditions should be satisfied:

$$\frac{\nu}{\nu'} = \begin{cases} > 1 & \text{(for vacancies),} \\ < 1 & \text{(for interstitials).} \end{cases} \qquad (3.101)$$

We shall use A to denote the factor in Eq. (3.75a) which is independent of temperature:

$$A = \exp(s_{20}/k). \qquad (3.102)$$

After the substitution of s_2 from Eq. (3.100), the factor A becomes:

$$A_V \approx (\nu/\nu')^q. \qquad (3.103)$$

We can see that for vacancies $A_V > 1$. In the case of the formation of the Frenkel defects, we are dealing with two quasiparticles: an interstitial atom and a vacancy. Consequently, the enthalpy of formation of a Frenkel defect h_0 is divided by 2 in Eq. (3.75a):

$$c_d = \exp[(s_{20}/k) - (h_0/2kT)]. \qquad (3.75b)$$

We shall use ν_1 to denote the frequency of an interstitial atom and ν'_1 to denote the frequency of q_1 of its nearest neighbors ($\nu'_1 > \nu$). We can easily see that A for the Frenkel defect is of the form [16]

$$A_F{}^2 = \frac{\nu^{q+q_1+3}}{\nu_1{}^3 \nu'_1{}^{q_1} \nu'{}^q}, \qquad (3.103a)$$

where — as before — ν' are the frequencies of the vibrations of the atoms next to a vacancy (these vibrations are directed along lines joining the vacancy with its neighboring atoms). Since $\nu_1 > \nu$, $\nu'_1 > \nu$, and $\nu' < \nu$, it follows that A_F can be less or greater than unity.

Let us now consider the dependence of the coefficient A on the volume of a crystal. According to [5], the frequency of vibrations in a solid depends exponentially on the volume:

$$\nu \propto \nu_0 \exp(-bx^{1/3}), \quad x = V/V_0, \quad b \sim 6\text{-}10$$

The interstitial atoms and their neighbors can be regarded as compressed somewhat compared with the atoms located at regular sites. For these atoms, $\nu_{01} > \nu_0$, $\nu'_{01} > \nu_0$, $b_1 < b$, $b'_1 < b$.

On the other hand, the atoms near a vacancy should be regarded as somewhat "elongated" compared with those at regular sites: $\nu'_0 < \nu'$, $b_0 > b$. Since this "compression" and "elongation" of the regions around defects is small, the quantities b_1, b'_1, and b' do not differ by more than a few percent from b. Consequently,

$$A_V \propto A_{V0} \exp[(b'-b)qx^{1/3}]$$

and the value of A_V also decreases with increasing pressure (decreasing x). We may assume that the dependence of A_V on x is much weaker than the corresponding dependence of the exponential factor $\exp(-h_0/kT)$, because

$$(b' - b)q \ll b.$$

In the case of Frenkel defects, we have

$$A_{\mathrm{F}}{}^2 \propto A_{\mathrm{F}0}{}^2 \exp\{[3b_1 + q_1 b_1' + q b' - (q + q_1 + 3)b]\, x^{1/3}\}.$$

Since the argument of the exponential function in A_{F}^2 can be positive or negative, the value of A_{F} may increase or decrease with decreasing x but the dependence on x is even weaker than in the case of vacancies.

We shall derive a formula by means of which the pressure dependence of h_0 can be deduced from the experimental data. To obtain this formula, we shall differentiate h_0 with respect to the pressure along the cold-compression isotherm

$$\frac{dh_0}{dP_0} = -\beta_{T0} h_0 \frac{d\ln h_0}{d\ln V_0}, \qquad \beta_{T0} = -\frac{1}{V_0}\frac{dV}{dP_0},$$

where the subscript "zero" is used for the cold-compression ($0°$K) isotherm. Integrating the above formula, we obtain the required relationship

$$\left.\begin{aligned}
h_0(P_0) &= h_{00} \exp \int_{P_{00}}^{P_0} \beta_{T0} \left| \frac{d\ln h_0}{d\ln V} \right| dP, \\
L_0 &= \frac{d\ln h_0}{d\ln V} = -\left| \frac{d\ln h_0}{d\ln V} \right|.
\end{aligned}\right\} \tag{3.104}$$

Here, h_{00} is the enthalpy under normal conditions, which can be determined from the experimental data obtained under such conditions; β_{T0} is the isothermal compressibility at $0°$K, which can be found using the equation of state. The value of the logarithmic derivative is approximately constant

$$L_0 \sim 1\text{-}3 \tag{3.105}$$

and it can be found from experiments designed to determine the influence of pressure on the concentration of the thermal defects. Consequently, Eq. (3.104) gives $h_0(P_0)$ in terms of relatively easily determined quantities h_{00} and β_{T0}, as well as in terms of the practically constant quantity L_0. The volume dependence of L_0 can be ignored in the case of small variations of pressure and this quantity can be taken outside the integral:

$$h_0(P_0) = h_{00}(\rho_0/\rho_{00})^{L_0} = h_{00}(1 + \beta_{T0} L_0 P_0), \qquad \beta_{T0} L_0 P_0 \ll 1. \tag{3.106}$$

Substituting h_0 from Eq. (3.106) into Eq. (3.75)

$$c_{\mathrm{d}}(P_0) = \exp(s_{20}/k)\exp[-(h_{00}/kT)(1 + \beta_{T0} L_0 P_0)], \tag{3.107}$$

we obtain a formula which can be used to determine the theoretical parameter L_0 by comparison with the experimental data obtained at moderate pressures. The volume (or pressure) dependence of h_0 can be found by a semiempirical approach, applying the method of potentials to the equation of state [19], as used in the present book. We shall apply this method to ionic crystals.

Determination of the Enthalpy of a Defect by the Method of Potentials

The energy of an ionic crystal in the method of potentials is selected in the form [20]

$$E_{\mathrm{p}} = A_1 \exp(-bx^{1/3}) - K x^{-1/3}, \qquad T = 0, \tag{3.108}$$

where A_1, b, K are constants found experimentally for each specific substance; $x = V/V_0$ is the dimensionless volume. According to Eq. (3.108), the cold-compression isotherm $P_0 \approx P_p$ is of the form

$$P_0 = -\frac{\partial E_p}{\partial V} = A_1 x^{-2/3} \exp(-bx^{1/3}) - K x^{-4/3}. \qquad (3.109)$$

Formulas (3.108) and (3.109) ignore the van der Waals attraction, which may be appreciable in some cases (for example, in the case of silver halides). However, we can easily show [21, 22] that these terms are included indirectly in the semiempirical equation by the experimental determination of the constants A_1 and b in the potential of the repulsive forces. The same is true of substances in which the attractive forces are partly covalent. The presence of such forces is effectively included by the selection of A_1, b, and K in Eq. (3.109) from the experimental data. If the electron structure of an ionic crystal is not greatly affected by the formation of a defect, the energy of the defect e can be represented in the form of an algebraic sum of terms representing changes in the Coulomb energy, the exchange repulsion energy, and the van der Waals energy. Since, in general, it is fairly difficult to calculate theoretically the coefficients in the expression for e, it seems reasonable to retain only the most important terms in the formula for e and to find the coefficients from the experimental data. At the present state of knowledge, we can determine two numerical coefficients using the experimental values of h_{00} and $(\partial c_d/\partial P)_T$ (fuller details are given in [19]).

Thus, the volume dependence of the enthalpy of formation of a Schottky defect h_v can be expressed as

$$h_v = a_1 x^{-1/3} - a_2 \exp(-bx^{1/3}) + P_0 v_{d0} x^n, \qquad n \sim 1, \qquad (3.110)$$

where P_0 is the pressure given by Eq. (3.109) and the volume dependence of v_d is taken in the form

$$v_d = v_{d0} x^n, \qquad n \sim 1. \qquad (3.111)$$

For NaCl-type crystals, we have $v_{d0} \sim 2a_0$, where a_0 is the lattice constant. The constants a_1 and a_2 in Eq. (3.110) can be determined from the experimental data and the coefficient b may be assumed to be equal to the corresponding coefficient occurring in the equation of state given by Eq. (3.109). It has been shown theoretically [23, 24] that $a_2 \exp(-b)/a_1 \approx 1/5$ for NaCl. We note also that the minus sign in front of a_2 in Eq. (3.110) is due to the fact that the absolute value of the repulsive (exponential) energy of a crystal decreases due to the formation of a Schottky defect. This minus sign is very important: because of it, vacancies do not collapse at high pressures since the rise in h_v due to the term $P_0 v_{d0} x^n$ is compensated to a great degree by the term $-a_2 \exp(-bx^{1/3})$. In the case of the Frenkel defects, the enthalpy of formation of a defect h_F should be used in the form

$$h_F = a_1 x^{-1/3} + a_2 \exp(-bx^{1/3}) + P_0 v_{d0} x^n, \qquad n \sim 1. \qquad (3.112)$$

Here, a_1 and a_2 are constants which are found from experimental data; b is assumed to be equal to the corresponding coefficient in the equation of state. The quantity v_{d0} in the case of the Frenkel defects is less than that in the case of vacancies (Schottky defects). No theoretical information can be obtained about the relative values of the coefficients a_1 and a_2 for the formation of a Frenkel defect. We can only assume that $a_1 \sim a_2 \exp(-b)$ and that $a_2 > 0$. The van der Waals energy must be included in the calculation of h_F. However, it is found that the net change in the van der Waals and exponential repulsion energies can be represented quite accurately by an expression of the type $a_2 \exp(-bx^{1/3})$ in a wide range of volumes $x \sim 1-0.5$ [21,22].

The term $P_0 v_{d0} x^n$, $n \sim 1$ in Eqs. (3.110) and (3.112) has the same structure as the first parts of these equations. Therefore, the following procedure should be used in the method of potentials. The enthalpy of a defect (a vacancy or a Frenkel defect) is assumed to be given by the expression

$$h_0 = a_1 x^{-1/3} + a_2 \exp(-bx^{1/3}),
\qquad (3.113)$$

where b is assumed to be equal to the corresponding coefficient in the equation of state and the coefficients a_1 and a_2 are determined for $x = 1$ from the following conditions:

$$h_{00} = a_1 + a_2 \exp(-b),$$

$$\left| \frac{d \ln h_0}{d \ln x} \right|_{x=1} = \frac{1}{3} \frac{a_1 + a_2 b \exp(-b)}{a_1 + a_2 \exp(-b)} = L_{00}.$$

Solving this system of equations for a_1 and a_2 and substituting the results obtained in Eq. (3.113), we obtain the volume dependence of h_0 in terms of the experimentally determined parameters h_{00} and L_{00}

$$h_0 = \frac{h_{00}}{(b-1)} \{ (b - 3L_{00}) x^{-1/3} + (3L_{00} - 1) \exp[b(1 - x^{1/3})] \}.
\qquad (3.113a)$$

Analysis of Eqs. (3.110)-(3.113) shows [19] that the quantity L_0 for vacancies (high values of v_{d0}) may reach 6-7, decreasing at high pressures to 3-2. However, if $L_{00} \sim 3$, this quantity remains constant or decreases slowly from 3 to 2.5 or 2 at pressures of the order of 10^6 atm. In the case of the Frenkel defects (low values of v_{d0}), a similar analysis shows that the value of L_0 varies slowly with the pressure and lies within the range 1-3.

We shall illustrate this method by considering NaCl. The theoretically calculated value of the energy of formation of a Schottky defect in NaCl under normal conditions, e_v, is in good agreement with the experimental values [23, 24]. Using in Eq. (3.109) the equation of state for NaCl, obtained in [20] from Bridgman's data ($A = 1.59 \cdot 10^9$ atm, $K = 1.07 \cdot 10^5$ atm, $b = 9.69$), bearing in mind the results given in [24], and assuming that $v_d = 2a_0^3$ (the volume per ion pair in NaCl), * we obtain $h_v = 2.26 x^{-1/3} - 0.83 \cdot 10^4 \exp(-9.69 x^{1/3}) + 2.8 \cdot 10^{-5} P_0 x$ eV. The dependences of h_v and L_0 on x are shown graphically in Fig. 3.4.

Continuum (Deformation) Theory of Defects [25]

The continuum (deformation) theory of defects is used widely. This theory was developed by Zener [26]. It is assumed that a solid can be described by the theory of elasticity and a defect is regarded as a stressed state in a continuous medium. The basic formulas of the continuum theory of defects are obtained as follows.

Let us consider a particle of an elastic isotropic continuous medium in the form of a cylinder, which is immersed in a liquid exerting a pressure P. Let us assume that a mechanical moment L is applied to the bases of the cylinder. The differential of the energy of this cylinder is

$$dE = -PdV + TdS + Ld\varphi,$$

---

*Due to relaxation effects, the value of v_{d0} for the formation of a pair of vacancies in NaCl is not exactly equal to $2a_0^3$, but is simply of that order. Consequently, the dependences of h_v and L_0 on the volume x are only qualitative representations of the variation of these quantities with the volume.

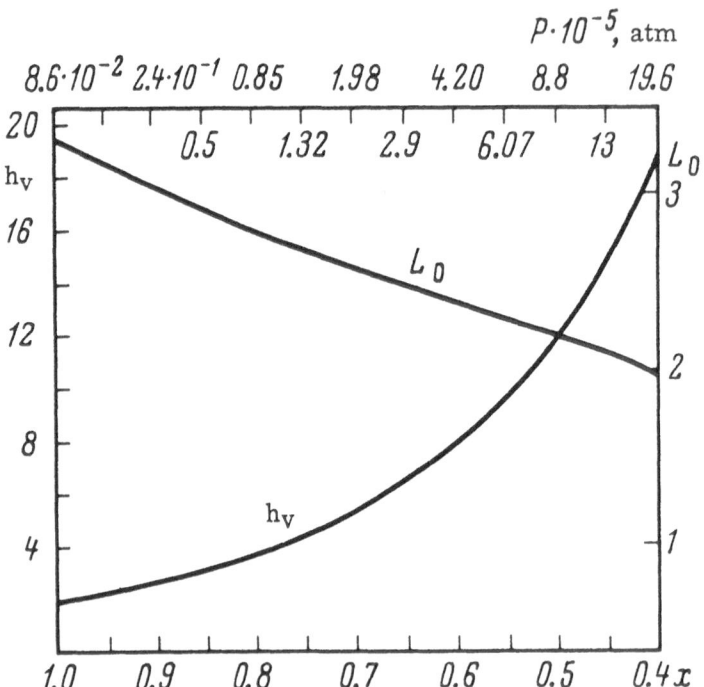

Fig. 3.4. Pressure and volume dependences of the energy
of formation of the Schottky defects.

where φ is the angular coordinate. The differential of the free Gibbs energy is

$$dG = d(E + PV - TS) = VdP - SdT + Ld\varphi.$$

The usual conditions which apply to the total differential yield the expressions

$$\left(\frac{\partial V}{\partial \varphi}\right)_{P,T} = \left(\frac{\partial L}{\partial P}\right)_{\varphi,T}, \tag{3.114}$$

$$\left(\frac{\partial S}{\partial \varphi}\right)_{T,P} = -\left(\frac{\partial L}{\partial T}\right)_{\varphi,P}. \tag{3.115}$$

Multiplying Eq. (3.114) by $1/L$, and using W to denote the work done in twisting the cylinder, we obtain

$$\left(\frac{\partial V}{\partial W}\right)_{P,T} = \left(\frac{\partial \ln L}{\partial P}\right)_{\varphi,T}.$$

For an isotropic elastic body, L is related to φ by

$$L = \frac{1}{2}\pi\mu R^4\varphi.$$

Consequently, Eq. (3.114) becomes

$$\left(\frac{\partial V}{\partial W}\right)_{P,T} = \left(\frac{\partial \ln \mu}{\partial P}\right)_T - \beta_T, \tag{3.116}$$

where μ is the shear modulus and β_T is the isothermal compressibility.

We shall now make a fundamental assumption on which the continuum (deformation) theory is based. This assumption postulates that the isothermal and isobaric work of deformation W is identical with the free energy of formation of a defect $g^* = h - s_2 T$ (the symbols used are explained at the beginning of the present section under the subheading Thermodynamic Formulas). Consequently, the change in volume associated with the deformation is identified with the activation volume of a defect v_d. As a result of this assumption, Eq. (3.116) assumes the required form:

$$\frac{v_d}{g^*} = \left(\frac{\partial \ln \mu}{\partial P}\right)_T - \beta_T = -\left[\left(\frac{\partial \ln \mu}{\partial \ln V}\right)_T + 1\right]\beta_T. \qquad (3.117)$$

Equation (3.117) can be simplified by expressing the logarithmic derivative of the shear modulus μ in terms of the Grüneisen parameter γ. We shall assume that the volume dependence is the same for all frequencies [27]:

$$\left(\frac{\partial \ln \mu}{\partial \ln V}\right)_T = -\left(2\gamma + \frac{1}{3}\right). \qquad (3.118)$$

It is known that the Grüneisen parameter can be expressed also in terms of the thermodynamic quantities

$$\gamma = \frac{\varkappa}{\beta_T c_V},$$

where c_V is the specific heat per unit volume. Substituting Eq. (3.118) in Eq. (3.117), we obtain

$$\frac{v_d}{g^*} = 2(\gamma - \tfrac{1}{3})\beta_T. \qquad (3.119)$$

Comparison of Eq. (3.119) with the experimental data [25] shows that the right hand of this equation is of the same order of magnitude as the left-hand side. Starting from Eq. (3.115) and making the assumptions used in the derivation of Eqs. (3.117) and (3.119), we find that

$$s_2/g^* = -[(\partial \ln \mu/\partial \ln V)_P + 1]\varkappa \qquad (3.120)$$

and

$$v_d/s_2 = \frac{\beta_T}{\varkappa}\frac{\left(\dfrac{\partial \ln \mu}{\partial \ln V}\right)_T + 1}{\left(\dfrac{\partial \ln \mu}{\partial \ln V}\right)_P + 1}. \qquad (3.121)$$

If we assume that μ depends on P and T only via the volume, it follows that

$$\left(\frac{\partial \ln \mu}{\partial \ln V}\right)_P = \left(\frac{\partial \ln \mu}{\partial \ln V}\right)_T.$$

Then, Eq. (3.121) assumes the very simple form

$$v_d/s_2 = \beta_T/\varkappa, \qquad (3.122)$$

and since

$$dg^* = v_d dP - s_2 dT, \qquad (3.121a)$$

it follows that

$$\frac{(\partial g^*/\partial P)_T}{(\partial g^*/\partial T)_P} = \frac{(\partial V/\partial P)_T}{(\partial V/\partial T)_P}. \tag{3.121b}$$

Comparison of Eqs. (3.120) and (3.121) with the experimental data shows that the two sides of these equations are of the same order of magnitude. It is reported in [25] that direct theoretical calculation of the ratios v_d/g^*, s_2/g^* using the continuum theory of defects can produce considerable errors. We can see that it is more reasonable to consider the formulas given by this theory as an attempt to establish some functional relationships between the thermodynamic parameters of defects and the thermodynamic properties of a body, which can be used to estimate the volume dependences of the ratios v_d/g^* and s_2/g^*.

Comparison with Experimental Data

It is not easy to obtain experimental data on the influence of pressure on the concentration of thermal defects because the number of these defects is small. Most of these data are deduced from measurements of the influence of the pressure on the diffusion or self-diffusion coefficients. The diffusion coefficient is defined by the well-known formula

$$D_i = \delta a^2 \bar{v} \exp[(s_a/k) - (h_a/kT)], \tag{3.123}$$

where δ is a numerical coefficient of the order of unity, whose exact value depends on the lattice geometry; a is the lattice constant; $\bar{v}$ is the frequency of the vibrations of a diffusing particle (this frequency is usually assumed to be equal to the Debye frequency, but also it can be taken as equal to the mean-square frequency). The quantities s_a and h_a denote, respectively, the entropy and enthalpy of activation of a particle in overcoming the potential barrier as it moves from one equilibrium position to another. Multiplying D_i by the concentration of the defects c_d, given by Eq. (3.75), we obtain the self-diffusion coefficient D

$$D = D_i c_d = \delta a^2 \bar{v} \exp(s_s/k) \exp(h_s/kT), \tag{3.124}$$

where

$$s_s = s_2 + s_a, \quad h_s = h_a + h. \tag{3.125}$$

Considering diffusion as an activation process, we can write

$$g_a^* = h - Ts_a,$$
$$dg_a^* = v_a dP - s_a dT; \tag{3.126}$$

$$v_s = \left[\frac{\partial(g^* + g_a^*)}{\partial P}\right]_T = v_a + v_d. \tag{3.127}$$

The self-diffusion activation volume v_s can be determined experimentally by investigating the pressure dependence of the self-diffusion under isothermal conditions. In this case, Eqs. (3.123)-(3.127) give

$$v_s = v_a + v_d = kT\left[\frac{\partial \ln D}{\partial P}\right]_T - kT\left(\frac{\partial \ln \delta a^2 v}{\partial P}\right)_T. \tag{3.128}$$

Usually, the contribution of the second term is small (less than 10% of v_s) compared with the first term. The diffusion coefficient of ionic crystals is related to the ion mobility μ_i by the Einstein relationship

Table 3.1

System	Structure	Measured quantity	v_{exper}, cm³/mole	Molar volume, cm³/mole
Li	bcc	$v_a + v_d$	3.4	13.1
Na	bcc	"	12.4	24
P	orthorh.	"	30	68
Ag	fcc	"	9.2	10.3
Au	fcc	v_d	1.5	20.2
Pb	fcc	$v_a + v_d$	13.0	18.2*
AgZn	fcc	"	5.4	10.0
O in V	bcc	v_d	1.7	2?
N in V	bcc	"	1.1	2?
C in Fe	bcc	"	0.0	2?
N in Fe	bcc	"	0.0	2?
AgBr	NaCl	v_a for Frenkel defects	16	9
	NaCl	v_a for Schottky defects	44	29
	NaCl	v_d for interstitial Ag ions	2.7	9
	NaCl	v_d for Ag ion vacancies	7.4	9
	NaCl	$v_a + v_d$ for Br ion vacancies	44	20
NaCl(Ca⁺⁺)	NaCl	v_d for Na ion vacancies	7.7	5.4
KCl(Sr⁺⁺)	NaCl	v_d for K ion vacancies	7.0	10.8
Au	fcc	$v_a + v_d$	7.2	10.2
	fcc	v_a	4.6-5.8	
Au in AuAg	fcc	$v_a + v_d$	7.5	10.3
Ag in AuAg	fcc	"	7.2	10.3

* Results of other workers.

$$D_i = \frac{kT}{e^*}\,\mu_i,$$

where e* is the ion charge (e* is used in order to avoid confusion with e employed for the energy of formation of a defect). Similarly, the self-diffusion coefficient is related to the electrical conductivity σ:

$$D = \sigma \frac{kT}{N(e^*)^2}.$$

Consequently, in the case of ionic crystals, the necessary information can be obtained by investigating the influence of the pressure on the ion mobility and the electrical conductivity. The experimental data on v_a and v_d [10, 11] are collected in Table 3.1 and are taken from Lazarus and Nachtrieb's paper [10], where a discussion of these results is given. We must point out that $v_d > 0$ and $v_a > 0$. Moreover, v_d is not equal to the atomic volume because of the relaxation of atoms around defects. At low pressures, at which these experimental data have been obtained, the pressure dependences of certain quantities can be expressed quite simply in terms of the activation volumes v_d and v_a. Expanding s_2 and h in terms of the pressure $(P - P_0)$, we obtain

$$s_2 = s_2(P_0, T) + \left(\frac{\partial s_2}{\partial P_0}\right)_T (P - P_0),$$

$$h = h(P_0, T) + \left(\frac{\partial h}{\partial P_0}\right)_T (P - P_0).$$

Substituting these expressions into Eq. (3.75) and using Eq. (3.81b), we find that

$$c_d(P) = c_{d0} \exp\left(-\frac{v_d \Delta P}{kT}\right), \qquad \Delta P = P - P_0, \tag{3.129}$$

where c_{d0} is the concentration of the defects at $P = P_0$. Similar formulas are obtained for the diffusion coefficient D_i of Eq. (3.123) and the self-diffusion coefficient D of Eq. (3.124):

$$D_i = D_{i0}\left(\frac{a^2 \bar{v}}{a_0{}^2 \bar{v}_0}\right) \exp\left(-\frac{v_a \Delta P}{kT}\right), \tag{3.123a}$$

$$D = D_0\left(\frac{a^2 \bar{v}}{a_0{}^2 \bar{v}_0}\right) \exp\left[-\frac{(v_d + v_a)\Delta P}{kT}\right], \tag{3.124a}$$

where the subscript zero refers to the values obtained at $P = P_0$. Assuming that the volume dependence of s_{20} in Eq. (3.107) for c_d can be neglected compared with the volume dependence of h, and comparing Eqs. (3.107) and (3.129), we obtain the following relationship between the logarithmic derivative L_0 and the experimentally determined value of v_d:

$$L_0 = \frac{v_d}{h_{00}\beta_{T0}}. \tag{3.130}$$

In those cases, when v_d is not known but only the sum $v_d + v_a$ is available, we can regard this sum as the upper limit of v_d:

$$(v_d)_{\max} = v_d + v_a. \tag{3.131}$$

Similarly, the upper limit of L_0 can be found from the formula

$$(L_0)_{\max} = \frac{v_d + v_a}{h_{00}\beta_{T0}}. \tag{3.132}$$

The relationship between v_a and the logarithmic derivative $L_{0a} = |(\partial \ln h_a/\partial \ln V)_T|$ can be obtained in a similar manner. We shall not consider this point in greater detail.

3.6. Electrons and Holes

in Insulators and Semiconductors

At first sight, it may seem that carriers in insulators and semiconductors cannot have any appreciable influence on the equation of state. This is because the number of thermally excited carriers in insulators and in semiconductors is so small, compared with the total number of atoms or ions, that their contribution to the thermodynamic functions of condensed matter (crystals and liquids) is negligible. However, when the pressure and temperature are increased, an insulator or a semiconductor may undergo a transition to the metallic state, i.e., the forbidden band (energy gap) may be closed in such a material. When this happens, thermally excited conduction electrons can make an appreciable contribution to the equation of state. It follows, therefore, that we must consider the problem of the pressure and temperature dependences of the forbidden band width (energy gap) in insulators and semiconductors.

These dependences are particularly important in shock-wave experiments. An analysis of the high-temperature part of a Hugoniot curve of an insulator or a semiconductor may give incorrect results if the possible metallization is ignored.

At sufficiently high temperatures, the thermally excited conduction electrons in the insulators and semiconductors can be described quite satisfactorily by the classical Boltzmann statistics. Consequently, we shall concentrate our attention on the classical approximation.

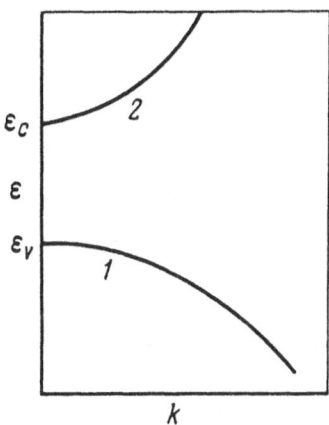

Fig. 3.5. Structure of the energy bands of a semiconductor (schematic representation). 1) Valence band; 2) conduction band.

Basic Relationships [28, 29]

Let us consider an insulator or a semiconductor. The energy band structure of such a solid is shown in Fig. 3.5. (We shall consider the properties of a semiconductor only in the intrinsic conduction region, where the influence of impurity electrons and holes can be neglected.) The energy of the upper edge of the valence band will be denoted by ε_v; ε_c will be used for the lower edge of the conduction band. In fact, the band structure of a semiconductor is much more complex than that shown in Fig. 3.5 (this point will be discussed later). The very simple energy band scheme depicted in Fig. 3.5 is used only when dealing with a little-known substance. Such a scheme is sufficient to obtain the relationships which are of interest to us.

Because of the presence of a forbidden band of width $\varepsilon_g = \varepsilon_c - \varepsilon_v$, the valence band of semiconductors and insulators is completely filled at $T = 0$ and the conduction band is completely empty. As the temperature is increased, electrons gradually jump the forbidden band to go over to the conduction band. Thus, the valence band becomes only partly filled. The unoccupied levels in the valence band behave effectively as positively charged quasiparticles, which are known as holes.

In the model of independent quasiparticles, holes behave as electrons with an effective mass m_p and a positive charge whose absolute value is equal to the electron charge. Holes, like electrons, obey the Fermi — Dirac statistics and — since they are complementary to electrons — their distribution function is not $n_F(\varepsilon)$ of Eq. (3.20) but $[1 - n_F(\varepsilon)]$. We shall use n and p to denote, respectively, the number of electrons and holes per unit volume (these quantities are known as electron and hole densities), so that $n_0 = nV$ and $p_0 = pV$. Then,

$$n = \int \nu_n(\nu) n_F(\varepsilon)\, d\varepsilon,$$
$$p = \int \nu_p(\varepsilon)[1 - n_F(\varepsilon)]\, d\varepsilon,$$

(3.133)

where $\nu(\varepsilon)$ is the density of states and the subscripts n and p refer to thermally excited electrons and holes. The position of the Fermi level ε_F in metals is given by Eq. (3.21), assuming that the number of particles N is known. In semiconductors, the electron and hole densities n and p are functions of the temperature and pressure. The position of the Fermi level ε_F is found from the condition for electrical neutrality:

$$n = p.$$

(3.134)

The solution of Eq. (3.134) requires the use of the Fermi functions and cannot be obtained in a finite form. These functions have been tabulated by McDougall and Stoner [30].

The calculations can be carried out quite simply in the classical limit. We assume that the densities of the electron and hole states are of the type given by Eq. (3.17):

$$\left.\begin{array}{l} \nu_n(\varepsilon) = \dfrac{(2m_n)^{3/2}}{2\pi^2\hbar^3}(\varepsilon - \varepsilon_c)^{1/2}, \qquad (\varepsilon > \varepsilon_c); \\[2mm] \nu_n(\varepsilon) = 0, \qquad (\varepsilon < \varepsilon_c) \end{array}\right\}$$

(3.135)

$$\left.\begin{aligned} v_p(\varepsilon) &= \frac{(2m_p)^{3/2}}{2\pi^2\hbar^3}\,(\varepsilon_v - \varepsilon)^{1/2}, \qquad (\varepsilon < \varepsilon_v); \\ v_p(\varepsilon) &= 0 \qquad (\varepsilon > \varepsilon_v). \end{aligned}\right\} \tag{3.136}$$

Substituting these expressions for $\nu(\varepsilon)$ in Eq. (3.134) and bearing in mind that $\exp[(\varepsilon - \varepsilon_F)/kT] \gg 1$ for electrons and $\exp[(\varepsilon - \varepsilon_F)/kT] \ll 1$ for holes, we easily obtain the required formulas

$$n = A_n(T)\exp[(\varepsilon_F - \varepsilon_c)/kT], \qquad A_n(T) = 2\left(\frac{m_n kT}{2\pi\hbar^2}\right)^{3/2}, \qquad n_0 = nV, \tag{3.137}$$

$$p = A_p(T)\exp[(\varepsilon_v - \varepsilon_F)/kT], \qquad A_p(T) = 2\left(\frac{m_p kT}{2\pi\hbar^2}\right)^{1/2}, \qquad p_0 = pV. \tag{3.138}$$

Using Eq. (3.134), we can find the Fermi energy of an intrinsic semiconductor in the classical limit:

$$\varepsilon_F = -\frac{3}{4}kT\ln\frac{m_n}{m_p} + \frac{\varepsilon_v + \varepsilon_c}{2}. \tag{3.139}$$

Using Eq. (3.134), the Fermi energy ε_F can be eliminated from the expressions (3.137) and (3.138)

$$np = n^2 = p^2 = A_n(T)A_p(T)\exp(-\varepsilon_G/kT), \tag{3.140}$$

or

$$n = [A_n(T)A_p(T)]^{1/2}\exp(-\varepsilon_G/2kT). \tag{3.141}$$

So far, we have considered the simplest band structure with energy extrema at the point $\mathbf{k} = 0$ ($\mathbf{k}$ is the wave vector) and we have assumed that the constant-energy surfaces in the wave-vector space are spheres: $\varepsilon = \hbar^2 k^2/2m^*$. In the case of more complex energy bands, the formulas we have derived still apply but the determination of the effective mass is different. Thus, the conduction bands of germanium and silicon have several minima of the energy ε along the $\langle 111 \rangle$ and $\langle 100 \rangle$ directions of the wave vector $\mathbf{k}$. The constant energy surfaces near the minima are prolate ellipsoids [see Eq. (3.18)]. In such cases, the quantity m_n must be replaced by $(m_1 m_2 m_3)^{1/3} N_n^{2/3}$, where m_1, m_2, m_3 are the effective masses along the three principal axes of the ellipsoids; N_n is the number of minima, which is four for germanium and six for silicon. The valence bands of germanium and silicon are degenerate at the point $\mathbf{k} = 0$. The constant-energy surfaces near this point are distorted spheres, and

$$\varepsilon = \frac{\hbar^2 k^2}{2m_p}\,\eta(\theta, \varphi),$$

where the function $\eta(\theta, \varphi)$ depends only on the direction of the wave vector. The density-of-states effective mass is now defined as

$$m_{\text{eff}} = m_p\left\{\frac{1}{4\pi}\sum_j\int_0^\pi\int_0^{2\pi}[\eta_j(\theta, \varphi)]^{-3/2}\sin\theta\,d\theta\,d\varphi\right\}^{2/3},$$

where the summation is extended over all the degenerate constant-energy surfaces.

We shall now consider qualitatively the transition from the classical to the quantum case, i.e., we shall deal with the degenerate case. We shall follow Shockley's treatment [28]. The quantities $A_n(T)$ and $A_p(T)$ in Eqs. (3.137)-(3.138) can be regarded as effective densities of states in the conduction and valence bands, respectively. Since the electron momentum in the classical limit is $p_n \propto (2m_n kT)^{\frac{1}{2}}$, the phase volume

$$\frac{2 \cdot 4\pi}{3} \frac{p_n^3}{\hbar^3}$$

corresponds approximately to the quantity $A_n(T)$. The same applies to $A_p(T)$. Thus, n and p are equal to the effective densities of states, multiplied by the probability of the existence of a state ε_c or ε_v, which is given by an exponential factor. Substituting numerical values of the parameters, we obtain

$$A_n(T) = 4.82 \cdot 10^{15} \left(\frac{m_n}{m_0}\right)^{3/2} T^{3/2},$$

$$A_p(T) = 4.82 \cdot 10^{15} \left(\frac{m_p}{m_0}\right)^{3/2} T^{3/2} \ (\mathrm{cm}^{-3}). \tag{3.142}$$

At room temperature $T \sim 300°K$ and for $m_n = m_p = m_0$, we obtain $A_n = A_p = 2.4 \cdot 10^{19}$ cm^{-3}, which is approximately $1/2000$ of the number of atoms in 1 cm^3. These formulas are not applicable at high electron (or hole) densities. In this case, as in the case of metals (Sec. 3.2), we must use quantum formulas. The treatment is different from that for metals because the energy is measured from the bottom of the conduction band ε_c. In the zeroth approximation [Eqs. (3.23) and (3.33)], we have

$$n = \int_{\varepsilon}^{\varepsilon_F} \nu_n(\varepsilon)\, d\varepsilon = \frac{8\pi}{3h^3} [2m_n(\varepsilon_F - \varepsilon_c)]^{3/2}$$

or

$$\varepsilon_F - \varepsilon_c = \varepsilon_m = \frac{h^2 (3n/8\pi)^{2/3}}{2m_n} = 21.6 \cdot 10^{-16} \left(\frac{m_0}{m_n}\right) n^{2/3}, \tag{3.143}$$

where ε_m is the maximum energy level to which electrons fill the conduction band. This case is governed by the inequality $kT \ll \varepsilon_m$. At any given temperature, we can encounter a transition from the classical to the quantum degeneracy case as the number of electrons in the conduction band increases. The boundary between the classical and quantum cases is arbitrarily set by the degeneracy carrier density n_{deg} or by the degeneracy temperature T_{deg}:

$$T_{deg} = \frac{\varepsilon_m}{k} = \left[\frac{3}{\pi}\right]^{2/3} \frac{h^2}{8km_n} n_{deg}^{2/3} = 4.2 \cdot 10^{-11} n_{deg}^{2/3}. \tag{3.144}$$

Equation (3.144) can be used to determine the degeneracy temperature at a given carrier density, and the carrier density at a given temperature. When $n \sim n_{deg}$, the dependence of n on T and $(\varepsilon_F - \varepsilon_c)$ is not described by a simple formula but can be determined only by the exact solution of Eq. (3.133). The dependence of n on $(\varepsilon_F - \varepsilon_c)$ and T, obtained by the exact integration of Eq. (3.133), is shown in Fig. 3.6, which includes also the curves representing the limiting cases of Eqs. (3.137) and (3.143). The point corresponding to $n = n_{deg}$ is found from $\varepsilon_m = kT$; on the curve representing the total degeneracy, it corresponds to the value of n for $\varepsilon_F - \varepsilon_c = kT$. It is found that this value is $n \approx 0.75 A_n(T)$. The exact solution for $n = 0.75 A_n(T)$ corresponds to a small negative value of the difference $(\varepsilon_F - \varepsilon_c)$, so that even states at the bottom of the band are less than half occupied by electrons and the distribution is not strongly degenerate. The distribution is completely degenerate when $(\varepsilon_F - \varepsilon_c) \geq 2kT$. This corresponds to values of n that are approximately three times larger than n_{deg} for a given temperature T, or to the temperatures T representing a fraction $3^{-2/3} = 0.47$ of T_{deg} for a given value of n. The case of a degenerate electron gas can be considered in the same way as the case of metals in Sec. 3.2.

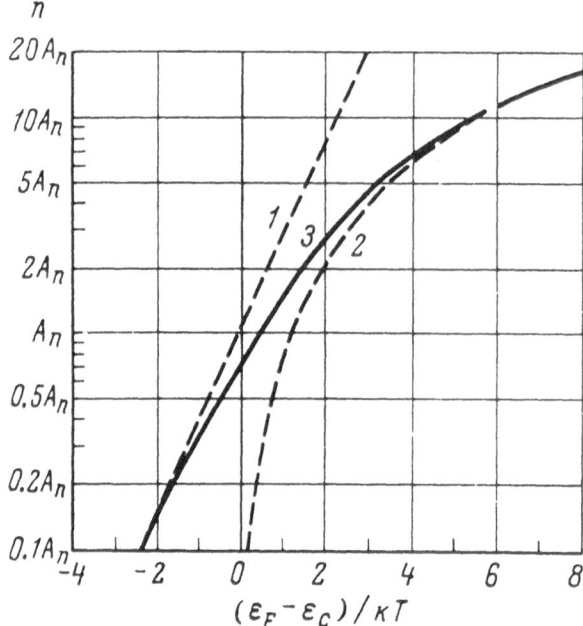

Fig. 3.6. Exact dependence of n on $(\varepsilon_F - \varepsilon_c)$ and T; this dependence is compared with the classical dependence and with the dependence in the case of total degeneracy (following Shockley). 1) Classical approximation: $n = A_n \cdot \exp [\varepsilon_F - \varepsilon_c)/kT]$; 2) approximation for the case of total degeneracy $n = \frac{8\pi}{3h^3} [2m (\varepsilon_F - \varepsilon_c)]^{3/2} = \frac{4A_n}{3\sqrt{\pi}} \left[\frac{\varepsilon_F - \varepsilon_c}{kT} \right]^{3/2}$; 3) exact value of n.

The relationships governing the behavior of holes can be analyzed in a similar manner. Equation (3.141) can be obtained by minimizing the free energy with respect to n_0 for constant volume and temperature. The change in the free energy of a semiconductor due to the formation of n_0 electrons (and, correspondingly, $p_0 = n_0$ holes) consists of the energy required for the transfer of n_0 electrons from the valence to the conduction band, which is equal to $n\varepsilon_G$ ($\varepsilon_G = \varepsilon_c - \varepsilon_v$), the free energy of a classical electron gas in the conduction band[*]

$$ F_n = - n_0 kT \ln \left[\frac{2eV}{n_0} \left(\frac{m_n kT}{2\pi\hbar^2} \right)^{3/2} \right], $$

and the free energy of a classical gas of holes

$$ F_p = - p_0 kT \ln \left[\frac{2eV}{p_0} \left(\frac{m_p kT}{2\pi\hbar^2} \right)^{3/2} \right]. $$

Using ΔF to represent the required change in the free energy, and employing the condition for electrical neutrality given by Eq. (3.134), we obtain

---

[*]To obtain F_n and F_p in the formula for the free energy of a classical gas, given by Eq. (C.7) in Appendix C, we must replace $\sum_k \exp (-\varepsilon_k'/kT)$ by 2 — which is the statistical weight of a particle of spin $\frac{1}{2}$ — and we must assume that $N = n_0$.

$$\Delta F = n_0 \left\{ \varepsilon_G - 2kT \ln\left[\frac{2eV}{n_0}\left(\frac{kT}{2\pi\hbar^2} \right)^{3/2} (m_n m_p)^{3/4} \right] \right\}. \tag{3.144a}$$

The number of electrons n_0 can be found from the condition

$$\left(\frac{\partial \Delta F}{\partial n_0} \right)_{V,T} = 0,$$

which gives the expression

$$\varepsilon_G - 2kT \ln\left[\frac{2V}{n_0}\left(\frac{kT}{2\pi\hbar^2} \right)^{3/2} (m_n m_p)^{3/4} \right] = 0,$$

identical with Eq. (3.141). Substituting into Eq. (3.144a) the value of n from Eq. (3.141), we obtain

$$\Delta F = -2n_0 kT.$$

Using the general thermodynamic formula (A.14) from Appendix A, we can determine the corresponding change in the energy of the semiconductor

$$\Delta E = -T^2\left(\frac{\partial}{\partial T}\frac{\Delta F}{T} \right)_V = 2kn_0 T\left\{ \frac{3}{2} + \frac{3}{4}\frac{\partial \ln(m_n m_p)}{\partial \ln T} + \frac{\varepsilon_G}{2kT}\left[1 - \left(\frac{\partial \ln \varepsilon_G}{\partial \ln T} \right)_V \right] \right\}. \tag{3.145}$$

The change in the specific heat of the semiconductor, Δc_V, is

$$\Delta c_V = \left(\frac{\partial \Delta E}{\partial T} \right)_V = 2kn_0 \left\{ \frac{3}{2} + \frac{3}{4}\frac{\partial \ln(m_n m_p)}{\partial \ln T} + \frac{\varepsilon_G}{2kT}\left[1 - \left(\frac{\partial \ln \varepsilon_G}{\partial \ln T} \right)_V \right] \right\} \times$$

$$\times \left\{ \frac{5}{2} + \frac{3}{4}\frac{\partial \ln(m_n m_p)}{\partial \ln T} + \frac{\varepsilon_G}{2kT}\left[1 - \left(\frac{\partial \ln \varepsilon_G}{\partial \ln T} \right)_V \right] - \frac{(\varepsilon_G/2kT)}{\frac{3}{2} + \frac{3}{4}\frac{\partial \ln(m_n m_p)}{\partial \ln T} + \frac{\varepsilon_G}{2kT}\left[1 - \left(\frac{\partial \ln \varepsilon_G}{\partial \ln T} \right)_V \right]} \right\}. \tag{3.146}$$

The contribution of the thermally excited electrons and holes to the thermal motion is

$$\Delta P = -\left(\frac{\partial \Delta F}{\partial V} \right)_T = 2kT\left(\frac{\partial n_0}{\partial V} \right)_T =$$

$$= 2kTn\left\{ 1 + \frac{3}{4}\left[\frac{\partial \ln(m_n m_p)}{\partial \ln V} \right]_T - \frac{\varepsilon_G}{2kT}\left(\frac{\partial \ln \varepsilon_G}{\partial \ln V} \right)_T \right\}. \tag{3.147}$$

We determine the "electron−hole analog" of the Grüneisen parameter using the standard formula

$$\gamma_{np} = \frac{V\Delta P}{\Delta E}. \tag{3.148}$$

Substituting ΔP of Eq. (3.147) and ΔE of Eq. (3.145), we obtain the required formula

$$\gamma_{np} = \frac{1 + \dfrac{3}{4}\left[\dfrac{\partial \ln(m_n m_p)}{\partial \ln V} \right]_T - \dfrac{\varepsilon_G}{2kT}\left(\dfrac{\partial \ln \varepsilon_G}{\partial \ln V} \right)_T}{\dfrac{3}{2} + \dfrac{3}{4}\left[\dfrac{\partial \ln(m_n m_p)}{\partial \ln T} \right]_V + \dfrac{\varepsilon_G}{2kT}\left[1 - \left(\dfrac{\partial \ln \varepsilon_G}{\partial \ln T} \right)_V \right]} \tag{3.148a}$$

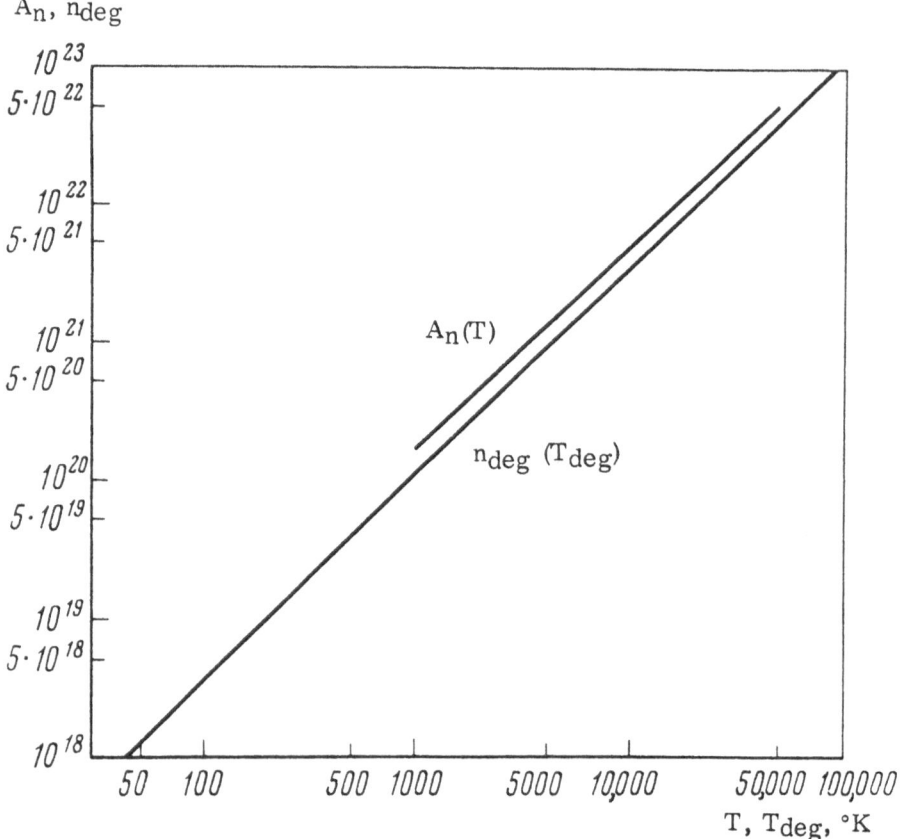

Fig. 3.7. Dependence of the density of states A_n on the temperature T and the dependence of the degenerate carrier density n_{deg} on the degeneracy temperature T_{deg}.

Finally, we obtain a correction to the isothermal bulk modulus

$$\Delta K_T = -V\left(\frac{\partial \Delta P}{\partial V}\right)_T = -2kTn\left\{\left[\frac{3}{4}\left(\frac{\partial \ln(m_n m_p)}{\partial \ln V}\right)_T - \frac{\varepsilon_G}{2kT}\left(\frac{\partial \ln \varepsilon}{\partial \ln V}\right)_T\right]\left[1 + \frac{3}{4}\left(\frac{\partial \ln(m_n m_p)}{\partial \ln V}\right)_T -\right.\right.$$

$$\left.\left.- \frac{\varepsilon_G}{2kT}\left(\frac{\partial \ln \varepsilon_G}{\partial \ln V}\right)_T\right] - \frac{\varepsilon_G}{2kT}\left[\left(\frac{\partial \ln \varepsilon_G}{\partial \ln V}\right)_T^2 + \left(\frac{\partial^2 \ln \varepsilon_G}{\partial (\ln V)^2}\right)_T\right] + \frac{3}{4}\left(\frac{\partial^2 \ln(m_n m_p)}{\partial (\ln V)^2}\right)_T\right\}. \quad (3.149)$$

We can see from these formulas that the thermally excited electrons and holes can make an appreciable contribution to the thermal energy and thermal motion in semiconductors and insulators only at sufficiently high temperatures of the order of 10^4–10^5 °K. This can be seen easily from Fig. (3.7), which shows, on a logarithmic scale, the dependences of $A_n(T)$ and of n_{deg} on, respectively, the temperature and the degeneracy temperature.

We must now determine the dependence of the width of the forbidden band (the energy gap) ε_G on the volume and temperature. It is very difficult to obtain these dependences theoretically and, therefore, we shall find them from the corresponding logarithmic derivatives and from the experimental data.

Differentiating ε_G with respect to the pressure at a constant temperature, we obtain

$$\left(-\frac{\partial \varepsilon_G}{\partial P}\right)_T = -\frac{\varepsilon_G}{K_T}\left(\frac{\partial \ln \varepsilon_G}{\partial \ln V}\right)_T. \tag{3.150}$$

Integrating Eq. (3.150) with respect to the pressure from a given value P_1 to some running value P, we obtain the required relationship

$$\varepsilon_G(P) = \varepsilon_G(P_1)\exp\left[-\int_{P_1}^{P}\left(\frac{\partial \ln \varepsilon_G}{\partial \ln V}\right)_T \frac{1}{K_T}dP\right], \quad T = T_1 = \text{const.} \tag{3.151}$$

Equation (3.151) gives the pressure dependence of the forbidden band width ε_G in terms of a practically constant value of the logarithmic derivative $(\partial \ln \varepsilon_G/\partial \ln V)_T$, which can be determined experimentally, and in terms of the isothermal bulk modulus K_T, which can be found easily from the equation of state.

In a narrow range of pressures, i.e., when the volume changes only slightly, we may regard both factors in the integrand of Eq. (3.151) as constants and we can take them outside the integral sign. This gives the formulas

$$\left.\begin{array}{c} \varepsilon_G(P) = \varepsilon_G(P_1)(V/V_1)^{L_{np}} \approx \varepsilon_G(P_1)[1 - L_{np}\beta_T(P - P_1)], \\[2mm] L_{np} \equiv \left(\dfrac{\partial \ln c_G}{\partial \ln V}\right)_T, \qquad \beta_T = \dfrac{1}{K_T}. \qquad L_{np}\beta_T(P - P_1) \ll 1, \end{array}\right\} \tag{3.151a}$$

which can be used to determine the logarithmic derivative L_{np} from the experimental data. We shall assume that the forbidden band width ε_G depends weakly on the temperature. Expanding $\varepsilon_G = \varepsilon_G(P, T)$ with respect to T and retaining only terms of the first order, we obtain

$$\varepsilon_G(P, T) = \varepsilon_{G0}(P) + \left(\frac{\partial \varepsilon_G}{\partial T}\right)_P T.$$

Since

$$\left(\frac{\partial \varepsilon_G}{\partial T}\right)_P = \left(\frac{\partial \varepsilon_G}{\partial T}\right)_V + \alpha\left(\frac{\partial \ln \varepsilon_G}{\partial \ln V}\right)_T \varepsilon_G, \tag{3.152}$$

it follows that

$$\varepsilon_G(P, T) = \varepsilon_{G0}(P) + \left[\left(\frac{\partial \varepsilon_G}{\partial T}\right)_V + \alpha\left(\frac{\partial \ln \varepsilon_G}{\partial \ln V}\right)_T \varepsilon_G\right]T = \varepsilon_{G0}(P) + \left[\left(\frac{\partial \varepsilon_G}{\partial T}\right)_V - \frac{\alpha}{\beta_T}\left(\frac{\partial \varepsilon_G}{\partial P}\right)_T\right]T.$$

In Eq. (3.152), the quantity α is the volume thermal expansion coefficient and the quantity $(\partial \varepsilon_G/\partial T)_V \neq 0$ is governed by the interaction of electrons and holes with phonons. Experiments show that the two terms enclosed in the square brackets are of the same order of magnitude:

$$\left(\frac{\partial \varepsilon_G}{\partial T}\right)_V \sim \alpha\left(\frac{\partial \ln \varepsilon_G}{\partial \ln V}\right)_T \varepsilon_G.$$

According to Eqs. (3.151) and (3.152), the dependences of ε_G on the pressure and temperature can be determined only if we know the experimental values of $(\partial \ln \varepsilon_G/\partial \ln V)_T$, $(\partial \varepsilon_G/\partial T)_V$, and α, as well as the equation of state $P = P(V, T)$.

Experimental Data

The energy spectrum of semiconductors is not as simple as that shown in Fig. (3.5). This means that the values of the logarithmic derivatives have a completely different meaning from that which would appear at first sight.

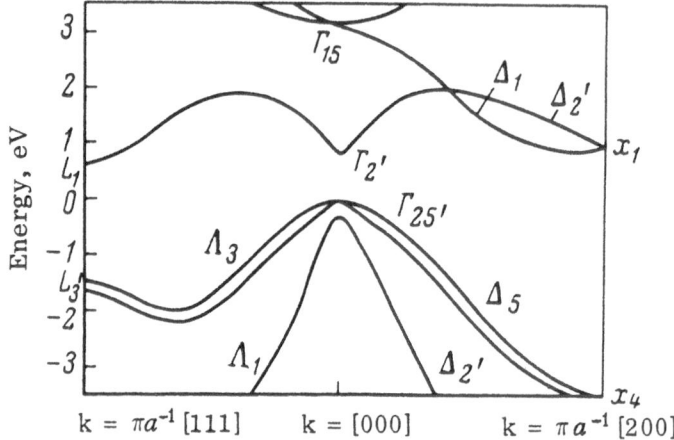

Fig. 3.8. Structure of the energy bands of germanium [35]. The abscissa represents the wave vectors along the [111] and [100] directions; the ordinate represents energy. The symbols used for the symmetry points are the same as those employed by Bouckaert, Smoluchowski, and Wigner [Phys. Rev., 50:58 (1936)]. The energy is measured from the valence band maximum $\Gamma_{25'}$. Approximate values of the energy gaps at 300°K are: $\Gamma_{25'} - L_1 = 0.65$ eV; $\Gamma_{25'} - \Delta_1$ (at its minimum) $= 0.85$ eV; $\Gamma_{25'} - \Gamma_{2'} = 0.80$ eV; $\Gamma_{25'} - \Gamma_{15} = 3.1$ eV; $L_1 - L_{3'} = 2.1$ eV; $X_1 - X_4 = 4.5$ eV.

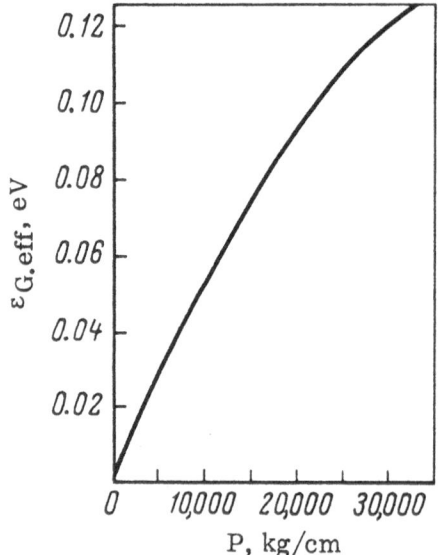

Fig. 3.9. Pressure dependence of the energy gap of germanium [33].

In order to analyze this problem in detail, we shall consider the best-known semiconductors, which are germanium and silicon. Figure 3.8 depicts the energy band structure of germanium [31]. It is shown in [32, 33] that the pressure dependence of the electrical conductivity of germanium can be explained in a natural manner in terms of a two-band model. When the pressure is increased, the conduction band edge along the [111] direction moves away from the valence band edge while the second conduction band edge along the [100] direction, which is initially further away, approaches the valence band. * The rates of shift of these bands are, respectively, $5 \cdot 10^{-6}$ eV /bar and $-2 \cdot 10^{-6}$ eV/bar. Thus, as the pressure increases, electrons are transferred from the [111] minimum to the [100] minimum of the conduction band. The effective forbidden band (energy gap) $\varepsilon_{G.eff}$ increases at first and then decreases. In the range of pressures used in investigations of germanium (30 kbar), we observe (Fig. 3.9) only the rising part of the dependence $\varepsilon_{G.eff}(P)$ and a reduction of the slope of this dependence. The falling part of the curve is reported in [34] on the basis of optical measurements (Fig. 3.10). A maximum is observed at $P \sim 5 \cdot 10^4$ atm.

---

*The existence of a second conduction band in germanium along the [100] direction has been demonstrated in studies of silicon – germanium alloys.

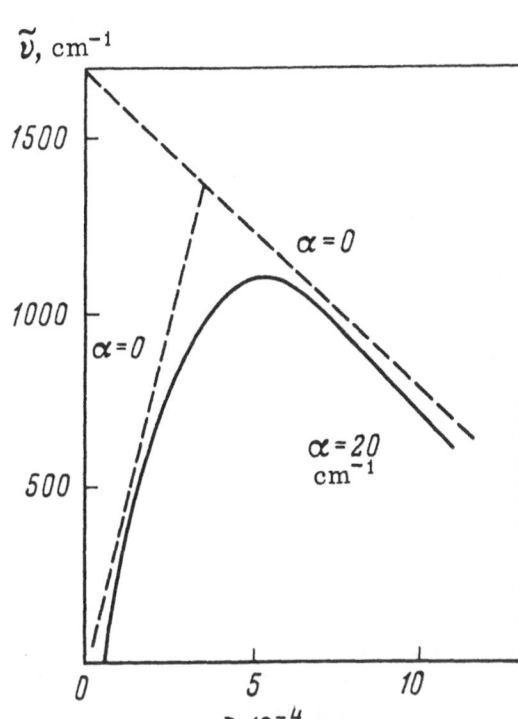

Fig. 3.10. Pressure dependence of the absorption edge of germanium for indirect transitions [34].

The expression for the effective forbidden band width $\varepsilon_{G.eff}$ is found as follows. According to Eqs. (3.137) and (3.138), the numbers of electrons in the first minimum n_1 ([111] band) and in the second minimum n_2 ([100] band) are given by

$$n_1 = A_{n_1}(T)\exp[\varepsilon_F - \varepsilon_c) / kT],$$

$$n_2 = A_{n_2}(T) \ \exp \ [(\varepsilon_F - \varepsilon_c - \ - \Delta\varepsilon) / kT],$$

where $\Delta\varepsilon$ is the difference between the energies of the minima of the first and second bands at $P = 0$. The number of holes is given by

$$p = A_P(T)\exp[(\varepsilon_v - \varepsilon_F)/ kT].$$

The effective forbidden band width (energy gap) $\varepsilon_{G.eff}$ is given by the formula

$$np = n^2 = (n_1 + n_2)p =$$

$$= A_{n_1}(T)A_P(T)\exp(-\varepsilon_{G.eff}/kT). \quad (3.153)$$

Substituting here n_1, n_2, p, and making certain simple transformations, we obtain

$$\varepsilon_{G.eff} = \varepsilon_c - \varepsilon_v - kT \ln\left[1 + \frac{A_{n_2}(T)}{A_{n_1}(T)} e^{-\Delta\varepsilon/kT}\right], \quad (3.154)$$

We can see that the shift of $\varepsilon_{G.eff}$ with increasing pressure may differ considerably from the required quantity $(\partial \ln \varepsilon_G / \partial \ln V)_T$, which represents the rate of closing of the energy gap of a semiconductor with increasing pressure. In the case of germanium, the quantity $L_{np} \equiv (\partial \ln \varepsilon_G / \partial \ln V)_T$ should be determined from the falling part of the dependence $\varepsilon_{G.eff} = \varepsilon_G(P)$ (see Fig. 3.10).* The falling part appears only at pressures $P > 50$ kbar. Consequently, the experimental data obtained at comparatively low pressures are frequently unsuitable for the determination of the range of temperatures and pressures in which the energy gap of insulators and semiconductors disappears. However, when the pressure is increased, we always reach a region in which the width of the energy gap (forbidden band) begins to decrease; only in this region need we determine the value of L_{np} which occurs in the formulas given in the present section.

The energy band scheme of silicon is shown in Fig. 3.11. In this case, the principal role is played by the [100] conduction band. Under pressure, this band approaches the valence band at a rate

$$\left(\frac{\partial\varepsilon_G}{\partial P}\right) = -1.5 \cdot 10^{-6} \text{ eV/bar,} \quad (3.155)$$

which is close to the rate for the [100] band of germanium. Figure 3.12 shows the shift of the absorption edge of silicon under the influence of pressure [34]. The reported measurements were carried out up to $1.4 \cdot 10^5$ atm. The rate of reduction of the gap was $-2 \cdot 10^{-6}$ eV/bar,

---

*The difference between the optical and thermal (electrical) values of the energy gap will be discussed later.

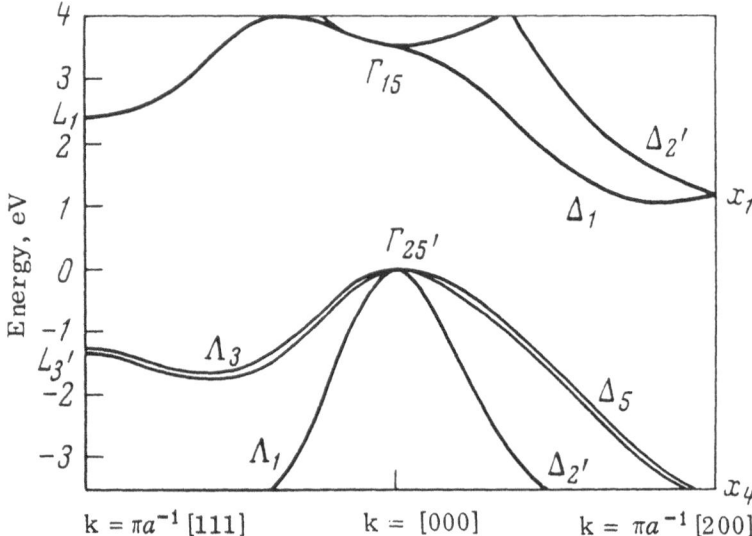

Fig. 3.11. Structure of the energy bands of silicon [35]. The abscissa represents the wave vectors along the [111] and [100] directions, the ordinate represents the energy. The symbols used for the characteristic symmetry points are the same as that employed by Bouckaert, Smoluchowski, and Wigner [Phys. Rev., 50:58 (1936)]. The energy is measured from the valence band maximum at $\Gamma_{25'}$. The approximate values of the energy gaps at 300°K are $\Gamma_{25'} - \Delta_1$ (at its minimum) = 1.1 eV; $\Gamma_{25'} - \Gamma_{15}$ = 3.5 eV; $L_1 - L_{2'}$ = 3.7 eV; $X_1 - X_4$ = 4.5 eV.

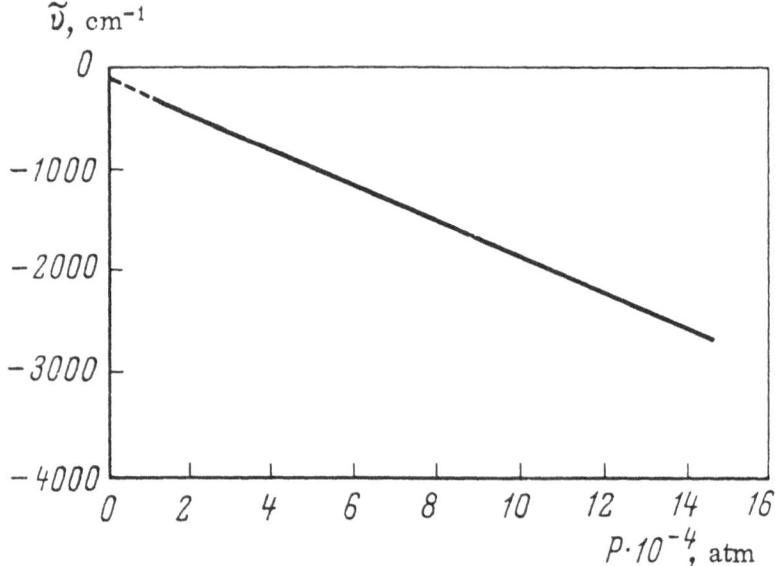

Fig. 3.12. Pressure dependence of the absorption edge of silicon [34].

TABLE 3.2

Substance	Lattice constant (25°C), Å	Energy gap		Conduction band min.	$\left(\dfrac{\partial \varepsilon_G}{\partial P}\right)_T \cdot 10^6$, eV·kgf^{-1}·cm^2	$\left(\dfrac{\partial \varepsilon_G}{\partial \ln V}\right)_T$, eV
		width, eV	temp., °K			
Diamond	3.567	5 3	300	$\Delta_{1\,opt}$		
Si	5.43	1.21	0	Δ_1	−1,5	+1.5
Ge	5.66	0.66	300	L_1	5	−3.8
		0.803	300	$\Gamma_{2'}$	12	−9
		0.85	300	Δ_1	0—−2	+1.5
Sn	6.489	0.08	0	L_1	5	
AlP	5.47	3.1	300			
AlAs	5.66	2.16	300			
AlSb	6.10	1.6	300	$\Delta_{1\,opt}$	−1.6	
GaP	5.47	{ 2.2	300	$\Delta_{1\,opt}$	−1.7	
		2.6	300	$\Gamma_{1\,opt}$	−1.8	
GaAs	5.66	1.53	300	Γ_1	{ 9.4	−7
					12	−9
		1.89	0	$\Delta_{1\,opt}$	<0	
GaSb	6.10	0.81	0	Γ_1	{ 16*	−9
					12*	−6.75
				L_1	5	−2.8
				Δ_1	<0	
InP	5.9	1.34	0	Γ_1	4.6	−6.15
				$\Delta_{1\,opt}$	−10	+7.45
InAs	6.07	0.36	300	Γ_1	{ 5.5 *	−3.3
					8.5 *	−5.1
					4.8 *	−2.9
InSb	6.49	0.27	0	Γ_1	{ 15.5 *	−6.7
					14.2 *	−6.1

* Results of different workers.

i.e., it was close to the rate found from electrical measurements and given by Eq. (3.155). Table 3.2, taken from the review of Paul and Wareschauer [35], presents the data on the influence of pressure on the energy gap in semiconductors. We can see that, with the exception of silicon, the results given in that table cannot be used to determine the logarithmic derivative L_{np}, which gives the rate of reduction of the gap with increasing pressure. The value of L_{np} for silicon is about 1.2.

Table 3.3 presents data on the influence of pressure on the optical value of the energy gap of insulators and conductors, obtained using the results of Drickamer [36].

Let us now consider the experimental data on the influence of temperature on the forbidden band width (energy gap). We have already considered some concepts relating to the optical energy gap of semiconductors [37]. In general, the band structure of semiconductors and insulators is very complex. This can be seen in the case of germanium and silicon (Figs. 3.8 and 3.9). The simplest optical transition is that in which an electron absorbs a photon and is transferred from the valence band to a state in the conduction band which has the same crystal momentum, for example, from $\Gamma_{25'}$ to $\Gamma_{2'}$ (Fig. 3.8). This type of transition may be accompanied by the formation of excitons (in an exciton, an electron and a hole combine to form a hydrogen-like quasiparticle), as well as by the absorption or emission of optical phonons with zero momentum, which establishes favorable conditions for transitions accompanied by the absorption of photons whose energies are less than the energy gap. Moreover, the dependence of the absorption coefficient on the photon energy is different for allowed and forbidden transitions.

TABLE 3.3

Substance	ν_0, cm^{-1}	ε_0, eV	$\dfrac{\Delta\nu}{\nu_0}$	$\dfrac{\Delta V}{V_0}$	ΔP, kbar	$\dfrac{d\ln\varepsilon_0}{d\ln V} = \dfrac{\Delta\nu}{\nu_0}\dfrac{V_0}{\Delta V}$	$\dfrac{d\ln\varepsilon_0}{dP}\cdot 10^3$, kbar^{-1}
TeJ	21 840	0.902	0.64	0.70	—	0.9	—
TeBr	23 950	0.989	0.63	0.375	—	2.8	—
TeCl	37 300	1.540	0.38	0.385	—	1.3	—
	29 300	1.210	0.61	0.32	—	1.9	—
PbCl$_2$	32 300	1.334	0.22	—	120	—	−1.8
	26 800	1.107	0.15	—	165	—	−0.9
PbBr$_2$	26 400	1.090	0.29	—	105	—	−2.7
	21 400	0.884	0.19	—	160	—	−1.2
HgCl$_2$	36 440	1.505	0.33	—	128	—	−2.4
	26 940	1.113	0.15	—	205	—	−0.7
HgI$_2$							
crystal	17 900	0.739	0.11	—	43	—	−2.5
	16 100	0.665	0.28	—	50	—	−5.6
compacted	18 450	0.762	0.11	—	43	—	−2.5
	16 650	0.688	0.27	—	50	—	−5.4
AgCl	24 300	1.004	0.09	—	450	—	−0.2
	23 300	0.962	0.09	—	250	—	−0.4
AgI	19 800	0.818	0.14	—	110	—	−1.3
	13 700	0.566	0.04	—	650	—	−0.0
Hg$_2$Cl$_2$	23 250	0.960	0.30	—	130	—	−2.3
	19 250	0.795	0.23	—	5.5	—	−4.2
Hg$_2$Br$_2$	21 250	0.878	0.33	—	90	—	−3.7
	15 750	0.650	0.48	—	68	—	−7.1
Hg$_2$I$_2$	19 400	0.811	0.36	—	70	—	−5.1
	17 200	0.710	0.44	—	54	—	−8.1
SnI$_4$	19 500	0.805	0.36	—	110	—	−3.3
	15 100	0.624	0.58	—	82	—	−7.0
CI$_4$	18 500	0.764	0.38	—	90	—	−4.2
	12 800	0.529	0.81	—	62	—	−13.1
I	12 500	0.516	0.08	—	100	—	−0.8
	8 000	0.330	0.69	—	100	—	−6.9
S	23 250	0.960	0.09	0.5	—	0.19	—
	17 750	0.733	1.05	0.33	—	3.3	—
Se	13 600	0.562	0.15	0.35	—	0.43	—
	5 600	0.231	3.4	0.88	—	3.86	—
As	7 250	0.299	0.27	0.23	—	1.17	—
	4 750	0.196	0.76	0.16	—	4.75	—
Red phosphorus	10 960	0.453	0.18	0.23	—	0.78	—
	8 460	0.349	0.425	0.16	—	2.66	—

TABLE 3.4

Substance	Melting point, °C	$\mathcal{E}_{G_0}$, eV	ε_0 (300° K), eV	$(\partial\varepsilon_0/\partial T)_P\cdot 10^4$, eV/deg K
PbS	1110	0.37	0.30	+4
PbSe	1065	—	0.22	+4
PbFe	904	—	0.27	+4
AlSb	1060	1.6	1.5	−3
GaP		—	2.4	−5.5
GaAs	1240		1.1—1.35	−5
GaSb	720	0.80	0.7	−3.5
InAs	940	0.47	0.35	−4
InSb	535	0.27	0.18	−3

TABLE 3.5

Substance	ε_{G0}, eV	$(\partial \varepsilon_G / \partial T)_P \cdot 10^4$, eV/deg K	ε_G (300°K), eV	ε_0 (300°K), eV	$(\partial \varepsilon_0 / \partial T)_P \cdot 10^4$, eV/deg K
Silicon	1.21	−4.2	1.09	1.05	−4
Germanium . . .	0.785	−4.0	0.65	0.62	−4.4

TABLE 3.6

Substance	$\left(\dfrac{\partial \varepsilon_G}{\partial T}\right)_P$ 10^4, eV/deg K	$-\dfrac{\alpha}{\beta_T}\left(\dfrac{\partial \varepsilon_G}{\partial P}\right) \cdot 10^4$, eV/deg K
Germanium:		
gap for [111] minimum of the conduction band.	−4	−0.7
gap for [100] minimum of the conduction band.	−4	−1.8
Silicon:		
gap for [100] minimum of the conduction band.	−2.5	+0.9

An optical transition without a change in the electron momentum is known as direct. Indirect transitions involve a change in the electron momentum and they can occur in insulators and semiconductors. In this case, an electron is transferred from a valence band maximum to the lowest minimum of the conduction band. The selection rules (the law of conservation of momentum) require an electron to emit or absorb a phonon whose momentum must be equal to the difference between the momenta of the electron in its initial and final states. Excitons may be formed (in various excited states) in such a transition. Other transitions and the corresponding absorption of light can take place in crystals with complex band structures [37]; we shall not consider this problem in detail. However, it is important to stress some physical causes responsible for the difference between the thermal (deduced from the temperature dependence of the electrical conductivity) and optical values of the energy gap. We shall denote the optical gap by ε_0 and the thermal gap by ε_G.

Table 3.4 gives information on the temperature dependences of the energy gap of various semiconductors. Table 3.5 lists the data for $(\partial \varepsilon_G / \partial T)_P$ and $(\partial \varepsilon_0 / \partial T)_P$ of silicon and germanium (the results given in Tables 3.4 and 3.5 are taken from the review by Burstein and Egli [38]). We can see that the values of these quantities are similar. More detailed data for germanium and silicon, reported in [38], are presented in Table 3.6. Following Burstein and Egli [38], we can obtain the general relationship:

$$\left(\frac{\partial \varepsilon}{\partial T}\right)_P \sim -(3-5) \cdot 10^{-4} \text{ eV/deg.} \tag{3.156}$$

The compounds PbS, PbSe, and PbTe are exceptions to this rule because for these compounds the quantity $(\partial \varepsilon / \partial T)_V$ is anomalously small and $(\partial \varepsilon / \partial T)_P$ is governed primarily by the thermal expansion.

3.7. Melting Curves

The locus of the melting points at various pressures, $T_{mp} = T_{mp}(P_{mp})$, is known as the melting curve and, by definition, it separates the solid state of a crystal from its liquid state. For this reason, the melting curve is important to the equations of state at high temperature.

Melting is a phase transition of the first kind and, in a rigorous treatment, the melting curve can be determined only if we know the chemical potential of the solid phase $\mu_s(P, T)$ as well as the chemical potential of the liquid phase $\mu_l(P, T)$. The melting curve is defined by the equality of the chemical potentials of the solid and liquid phases:

$$\mu_s(P, T) = \mu_l(P, T). \tag{3.157}$$

Solving Eq. (3.157) for temperature, we obtain the required equation $T_{mp} = T_{mp}(P_{mp})$. Melting is accompanied by the absorption of some heat, which is known as the latent heat of fusion. The process of melting (fusion) takes place at constant P and T. Consequently, the amount of heat absorbed during melting is equal to the change in the enthalpy of the substance under investigation. Using q to denote the heat of fusion per molecule, we obtain

$$q = w_l - w_s.$$

Using Eq. (3.157), we can express q in terms of a change in the entropy at the melting point:

$$\varepsilon_l - \varepsilon_s - T_{mp}(s_l - s_s) + P_{mp}(v_l - v_s) = w_l - w_s - T_{mp}(s_l - s_s) = 0.$$

Consequently,

$$q = T_{mp}(s_l - s_s) \tag{3.158}$$

(where ε, s, and v are the molecular energy, entropy, and volume). The value of q is only a few percent of the values of ε or w:

$$q \ll \varepsilon, w.$$

We cannot calculate (ε_s, w_s) and (ε_l, w_l) with the required accuracy. Therefore, Eq. (3.157) cannot be used to determine the melting curve and we must use semiempirical methods.

The slope of the initial part of the melting curve can be obtained using experimental data obtained at atmospheric pressure, as well as the Clapeyron — Clausius equation. This equation is obtained by differentiating Eq. (3.157) with respect to the temperature, as shown in Appendix A:

$$\frac{\partial \mu_s}{\partial T_{mp}} + \frac{\partial \mu_s}{\partial P_{mp}} \frac{dP_{mp}}{dT_{mp}} = \frac{\partial \mu_l}{\partial T_{mp}} + \frac{\partial \mu_l}{\partial P_{mp}} \frac{dP_{mp}}{dT_{mp}}. \tag{A.41}$$

Using Eqs. (A.41) and (3.158), we obtain

$$\frac{dP_{mp}}{dT_{mp}} = \frac{s_s - s_l}{v_s - v_l} = \frac{q}{T_{mp}(v_l - v_s)} = \frac{q}{T_{mp}\Delta v_{mp}} \tag{3.159}$$

or

$$\frac{dT_{mp}}{dP_{mp}} = \frac{T_{mp}\Delta v_{mp}}{q} \tag{3.159a}$$

where $\Delta v_{mp} = v_l - v_s$ is the change in the volume at the melting point, per molecule. Table 3.7, taken from Kaufman's review [39], compares the slopes of the melting curves (dT_{mp}/dP_{mp}) found experimentally and those deduced from the Clapeyron — Clausius formula (3.159a) using the measured values of q and Δv_{mp} at atmospheric pressure. The agreement between the two slopes is in general good and the discrepancies are simply due to the inaccuracy of the experimental measurements.

We shall now consider the semiempirical methods for the determination of the melting curves. All these methods are based on the properties of solids since the latter can be de-

TABLE 3.7

Metal	Structure	Volume at T_{mp}, cm^3/mole	$\Delta s = s_l - s_s$, cal·mole^{-1}·deg^{-1}	$\Delta v_{mp}/v$	dT_{mp}/dP_{mp} (calc), deg/kbar	dT_{mp}/dP_{mp} (exptl) deg/kbar
Li	bcc	13.3	1.59	0.0165	3.3	3.3 *
						1.5 *
Na	bcc	24.1	1.68	0.025	8.6	7.8 *
						6.5 *
K	bcc	46.0	1.65	0.0255	16.9	13.3 *
						8.5 *
Rb	bcc	56.1	1.79	0.025	18.7	18.0
Cs	bcc	70.0	1.69	0.026	25.7	20.0 *
						19.0 *
Al	fcc	10.5	2.74	0.060	5.5	6.4
Cu	fcc	7.6	2.30	0.0415	3.3	4.2
Ag	fcc	10.9	2.19	0.038	4.5	5.5
Au	fcc	10.7	2.21	0.051	5.9	
Ni	fcc	7.1	2.44	0.037	2 6	3.7
Pt	fcc	9.5	2.30	(0.038)	(3.8)	5.0
Rn	fcc	8.7	(2.32)	(0.039)	(3.5)	5.9
Pb	fcc	18.9	1.90	0.035	8.3	10.0
						6.6
Fe	fcc	7.7	2.20	(0.032)	2.7	3.0
Fe	bcc	7.7	2.03	0.030	2.7	
Tl	bcc	17.8	1.77	0.022	5.3	9.0
						5.0
Tl	fcc	17.8	(1.98)	(0.023)	4.9	
Mg	hcp	14.8	2.31	0.041	6.3	7.5
Zn	hcp	9.5	2.55	0.042	3.7	4.5
						4.8
Cd	hcp	13.4	2.44	0.040	5.3	9.0
						5.6
In	fc. tetra.	16.2	1.82	0.020	4.3	4.8
						5.6
Sn	bc. tetra.	16.5	3.41	0.028	3.2	2.8
						4.3
						2.7
Te	hexagonal	21.0	5.78	(0.020)	1.7	0.1
Sb	trigonal	18.7	5.25	−0.0095	−0.8	−0.50
						−0.44
Bi	trigonal	21.5	4.78	−0.0335	−3.6	−3.9
Ge	diamondlike cubic	13.9	6.28	−0.050	−2.7	−3.3
Ga	orthorhom.	11.8	4.41	−0.032	−2.0	−2 1

* According to different authors.

scribed theoretically much more simply than liquids. We shall now describe Lindemann's method and the method of the critical concentration of thermal defects. We shall also consider the empirical formula for the melting curve, derived by Simon.

Lindemann Melting Formula

In 1910, Lindemann [40] suggested that the ratio of the average amplitude of the thermal vibrations to the lattice parameter remained constant along the melting curve. Recently, this approach was considered in detail by Gilvarry [41]. The average amplitude of the thermal vibrations without allowance for the anharmonicity is equal to zero. Therefore, the currently used formulation of the Lindemann assumption reads as follows: "The ratio of the mean-square amplitude of the thermal vibrations $\langle u^2 \rangle$ to the square of the lattice parameter a^2 is constant along the melting curve":

$$\frac{\langle u^2 \rangle}{a^2} = \chi^2 = \text{const} \qquad \text{for} \qquad T_{\mathrm{mp}} = T_{\mathrm{mp}}(P_{\mathrm{mp}}). \qquad (3.160)$$

According to Gilvarry [41], $\chi \sim 0.075$.

Let us now consider this problem in a quantitative manner. The energy of thermal vibrations E_T, expressed in normal coordinates (p_α is the momentum and q_α is the coordinate) can be represented in the form:

$$E_T = \frac{1}{2m} \sum_{\alpha=1}^{3N} (p_\alpha{}^2 + m^2 \omega_\alpha{}^2 q_\alpha{}^2), \qquad (3.161)$$

where m is the mass of an atom in a given crystal.* The average value of the square of the normal coordinate α_1 is given by the standard formula

$$\langle q_{\alpha_1}^2 \rangle = \frac{\int\limits_{-\infty}^{\infty} q_{\alpha_1}^2 \exp(-\beta_1 q_{\alpha_1}^2)\, dq_{\alpha_1}}{\int\limits_{-\infty}^{\infty} \exp(-\beta_1 q_{\alpha_1}^2)\, dq_{\alpha_1}}, \qquad \beta_1 = \frac{m \omega_{\alpha_1}^2}{2kT}.$$

The integrals which occur in this formula have been encountered before in our discussion of the anharmonicity (Sec. 3.2). We thus obtain

$$\langle q_{\alpha_1}^2 \rangle = \frac{kT}{m \omega_{\alpha_1}^2}. \qquad (3.162)$$

We shall average $\langle q_{\alpha_1}^2 \rangle$

$$\langle q^2 \rangle = \frac{1}{3N} \sum_{\alpha=1}^{3N} \langle q_\alpha{}^2 \rangle$$

for the Debye model of a solid. The frequency distribution for this model (Chapter 2) is given by the expression

$$\rho(\omega)\, d\omega = \begin{cases} 9N \dfrac{\omega^2 d\omega}{\omega_m{}^3} & \text{for} \qquad \omega \leqslant \omega_m = \bar{c} \left(\dfrac{6\pi^2}{v} \right)^{1/3}, \\ 0 & \text{for} \qquad \omega > \omega_m, \end{cases}$$

where $\bar{c}$ is the average velocity of sound and v is the volume of one atom. Replacing summation with integration over frequencies, we obtain

$$\langle q^2 \rangle = \frac{1}{3N} \int\limits_{0}^{\omega_m} \langle q_\alpha{}^2 \rangle \, \rho(\omega)\, d\omega = \frac{3kT}{m\omega_m{}^2}.$$

Identifying the mean-square amplitude of the thermal vibrations $\langle u^2 \rangle$ with $\langle q^2 \rangle$, we obtain

---

*The application of the Lindemann approach to multiatomic solids is less justified [42]. This will be discussed in detail later.

$$\langle u^2 \rangle = \frac{3kT}{m\omega_m^2} = \frac{3kT}{m\theta^2}\left(\frac{\hbar}{k}\right)^2, \tag{3.163}$$

in which we have used the definition of the Debye temperature $k\theta = \hbar\omega_m$. The Debye temperature θ is related to the average velocity of sound $\bar{c}$, the bulk modulus K_S, and Poisson's ratio σ by the well-known expression [13]

$$\theta = \frac{\hbar}{k}\left(\frac{6\pi^2\rho_0}{m}\right)^{1/3} x^{-1/3}\bar{c},$$

$$\bar{c} = 3^{5/6}\left(\frac{K_S}{\rho}\right)^{1/2}\delta(\sigma),$$

$$\delta(\sigma) = \sqrt{\frac{1-\sigma}{1+\sigma}}\left[1 + 2^{5/2}\left(\frac{1-\sigma}{1-2\sigma}\right)^{3/2}\right]^{-1/3}.$$

Substituting in Eq. (3.160) the value of $\langle u^2 \rangle$ from Eq. (3.163), as well as the values of θ and $\bar{c}$, we obtain

$$\frac{\langle u^2 \rangle}{a^2} = \left(\frac{\xi_0}{18\pi^2}\right)^{1/3}\frac{kT\rho}{mK_S\delta^2(\sigma)} = \chi^2 = \text{const}, \tag{3.164}$$

where ξ_0 is a constant which relates the volume of one atom at atmospheric pressure v_0 with the cube of the lattice constant a_0^3 ($v_0 = \xi_0 a_0^3$). Assigning the index zero to all quantities at the beginning of the melting curve [T_{mp0}, $K_{S,mp0}$, ρ_{mp0}, $\delta^2(\sigma_{mp0})$ are known quantities], we obtain the equation for the melting curve in the Lindemann approach based on the Debye model of a solid:

$$T_{mp} = T_{mp0}\left[\left(\frac{\rho_{mp0}}{\rho_{mp}}\right)\left(\frac{K_{s,mp}}{K_{s\cdot mp0}}\right)\left(\frac{\delta^2(\sigma_{mp})}{\delta^2(\sigma_{mp0})}\right)\right]. \tag{3.165}$$

Using Eq. (3.165), we can usually assume that the factor depending on Poisson's ratio is equal to unity:

$$T_{mp} = T_{mp0}\left[\frac{\rho_{mp\,0}K_{s.mp}}{\rho_{mp}K_{s\,mp\,0}}\right]. \tag{3.165a}$$

In this case, the melting curve can be written in the analytic form if the equation of state is known. Thus, in the case of the Born—Mayer equation of state given by Eq. (3.50a):

$$P = A_1 x^{-2/3}\exp(-bx^{1/3}) - Kx^{-4/3} + \frac{3RT\gamma\rho_0}{xM} + (A_a\gamma_a + A_e\gamma_e)x^{-1}T^2. \tag{3.166}$$

The equation for the melting curve then assumes the form

$$T_{mp} = T_{mp0}\frac{A_1(2x_{mp}^{-2/3} + bx_{mp}^{-1/3})\exp(-bx_{mp}^{1/3}) - 4K\,x_{mp}^{-4/3} + 3f(x_{mp}T_{mp})}{3K_{s\,mp0}}\frac{x_{mp}}{x_{mp0}}, \tag{3.167}$$

where

$$f(x_{mp}T_{mp}) = \frac{3R\rho_0\gamma_{mp}T_{mp}}{Mx_{mp}}\left(1 - \frac{d\ln\gamma_{mp}}{d\ln x_{mp}}\right) - x_{mp}(A_a'' + A_e'')_{mp}T_{mp}^2 +$$

$$+ \frac{x_{mp}T_{mp}\left[\frac{3R\rho_0\gamma_{mp}}{Mx_{mp}} + 2x_{mp}^{-1}(A_a\gamma_a + A_e\gamma_e)_{mp}T_{mp}\right]^2}{\frac{3R\rho_0}{M}\left[1 + \frac{2(A_a + A_e)_{mp}T_{mp}M}{3R\rho_0}\right]} \tag{3.168}$$

The thermal components in $K_{S\,mp0}$ and $K_{S\,mp}$ are generally of little importance and in those cases where they are not known accurately, we can neglect them quite safely.

A difficulty is encountered in the determination of the mean-square displacement $\langle u^2 \rangle$ when the Lindemann approach is applied to crystals with complex unit cells. In this case, in addition to the three acoustical vibration modes, we have several optical modes. The averaging for the acoustical modes can be carried out exactly in the same way as for a monatomic Debye solid. Denoting the mean-square amplitude in the acoustical case by $\langle u^2 \rangle_{ac}$, we obtain for this amplitude a formula which is identical with Eq. (3.163),

$$\langle u^2 \rangle_{ac} = \frac{3kT}{m_a \omega_m^2} = \frac{3kT}{m_a \theta^2} \left(\frac{\hbar}{k} \right)^2 , \tag{3.163a}$$

where m_a is the average atomic mass. Since the optical frequencies belonging to a single optical mode are similar, it follows that we should use the formula (3.162) for the optical mean square amplitudes of the vibrations $\langle u^2 \rangle_{0i}$ (i is the number of the optical mode):

$$\langle u^2 \rangle_{0i} \approx \frac{kT}{m_1 \omega_{0i}^2}, \tag{3.162a}$$

where m_1 is the reduced atomic mass and ω_{0i} is the average optical frequency for the i-th optical mode.

Usually, the optical frequencies are of the same order but higher than the Debye frequency. Since, according to Eqs. (3.162) and (3.163), the melting is governed by the lowest of the characteristic frequencies and $\omega_m < \omega_{0i}$, it follows that, in order to obtain the melting curve for complex crystals, we can base our treatment on Eq. (3.163a) and revert to the original equation Eq. (3.165a). It follows that the Debye averaging of the initial formula (3.162) is not unique. In particular, in the Einstein model of a solid, we have $\omega_\alpha = (k/\hbar)\theta_E$ and the equation for the melting curve reduces to the form

$$T_{mp} = T_{mp0} (\rho_{mp0}/\rho_{mp})^{1/3} \left(\frac{\theta_{E\,mp}}{\theta_{E\,mp0}} \right)^2 . \tag{3.169}$$

Another important special case of Eq. (3.162) is obtained by replacing $\omega_{\alpha_1}^2$ by the mean-square frequency

$$\langle u^2 \rangle_{ms} = \frac{kT}{m \langle \omega^2 \rangle}. \tag{3.162b}$$

In this case, the equation for the melting curve is of the form

$$T_{mp} = T_{mp0} \left(\frac{\rho_{mp0}}{\rho_{mp}} \right)^{1/3} \frac{\langle \omega^2 \rangle_{mp}}{\langle \omega^2 \rangle_{mp0}}. \tag{3.170}$$

It is known [1, 3, 5] that in the case of central forces the mean-square frequency can be calculated from the potential of the interaction between the atoms. This approach also gives the analytic form of the melting curve.

In the case of van der Waals crystals, the expression for $\langle \omega^2 \rangle$, obtained in Sec. 6.3 (where a fuller explanation is given), is of the form

$$\langle \omega^2 \rangle = \frac{zx^2}{3ml_0^2} \left\{ \left(1 - \frac{2}{xy} \right) A_x e^{-xy} - \frac{n_1(n_1+1)}{x^2} \left(1 - \frac{2}{(n_1+1)y} \right) A_{n_1} y^{-(n_1+2)} \right\}.$$

Substituting this expression for $\langle \omega^2 \rangle$ into Eq. (3.170), we obtain

$$T_{mp} = T_{mp0} \left(\frac{\rho_{mp0}}{\rho_{mp}} \right)^{1/3} \frac{1}{\langle \omega^2 \rangle_{mp0}} \frac{zx^2}{3ml_0^2} \left[\left(1 - \frac{2}{xy} \right) A_x e^{-xy} - \frac{n_1(n_1+1)}{x^2} \left(1 - \frac{2}{(n_1+1)y} \right) A_{n_1} y^{-(n_1+2)} \right]. \tag{3.171}$$

Equation (3.171) can be used to determine the equation of state for van der Waals crystals as well as the melting curve $T_{mp} = T_{mp}(P_{mp})$. In the case of ionic crystals of the NaCl or CsCl type, the expression for $\langle \omega^2 \rangle$ is of the form (Sec. 6.3)

$$\langle \omega^2 \rangle = \frac{1}{6} \frac{z \varkappa^2}{l_0^2} \left[\left(1 - \frac{2}{\varkappa y} \right) \frac{A_\varkappa}{M} e^{-\varkappa y} + \frac{\bar{z}}{z} \left(1 - \frac{2}{\varkappa_1 y} \right) A_M e^{-\varkappa_1 y} \right]$$

and, consequently, $T_{mp} = T_{mp}(x_{mp})$ is given by

$$T_{mp} = T_{mp0} \left(\rho_{mp0}/\rho_{mp} \right)^{\frac{1}{3}} \frac{1}{\langle \omega^2 \rangle_{mp0}} \frac{z \varkappa^2}{6 l_0^2} \left[\left(1 - \frac{2}{\varkappa y} \right) \frac{A_\varkappa}{M} e^{-\varkappa y} + \frac{\bar{z}}{z} \left(1 - \frac{2}{\varkappa_1 y} \right) A_M e^{-\varkappa_1 y} \right]. \qquad (3.172)$$

Using Eq. (3.172) and the corresponding equation of state, we can easily calculate the melting curve $T_{mp} = T_{mp}(P_{mp})$ for ionic crystals.

Comparing Eqs. (3.172) and (3.167), we find that the dependence $T_{mp}(x_{mp})$ is primarily an exponential dependence on the reduced volume x_{mp}. However, the exact nature of the melting curve $T_{mp} = T_{mp}(x_{mp})$ obtained by the Lindemann approach depends on the method used to average Eq. (3.162) over the frequencies.

Method of the Critical Concentration of Thermal Defects [42]

In this, the dependence of the melting temperature T_{mp} on the pressure P_{mp} is determined assuming that the melting begins when the concentration of the thermal defects $c_d = c_d(T, P)$ (these may be the Frenkel or the Schottky defects) reaches a critical value $c_{d.cr}$:

$$c_{d.cr} = c_d(T_{mp0}, P_{mp0}) = c_d(T_{mp}, P_{mp}) = \text{const.} \qquad (3.173)$$

The concentration of the thermal defects near the melting point is of the order of 10^{-3}–10^{-4} and it increases exponentially with the temperature and pressure (the argument of the exponential function at the melting point is slightly higher than 10).

The concentration of such defects cannot increase appreciably along the melting curve because the activation energy for the formation of defects decreases considerably (due to the interaction between defects) at defect concentrations $c_d > 10^{-3}$–10^{-4}. Therefore, the melting curves based on the hypothesis represented by Eq. (3.173) are likely to overestimate the melting point. However, an important advantage of this method, compared with the Lindemann approach, is the fact that the defect concentration method is physically valid for monatomic as well as multiatomic unit cells.

We shall use the general formula (3.75a), which gives the dependence of the defect concentration c_d on the pressure and temperature. The condition (3.173) then assumes the form

$$T_{mp} = T_{mp0} \frac{h_0(P_0)}{h_{00} + T_{mp0}[s_{20}(P_0) - s_{200}]},$$

where $h_0(P_0)$ is the enthalpy of formation of a defect on the cold-compression isotherm P_0; s_{20} is the thermal entropy of formation of one defect on the cold-compression isotherm; h_{00} and s_{200} are the corresponding enthalpy and entropy at $P_0 = P_{00}$. Expressing h_0 in terms of the compressibility and the logarithmic derivative of h_0 with respect to the volume, which is denoted by L_0, we can use Eq. (3.104) to obtain

$$T_{mp} = T_{mp0} \frac{\exp \left\{ \int_{P_{00}}^{P_0} \beta_{T0} L_0 dP \right\}}{1 + \dfrac{T_{mp0}}{h_{00}} [s_{20}(P_0) - s_{200}]}, \qquad (3.174)$$

and since

$$\frac{T_{mp0}}{h_{00}}[s_{20}(P_0) - s_{200}] \ll 1, \tag{3.175}$$

the equation for the melting curve (3.174) simplifies considerably to

$$T_{mp} = T_{mp0} \exp\left\{\int_{P_{00}}^{P_0} \beta_{T0} L_0 dP\right\}. \tag{3.176}$$

The logarithmic derivative L_0 is a weak function of the volume. For the sake of simplicity, this derivative can be expressed in the form

$$L_0(x) = L_{00} x^n. \tag{3.177}$$

At high pressures, $n = 1/3$. When the dependence of L_0 on the volume is unknown, we may assume that $n = 1/3$ throughout the investigated range of pressures:

$$T_{mp} = T_{mp0} \exp\left\{L_{00} \int_{P_{00}}^{P_0} \beta_{T0} x^{1/3} dP\right\}. \tag{3.176a}$$

In the limiting case of low pressures

$$\int_{P_{00}}^{P_0} \beta_{T0} L_0 dP \ll 1$$

the logarithmic derivative L_0 can be replaced by its value under normal conditions L_{00}. Consequently, we obtain the following relationships:

$$T_{mp} = T_{mp0}\left\{1 + L_{00}\ln\frac{\rho_{mp}}{\rho_{mp0}}\right\} = T_{mp0}\{1 + L_{00}\beta_{T00}(P_0 - P_{00})\}, \quad \frac{dT_{mp}}{dP} = T_{mp0} L_{00}\beta_{T0}. \tag{3.178}$$

The value of L_{00} can be found from Eq. (3.178) and the initial slope of the experimentally determined melting curve; alternatively, it can be found from the Clapeyron–Clausius equation

$$L_{00}\beta_{T0} = \frac{\Delta v_{mp}}{q}, \tag{3.179}$$

if Δv_{mp}, q, and β_{T0} are known.

Simon's Equation

Simon's equation [43] is used widely; it is given by

$$\frac{P_{mp} - P_{mp0}}{a} = \left(\frac{T_{mp}}{T_{mp0}}\right)^c - 1, \tag{3.180}$$

where P_{mp0}, T_{mp0} are the coordinates of the triple point of the substance under consideration; a and c are constants which must be found from the experimental data for each specific substance. In the majority of cases, P_{mp} is of the order of a few kilobars or more, while the value of P_{mp0} is of the order of 1 bar. Therefore, we can drop P_{mp0} from Eq. (3.180) to obtain

$$P_{mp} = a[(T_{mp}/T_{mp0})^c - 1]. \tag{3.180a}$$

Calculating the derivative $(dP_{mp}/dT_{mp})_{mp0}$ and comparing it with the Clapeyron—Clausius formula (3.159) and with the formula (3.178), we obtain

$$\frac{dP_{mp}}{dT_{mp}} = \frac{ac}{T_{mp0}}\left(\frac{T_{mp}}{T_{mp0}}\right)^{c-1},$$ (3.181)

$$\frac{\Delta v_{mp}}{2} = L_{00}\beta_{T0} = \frac{1}{ac}.$$ (3.182)

Simon's equation is empirical.

Extrapolation of this equation to wider ranges of the parameters (which is sometimes done in geophysics) can give rise to large unknown errors. The usefulness of Simon's equation lies in the fact that experimentalists have become accustomed to analyzing their results by means of this equation. Extensive experimental data on the melting curves are analyzed systematically in [44] on the basis of Simon's equation. Some of the data in [44] are included in Table 3.8.

The success of Simon's equation in describing the experimental data is simply due to the fact that the two–parameter equation (3.180a) describes well that small part of the melting curve, which is usually investigated experimentally. Nevertheless, attempts have been made to justify this equation theoretically. Thus, Eq. (3.180a) is derived in [44] from the Mie—Grüneisen equation of state, from the Lindemann equation in the form given by Eq. (3.169), and from the Grüneisen parameter $\gamma = -d\ln\theta/d\ln V$. However, this derivation is based on an unjustified assumption, according to which the parameter a in Eq. (3.180a) is equal to the pressure on the cold-compression isotherm at $T = 0°K$, $V = V_{0°K}$. This assumption is unacceptable and, therefore, the theoretical justification of Eq. (3.180a), as given in [44], is unsatisfactory.

We have already mentioned that the parameters in Eq. (3.180a) are selected using a small part of the melting curve corresponding to a pressure range of the order of ten kilobars. In this case, the change in the volume of a solid on the melting curve is small and it can be expanded as a series in terms of the corresponding changes

$$\Delta P_{mp} = P_{mp} - P_{mp0} \text{ and } \Delta T_{mp} = T_{mp} - T_{mp0},$$
$$V_{mp} = V_{mp0}[1 - (\beta_{T0})_{mp0}\Delta P_{mp} + (\alpha)_{mp0}\Delta T_{mp}],$$

where $(\beta_{T0})_{mp0}$ and $(\alpha)_{mp0}$ are, respectively, the compressibility and the thermal expansion coefficient of a solid at the beginning of the melting curve. Next, we shall express the volume dependence of the characteristic temperature θ of Eq. (3.169) in terms of the Grüneisen parameter $\gamma = -d\ln\theta/d\ln V$. The Grüneisen parameter itself can be regarded as a function of the volume and it can be represented by the simple expression

$$\gamma_{mp} = \gamma_{mp0}\left(\frac{V_{mp}}{V_{mp0}}\right)^{x}$$

where χ is some constant. Then, after certain elementary transformations have been made, the Lindemann formula (3.169) yields Simon's equation:

$$\frac{\Delta P_{mp}}{(K_T)_{mp0}} = \left(\frac{T_{mp}}{T_{mp0}}\right)^{\frac{1}{2(\gamma_{mp0}-1/3)} +(\alpha)_{mp0}T_{mp0} \mid} -1.$$ (3.183)

Since the derivation is valid with an accuracy to terms linear in ΔP_{mp} and ΔT_{mp}, the linearization of Eq. (3.183) and the comparison of the resultant expression with Eq. (3.81) gives

TABLE 3.8

Substance	$T_{mp0},$ °K	a, bar	c or A, bar/deg C*
Normal melting curves			
Aluminum	933.3		$A = 156$
Argon	83.812	2114	1.593
Bismuth VI	558	6600	11.1
Bismuth VII	462	6600	6.2
Cadmium	594.1	45000	2.4
Cesium	302.9	2674	4.49
Copper	1356.2		$A = 213$
Gallium II	276		$A = 437$
Gallium III	318	50000	2.4
Helium He^3	3.252	117.60	1.5178
Helium He^4	2.046	50.96	1.5602
Hydrogen H_2	14.155	274.22	1.744070
Deuterium D_2	18.811	429.67	1.787213
Tritium T_2	20.681	529.94	1.764179
Hydrodeuterium HD	16.6	259.6	2.23
Indium	429.76	35800	2.30
Iron	1805	1070000	1.76
Krypton	115.745	2376	1.6169
Lead	600.2	32300	2.405
Lithium	453.7	9000	14.8
Magnesium	923	6400	5.8
Mercury	234.29	38215	1.177
Neon	24.617	1037.53	1.599916
Nickel	1726	1020000	2.2
Nitrogen	63.1604	1606.56	1.79100
Oxygen	54.383	2732.95	1.742594
Phosphorus (white)	317.4	5525	1.92
Platinum	2046	1020000	2.0
Potassium	335.7	4270	4.44
Rhodium	2253	500000	1.30
Rubidium	311.9	3951	3.74
Silver	1234	9000	7.6
Sodium	370.78	11971	3.533
Selenium	495.5	11700	2.04
Sulfur (L-monoclinic)	387.2	2795	4.18
Sulfur (L-orthorhombic)	393.17	2914	4.31
Tellurium II	716		$A = 336 \pm 7$
Thallium	576.8	64000	1.38
Tin I	505.05	57000	3.4
Tin II	591	14700	5.2
Xenon	161.364	2610	1.5892
Zinc	692.7	60000	2.4
Cesium chloride	933	8400	2.3
Indium antimonide II	608		$A = 437$
Lithium chloride	878	14500	2.5
Potassium chloride I	1043	6900	5.7
Potassium chloride II	1315	12100	4
Rubidium chloride I	990.5	6600	6
Rubidium chloride II	1127	7200	4
Sodium bromide	1014	12200	2.9
Sodium chloride	1073.5	16700	2.7
Sodium fluoride	1265	14300	5.5
Sodium iodide	928	10100	2.8

*For substances whose parameters a and c are unknown, we give the quantity A, which is the initial slope of the melting curve.

TABLE 3.8 (continued)

Substance	T_{mp_0}, °K	a, bar	c or A, bar/deg.C
Anomalous melting curves			
Antimony.	903.7	—29 400	60
	903.7	—91 000	17
Bismuth I.	544.2	—27 250	5.60
Cesium II	?	?	7.7
Gallium I	303.01	—57 500	2.5
Germanium.	1213		$A = -263$
Silicon.	1693		$A = -172$
Aluminum antimonide	1333		$A = -145$
Cadmium telluride I	1318		$A = -50$
Cadmium telluride II	1041	—3400	16
Copper chloride	703	—55 000	11
Gallium antimonide	973		$A = -200$
Gallium arsenide	1411		$A = -294$
Indium antimonide I	1213		$A = -233$
Indium arsenide	1298		$A = -100$
Indium phosphide.	1333		$A = -344$
Potassium tetrasilicate	1038		$A = -166$
Water H_2O I . . . . . . : . . .	273.15	—3952	9.0
Heavy water D_2O II	276.97	—4596	7.75

* For substances whose parameters a and c are unknown, we give the quantity A, which is the initial slope of the melting curve.

$$ac = (K_T)_{mp_0} \left[\frac{1}{2(\gamma_{mp_0} - 1/3)} + (a)_{mp_0} T_{mp} \right]. \tag{3.184}$$

We have thus found a relationship between the product of the constants in Simon's equation on the one hand and the bulk modulus $(K_T)_{mp_0}$ as well as the Grüneisen parameter γ_{mp_0} on the other, since

$$(a)_{mp_0} T_{mp_0} \ll \frac{1}{2(\gamma_{mp_0} - 1/3)}.$$

The values of $(K_T)_{mp}$ and γ_{mp} vary along the melting curve and the "constants" a and c in Eq. (3.180a) should themselves depend on P_{mp} and T_{mp}.

It would be interesting to estimate the change in the latent heat of fusion q_{mp} along the melting curve. It is reasonable to assume that the work done in the breakup of the crystal lattice in the melting process (the process at given values of P_{mp} and T_{mp}) is proportional to the enthalpy of the formation of thermal defects h_{mp}. Then, we can easily write the relationship

$$q_{mp} \sim q_{mp_0}(T_{mp}/T_{mp_0}), \tag{3.185}$$

which is the analog of Trouton's rule at high pressures (the assumption that the latent heat of fusion is proportional to the melting point is the essence of Trouton's rule).

Many anomalies have recently been discovered in the behavior of the melting curves. Physically, these anomalies are always due to the fact that the solid and liquid phases at the melting point may have different structures. For example, the solid may have the bcc structure, whereas the number of nearest neighbors in the liquid may correspond to the fcc structure. This difference between the structures may be even more complex. At sufficiently high pressures, when all the polymorphic transitions are completed, the solid phase always transforms to the corresponding liquid phase and the melting curves are "normal." In this case,

T_{mp} is a smooth function of P_{mp} with a slowly decreasing slope ($d^2T_{mp}/dP^2_{mp} < 0$). If the slope of the melting curve is known, we can assume that it is constant and extrapolate the curve to obtain the melting points at higher pressures.

Literature Cited

1. R. E. Peierls, Quantum Theory of Solids, Clarendon Press, Oxford (1955).
2. L. D. Landau and E. M. Lifshitz, Statistical Physics, 2nd ed., Pergamon Press, Oxford (1968).
3. G. Leibfried and W. Ludwig, "Theory of anharmonic effects in crystals," Solid State Phys., 12:276 (1961).
4. V. N. Zharkov, Dokl. Akad. Nauk SSSR, 154:302 (1964).
5. V. N. Zharkov, in: Solids under Pressures and Temperatures in the Interior of the Earth [in Russian], Nauka, Moscow (1964), p. 41.
6. I. S. Gradshtein and I. M. Ryzhik, Table of Integrals, Series, and Products, Academic Press, New York (1965).
7. A. Sommerfeld and H. A. Bethe, "Elektronentheorie der Metalle," in: Handbuch der Physik, Vol. 24, Part 2, Springer-Verlag, Berlin (1933), p. 333.
8. R. E. Peierls, "Elektronentheorie der Metalle," Ergeb. Exakt. Naturwiss, 11:264 (1932).
9. I. M. Lifshits and M. I. Kaganov, Usp. Fiz. Nauk, 78:411 (1962).
10. D. Lazarus and N. H. Nachtrieb, "Effect of high pressure on diffusion," in: Solids under Pressure (ed. by W. Paul and D. M. Wareschauer), McGraw-Hill, New York (1963), p. 43.
11. R. H. Dickerson, R. C. Lowell, and C. T. Tomizuka, Phys. Rev., 137:A613 (1965).
12. V. N. Zharkov and V. A. Kalinin, Dokl. Akad. Nauk SSSR, 135:811 (1960).
13. V. N. Zharkov, Trudy Inst. Fiziki Zemli Akad. Nauk SSSR, No. 20(187), p. 3 (1962).
14. J. Frenkel, Z. Physik, 35:652 (1926).
15. J. Frenkel, Kinetic Theory of Liquids, Oxford University Press (1946).
16. N. F. Mott and R. W. Gurney, Electronic Processes in Ionic Crystals, Clarendon Press, Oxford (1948).
17. A. B. Lidiard, "Ionic conductivity," in: Handbuch der Physik (ed. by S. Flügge), Vol. 20, Springer-Verlag, Berlin (1957), pp. 246-349.
18. H. G. van Bueren, Imperfections in Crystals, 2nd ed., North-Holland, Amsterdam (1961).
19. V. N. Zharkov, Trudy Inst. Fiziki Zemli Akad. Nauk SSSR, No. 11(178), p. 14 (1960).
20. B. I. Davydov, Izv. Akad. Nauk Ser. Geofiz., No. 12, p. 1411 (1956).
21. J. Yamashita and T. Kurosawa, J. Phys. Soc. Japan, 10:610 (1955).
22. T. Kurosawa, J. Phys. Soc. Japan, 13:153 (1958).
23. N. F. Mott and M. J. Littleton, Trans. Faraday Soc., 34:485 (1938).
24. F. G. Fumi and M. R. Tosi, Discussions Faraday Soc., 23:92 (1957).
25. R. W. Keyes, "Continuum models of the effect of pressure on activated processes," in: Solids under Pressure (ed. by W. Paul and D. M. Wareschauer), McGraw-Hill, New York (1963), p. 71.
26. C. Zener, Trans. AIME, 147:361 (1942).
27. J. C. Slater, Introduction to Chemical Physics, McGraw-Hill, New York (1969), p. 239.
28. W. Shockley, Electrons and Holes in Semiconductors, Van Nostrand, New York (1950).
29. F. J. Blatt, "Theory of mobility of electrons in solids," Solid State Phys., 4:200 (1957).
30. J. McDougall and E. C. Stoner, Phil. Trans. Roy. Soc. London, A237:67 (1938).
31. W. Paul, in: High Pressure Physics and Chemistry (ed. by R. S. Bradley), Vol. 1, Academic Press, New York (1963), p. 299.
32. H. Brooks and W. Paul, Bull. Am. Phys. Soc., 1:48 (1956).
33. M. I. Nathan, W. Paul, and H. Brooks, Phys. Rev., 124:391 (1961).
34. T. E. Slykhouse and H. G. Drickamer, J. Phys. Chem. Solids, 7:210 (1958).
35. W. Paul and D. M. Wareschauer, "The role of pressure in semiconductor research," in: Solids under Pressure (ed. by W. Paul and D. M. Wareschauer), McGraw-Hill, New York (1963), p. 179.

36. D. H. Drickamer, "The electronic structure of solids under pressure," in: Solids under Pressure (ed. by W. Paul and D. M. Wareschauer), McGraw-Hill, New York (1963), p. 357.
37. W. Paul and H. Brooks, Progress in Semiconductors, 7:135 (1963).
38. E. Burstein and P. H. Egli, Adv. Electronics Electron Phys., 7:1 (1955).
39. L. Kaufman, "Phase equilibria and transformations in metals under pressure," in: Solids under Pressure (ed. by W. Paul and D. M. Wareschauer), McGraw-Hill, New York (1963), p. 303.
40. F. A. Lindemann, Phys. Z., 11:609 (1910).
41. J. J. Gilvarry, Phys. Rev., 102:308, 317, 325 (1956); 104:908 (1956).
42. V. N. Zharkov, Izv. Akad. Nauk Ser. Geofiz., No. 3, p. 465 (1959).
43. F. E. Simon and G. Glatzel, Z. Anorg. Allgem. Chem., 178:309 (1929).
44. S. E. Babb, Jr., Rev. Mod. Phys., 35:400 (1963).

CHAPTER 4

EQUATIONS OF STATE DEDUCED FROM EXPERIMENTAL DATA

4.1. High-Pressure Techniques

The last decade has seen considerable progress in experimental investigations of solids at high pressures. This has been stimulated by the high-pressure synthesis of materials which are of special interest technologically and geophysically. Recently, many laboratories began high-pressure studies not only to synthesize new materials at high pressures, but also to investigate the properties of matter at these pressures.

The generation of high static pressures under laboratory conditions and the making of measurements at these pressures are fairly difficult. The experimental techniques grow increasingly complex as the pressures get higher. A major problem is that the useful volume available in piezometers decreases rapidly with increasing pressure.

We shall describe the most important techniques used at high pressures. Details of the apparatus described here, as well as of apparatus designed for specific purposes, can be found in [1-5]. In most cases, high pressures are generated using multiton presses. The force generated by such a press is transmitted by a piston to a high-pressure chamber where the sample under investigation is compressed.

The apparatus shown in Fig. 4.1a is used most widely in the pressure range up to a few tens of kilobars. The substance being studied must be a nonviscous liquid in order to obtain a true hydrostatic pressure in its interior. At room temperature, the upper limit of purely hydrostatic pressures lies between 25 and 30 kbar, because at these pressures most liquids either solidify or become very viscous. At higher pressures, liquids cannot exist and, therefore, the pressures obtained are only approximately hydrostatic. If the substance under investigation is sufficiently rigid, it can be compressed hydrostatically using a pressure-transmitting medium which (depending on the experimental conditions) may be a liquid or a sufficiently plastic solid (Fig. 4.1b). Sealing rings, which enclose completely the compressed substance within the high-pressure chamber, are used to prevent leakage of the substance from the chamber. The change in volume due to compression is deduced from the displacement of the piston, after correction for the deformation of the pistons, cylinder walls, and of the pressure-transmitting medium [6].

The plastic deformation of pistons and cylinders begins from about 15-20 kbar and, if suitable measures are not taken, such deformation eventually leads to collapse of the apparatus. This is prevented by various supporting devices.

A cylinder is usually supported as follows. Under an appropriate load, the cylinder is stressed so that its inner wall is in a state of high compression and the outer wall is in a state of high tension. When the maximum pressure is applied, the cylinder is stressed to a limit close to the yield point. This can be achieved, for example, by compressing the cylinder with

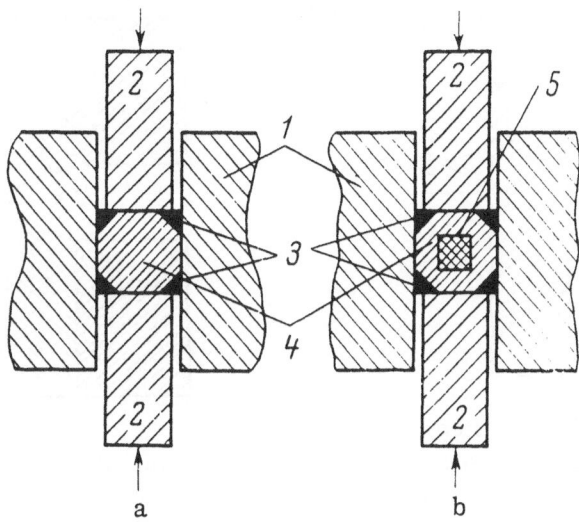

Fig. 4.1. Simple cylinder−piston apparatus. 1) Cylinder;
2) pistons; 3) sealing rings; 4) liquid or plastic solid; 5)
sample.

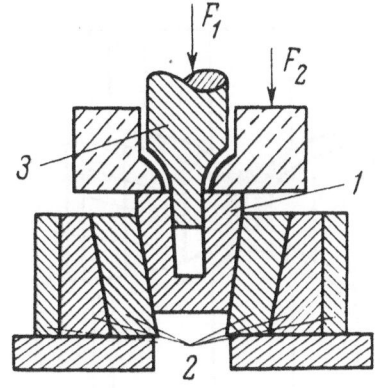

Fig. 4.2. Supported-cylinder apparatus. 1) Cylinder; 2) supporting
rings; 3) tapered piston.

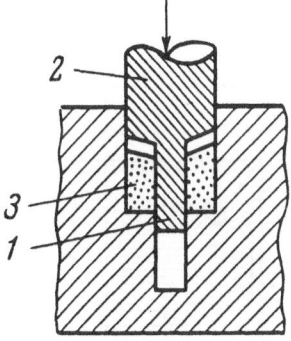

Fig. 4.3. Piston supported by hydrostatic pressure. 1) High-pressure piston;
2) low-pressure piston; 3) supporting
sleeve kept at hydrostatic pressure.

supporting rings which are heated and shrunk onto the cylinder. Another method for supporting
the cylinder is shown in Fig. 4.2. In this case, the support is increased gradually. A sample,
placed in the high-temperature chamber, is compressed by a force F_1 applied to the piston. A
force F_2, which is applied simultaneously to the top end of the chamber, pushes it into a system
of supporting rings. The forces F_1 and F_2 are applied in such a way as to ensure that the
stresses in the outer cylindrical channel do not exceed the tensile strength of the materials
used in the apparatus. This type of apparatus withstands pressures up to 50 kbar.

Cylindrical pistons made of high-quality steels or of hard alloys are capable of supporting an axial load up to 50 kbar. The fracture of a piston can be prevented in two ways. The
simplest way is to apply radial compression to the piston by supporting rings or by a medium
which is under hydrostatic pressure (Fig. 4.3). In the second case, a cylindrical piston is replaced by a tapered one such as that shown in Fig. 4.2. Using a piston of this shape, we can obtain pressures considerably higher than the yield stress of the piston material. In both cases,
the insertion of the rod into a cylinder is a fairly difficult process.

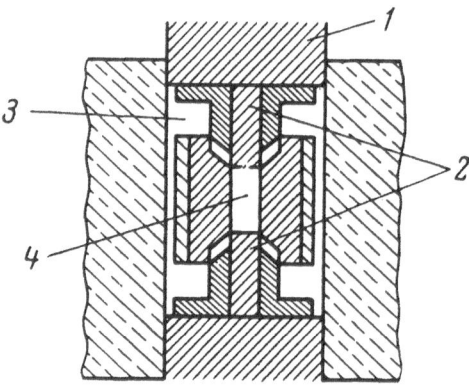

Fig. 4.4. Basic parts of the two-step Bridgman apparatus. 1) Low-pressure piston; 2) high-pressure pistons; 3) low-pressure chamber filled with pressure-transmitting medium; 4) high-pressure chamber.

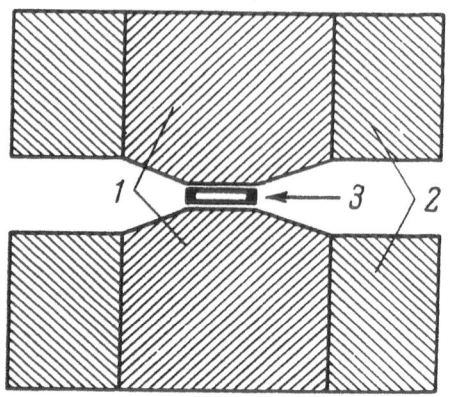

Fig. 4.5. Bridgman anvils. 1) Anvils; 2) supporting rings; 3) thin-disk sample, surrounded by a sealant.

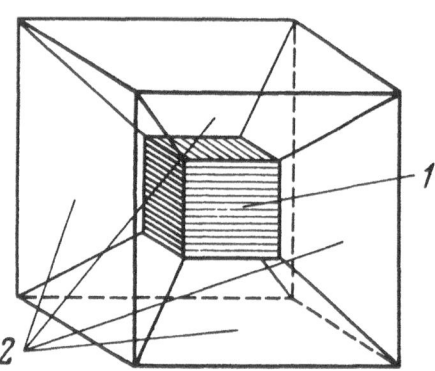

Fig. 4.6. Cubic anvil apparatus. 1) High-pressure chamber; 2) truncated-pyramid anvils.

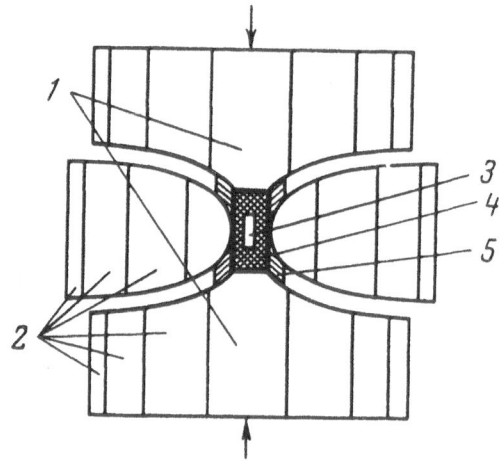

Fig. 4.7. Belt apparatus: 1) Anvils; 2) supporting rings; 3) sample; 4) pressure-transmitting medium; 5) seals.

Bridgman investigated the compressibility of various substances up to 100 kbar using a multistep method in which a high-pressure chamber was placed within a low-pressure chamber. The low-pressure chamber in Bridgman's apparatus is similar to that shown in Fig. 4.2. The high-pressure chamber is a cylinder with two pistons and this cylinder is located in the upper part of the low-pressure chamber (Fig. 4.4). The instruments for recording the displacement of the high-pressure piston and the force applied to it are placed elsewhere in the low-pressure chamber.

A characteristic feature of recently developed apparatus for pressures of the order of 100 kbar or more is a piston in the form of a truncated cone or pyramid (such pistons are known as anvils). In this apparatus, a cylinder is dispensed with or is replaced by pistons themselves (multipiston apparatus); sometimes the cylinder is modified considerably (belt apparatus).

One variant is known as the Bridgman anvil apparatus. It has two opposed pistons (anvils) whose tops are truncated cones (Fig. 4.5). High pressures are generated between the opposed tops of the two pistons. The sample, thin and cylindrical, is placed between the tops of these pistons. The hard steel or alloy pistons are surrounded by strong steel rings which are in a state of high tension and therefore compress the pistons diametrically and elongate them axially. When an axial force is applied to the pistons during the compression of a sample, the distribution of stresses in the pistons is reversed and the pistons are capable of withstanding high loads.

In multipiston apparatus, the substance under study is compressed simultaneously from several directions. The piston tops in such apparatus are in the form of truncated pyramids. These pyramidal tops are different in different variants and they depend on the number of pistons.

The assembly and operation of such apparatus can be described as follows. Let us inscribe a regular polyhedron into a sphere and then let us cut it into equal pyramids whose vertices are located at the center of the sphere and whose bases are the polyhedron faces. By cutting off the vertices of the pyramids, we obtain a high-pressure chamber in the form of a regular polyhedron, similar to the large polyhedron from which we have started (Fig. 4.6). The pyramidal pistons are pushed apart somewhat at the center of the sphere and the resulting central chamber is filled with a pressure-transmitting medium. Finally, the sample is placed in this medium and the pyramidal pistons (anvils) are pushed by external forces toward the center of the sphere.

Two forms of this apparatus have been developed: tetrahedral anvil apparatus (four anvils) and cubic apparatus (six anvils). Both types of apparatus have been used successfully up to pressures of the order of 100 kbar.

The belt apparatus consists of two truncated-cone anvils which are inserted from opposite sides into an approximately cylindrical chamber surrounded by a toroidal "belt" of supporting rings (Fig. 4.7). The sample is surrounded by the pressure-transmitting medium and is located at the center of the high-pressure chamber. Funnel-shaped spacers separate the anvils and the chamber. These spacers act as seals preventing leakage of the substance under study from the high-pressure chamber. Such apparatus can be used to obtain pressures above 100 kbar.

The pressure calibration of such apparatus is a difficult problem. The calibration procedure is simplest for the cylinder-piston type of apparatus. This is because a relatively small fraction of applied forces is lost in friction and because the frictional forces are approximately equal during the forward and reverse stroke of a piston (this makes it possible to determine the friction forces from the difference of pressures during the forward and reverse stroke of the piston). In such apparatus, the pressure is equal to the force acting in the piston divided by the piston's cross-sectional area.

The principal difficulty in the pressure calibration of other apparatus is the unknown fraction of the applied force lost due to friction and the pressure within the spacers. Since the true pressure in the chamber is not given by the ratio of the force to the piston area, one has to use other methods for the determination of pressure. In most cases, a chamber is calibrated using some reference points. This is done by placing, in the high-pressure chamber, various substances whose electrical resistivity or volume changes suddenly at definite pressures. Knowing the force applied by a press and the pressures at which the transitions occur in such substances, we can plot a calibration curve of the type shown in Fig. 4.8 for each apparatus.

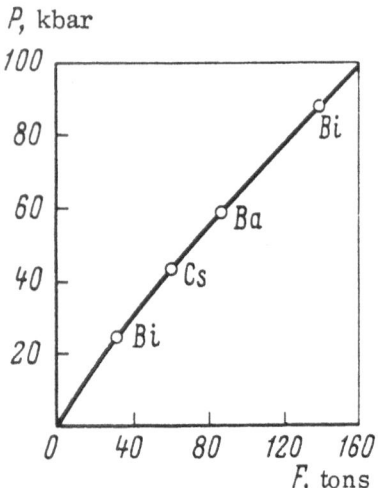

Fig. 4.8. Calibration curve for the belt-type apparatus. F is the force applied by a press; P is the pressure in the chamber.

The pressure calibration procedure is usually carried out at room temperature because the calibration data have been originally obtained at this temperature. In some cases, material under investigation is subjected simultaneously to high pressure and temperature. The inner part of the high-pressure chamber is heated after the establishment of a pressure at the reference temperature. Such heating unavoidably changes the pressure in the chamber. Measurements of pressure at high temperatures are very difficult. Moreover, it is difficult to measure temperature in a high-pressure chamber. Various workers have suggested very ingenious ways of overcoming these difficulties but the problems have not yet been solved completely.

4.2. Review of Experimental Data

The published experimental data on the static compressibility of solids differ greatly both in respect of the methods by which they were obtained, and the accuracy and the range of pressures covered. Therefore, only some data may be used to determine reliably the equation of state of a solid suitable for extrapolation to other pressures.

We shall divide the available experimental data according to their pressure range. We shall consider all the data obtained at pressures not higher than 10-20 kbar as one group, and those obtained at pressures higher than 20 kbar as the second group. The first group can be divided into three subgroups according to the method used to obtain the data.

TABLE 4.1

Range of pressures, kbar	Ref.	Substance
0—100	[7]	Li, Na, K, Rb, Ca, Sr, Ba, Zn, Cd, In, Tl, Sn, As, Sb, Bi, Se, Te
0—100	[8]	Pb, In, S, AgBrO$_3$; A'Cl, A'Br, A'I, A'NO$_3$ (A' = Na, K, Rb, Cs, Ag, Tl, NH$_3$)
0—25	[9]	Tl, As, Cd, In; Pb, NaCl; A'NO$_3$ (A' = Na, K, Rb, Cs, Ag, Tl); B'ClO$_3$ (B' = Na, K, Cs); C'BrO$_3$ (C' = Na, Ag); D'IO$_3$ (D' = Na, K, NH$_4$); E'ClO$_4$ (E' = Li, Na, K, Rb, Cs, NH$_4$); F'IO$_4$ (F' = Na, K, Rb, NH$_4$)
0—100	[10]	Li, Na, K, Rb, Cs, Be, C (graphite), Mg, Al, Si, P (white and black), Ti, U, Mn, Ge, As, Zr, La, Ce, Pr, Nd, Th; ZnA', HgA', PbA' (A' = S, Se, Te), SiO$_2$
0—40	[19]	Li, Na, K, Rb, Cs, Be, Ti, Zn, Ge, Zr, La, Ce, Pr, Nd, U, quartz, orthoclase, labradorite, calcite, olivine, diopside, grossularite, garnet, hypersthene, hallite
0—30	[20]	Mg
0—30	[21]	Ir, W, Ru, Pt, Mo, Rh, Ta, Co, Ni, Pd, Au, Nb, Fe, Cu, Ti, Ag, Si, Ge, Al, Sb, Zn, Th, Sn, Te: Cu$_3$Au, AuZn, Cu$_{31}$Sn$_8$, CuZn, Ag$_2$Al, Cu$_5$Zn$_8$, AgZn, Ag$_5$Zn$_8$, Cu$_5$Cd$_8$, SbSn, Sb$_2$Tl$_7$, 35% Ni—65% Fe; garnet, MgO, Fe$_3$O$_4$, andradite, FeS$_2$, cobaltite, fluorspar, LiF, NaCl, Al$_2$O$_3$, beryl, topaz, apatite, tourmaline, spodumene, barite, quartz
0—40	[22]	La, Ce, Pr, Nd, Sm, Gd, Dy, Ho, Er, Yb, Lu

Let us consider first the compressibilities determined by the classical cylinder-and-piston method. The available data are very extensive. They have been analyzed systematically by Birch, tabulated, and published in [11]. An important defect of these data is that the volume of most substances changes only slightly at pressures up to 10-20 kbar; therefore, the accuracy of the measurements of the changes in volume by the cylinder-and-piston method is not high. Moreover, various properties of particular samples (such as porosity) may affect the results obtained at these pressures, making these results irrelevant at high pressures. All this means that such results cannot be used to determine reliably the equation of state. However, we must mention specifically a series of measurements carried out at low temperatures on various highly compressible substances [4, 12]. The experimental techniques used in this series are complex but promising because they make it possible to eliminate the thermal component from the equation of state.

Another batch of data has been obtained by the ultrasonic method. This method yields, in principle, all the elastic constants of a crystal [13]. However, in practice, it is usual to measure the velocities of longitudinal and transverse elastic waves in polycrystalline samples [14-16]. Such measurements are usually carried out at relatively low pressures and they can be affected appreciably by the specific properties of particular samples, as mentioned in the preceding paragraph.

The equation of state can definitely be refined by raising the upper limit of the pressures employed [17].

Compressibilities at low pressures have been measured also by the x-ray diffraction method. This method makes it possible to determine directly the pressure dependences of the interatomic distances. The experimental technique and the results of measurements are described in detail in [18]. Unfortunately, very few substances have been investigated by this method.

Let us now consider the static data on the compressibility of solids obtained at pressures above 20 kbar. The fullest and most extensive investigations at these pressures were carried out by Bridgman [7-10, 19-22]. In spite of the fact that the results given by Bridgman in different papers do not always agree well, we shall use them to derive equations of state. Bridgman's measurements were carried out mainly at room temperature. For the sake of convenience, Bridgman's values of the compressibility of solids are summarized in Table 4.1. The numerical values for many substances are given in [11].

4.3. Determination of Equations of State
from Experimental Data

The measurements of the isothermal compressibility described in the preceding section can be used to determine fairly reliably the interaction potentials of solids. However, it must be pointed out that the reliability of extrapolation decreases with increasing pressure. This is due to the unavoidable experimental errors in the measurement of $x = V/V_0$. It must be stressed that the reliability is a function not only of the experimental accuracy but also of the range of x in which measurements are carried out.

Static measurements of the compressibility are usually carried out at low or moderate temperatures. This makes it possible to drop from the equation of state the terms proportional to T^2, because they are small.

The equation of state now consists of two terms which represent, respectively, the interaction potential and the contribution from phonons. The potential term depends on the type of solid (see Chapter 1). We shall describe the phonon component using the Debye approximation (see Chapter 2). This approach yields the following theoretical equation of state, as given by Eq. (2.25):

$$P = P_{\mathrm{p}}(x) + \frac{\rho_0 \gamma}{x} \frac{R}{\mu} \left[\frac{9}{8} \theta + 3TD \left(\frac{\theta}{T} \right) \right]. \tag{4.1}$$

This equation includes two unknown functions P_{p} and γ. According to Eq. (2.26), the Debye temperature θ can be expressed in terms of γ:

$$\theta = \theta_0 \exp \int_x^1 \frac{\gamma}{x} \, dx, \tag{4.2}$$

where θ_0 is the Debye temperature under normal conditions $(x = 1)$. A single experimental curve $P(x)$ at $T = $ const is insufficient for the determination of the two functions P_{p} and γ. Therefore, we must make additional assumptions about a functional relationship between P_{p} and γ. We shall assume that this functional relationship is given by Eq. (2.44)

$$\gamma = -\frac{x}{2} \frac{\partial^2 (P_{\mathrm{p}} x^{2m/3})/\partial x^2}{\partial (P_{\mathrm{p}} x^{2m/3})/\partial x} + \frac{1}{3}(m-2) + \delta, \tag{4.3}$$

which contains the same unknown parameters A, b, and K as the function P_{p}. The additional parameter δ is a normalization constant which is found by postulating that the value of γ given by Eq. (4.3) for $x = 1$ is identical with the thermodynamic value γ_0 under normal conditions, calculated from Eq. (1.35). One of the three parameters A, b, and K can be eliminated using the condition that $P = 0$ at $x = 1$. In practice, it is convenient to eliminate the parameter K; then, Eqs. (4.1) and (1.9) yield

$$K = A + \rho_0 \gamma_0 \frac{R}{\mu} \left[\frac{9}{8} \theta_0 + 3T_0 D \left(\frac{\theta_0}{T_0} \right) \right]. \tag{4.4}$$

The second term in Eq. (4.4) is free of unknown parameters. The values of ρ_0, γ_0, and θ_0 are assumed to be known at $P = 0$ and T_0, which is the temperature at which the published experiments were carried out.

The remaining two unknown parameters A and b can be found from Eq. (4.1) by substituting into the left-hand side of this equation the experimentally determined values of P corresponding to various values of x. In this operation, it is formally sufficient to know two accurately determined pairs of values of P and x. In practice, because of the unavoidable experimental errors, two such pairs are insufficient to determine A and b with the required accuracy. It is usual to employ more than two pairs of experimental values of P and x and, consequently, it is convenient to use the least-squares method. In this method, we calculate the parameters A and b from the condition for a minimum of the function

$$M = \sum_{i=1}^{n} [P_{ei} - P(x_i)]^2 \frac{1}{\sigma_i{}^2}, \tag{4.5}$$

where $P(x_i)$ is the theoretical value of the pressure at a point x_i, calculated using Eq. (4.1); P_{ei} is the corresponding experimental value of the pressure at the same point; n is the number of experimental points; $1/\sigma_i^2$ is the weight of the point i; σ_i is the absolute value of the experimental error at this point. If all the points are obtained with the same relative error ξ, it follows that $\sigma_i = P_{ei}\xi$ and the function (4.5) transforms to

$$M = \frac{1}{\xi^2} \sum_{i=1}^{n} \frac{[P_{ei} - P(x_i)]^2}{P_{ei}^2}. \tag{4.5a}$$

The equations for the determination of A and b are found from the condition for a minimum of M, i.e., from

$$\frac{\partial M}{\partial A} = 0, \qquad \frac{\partial M}{\partial b} = 0.$$

When these equations are written out in full, they are of the form

$$\sum_{i=1}^{n} \frac{1}{\sigma_i^2} [P_{ei} - P(x_i)] \frac{\partial P(x_i)}{\partial A} = 0,$$

$$\sum_{i=1}^{n} \frac{1}{\sigma_i^2} [P_{ei} - P(x_i)] \frac{\partial P(x_i)}{\partial b} = 0.$$

(4.6)

These two equations are evidently nonlinear in A and b or, more exactly, they are trans-cendental in respect of b. Therefore, the system (4.6) must be solved by the method of succes-sive approximations, assuming that

$$A = A_0(1 + a), \quad b = b_0(1 + \beta),$$

(4.7)

where α and β are small corrections; A_0 and b_0 are the values of A and b in the zeroth approxi-mation. The procedure to be followed in the calculation of A and b is described in detail in Appendix D (this Appendix also gives all the formulas required in the calculation). The least-squares method and its application to nonlinear functions are described in [23].

The parameters A, b, K, and δ calculated by this method are listed in Table 4.2. This table gives, in the first column, the source of the experimental data used in the calculations of the equation of state. Three rows of figures are given for each substance. These figures dif-fer in respect of the formula for γ used in the calculations. The upper row corresponds to m = 0 in Eq. (4.3), the middle row to m = 1, and the lower row to m = 2.

Using the tabulated parameters (the values of ρ_0, γ_0, and θ_0 are listed separately in Table 4.3) and Eq. (4.1), we can calculate any thermodynamic property of a solid in its equilib-rium state described by a given set of values $P - x - T$.

Figure 4.9 shows the isotherm of iron before a phase transition. This isotherm is cal-culated using Eq. (4.1), the parameters listed in Table 4.2, and m = 1. It is evident from this figure that the agreement between the theoretical curve and experimental data obtained by the x-ray diffraction method is very good. However, it must be pointed out that iron was used by Bridgman as a standard substance and, therefore, the static data on the compressibility of iron were determined by Bridgman more accurately than for other substances. Unfortunately, the corresponding data for other substances are often of poorer quality. This will be discussed in detail in Chapter 5.

Figures 4.10-4.25 show the equations of state for various substances. The values given in these figures are calculated using the parameters A, b, K, and δ taken from the middle rows of Table 4.2, i.e., these values correspond to m = 1. The potential pressure P_p is calcu-lated using Eq. (1.9), the Grüneisen parameter γ is evaluated using Eq. (4.3) and assuming that m = 1, and the quantity $\Phi_p = K_p/\rho$ (K_p is the potential component of the bulk modulus) is cal-culated using Eq. (1.11). The Hugoniot curve P_H and the temperature along this curve T_H, shown in these figures and discussed in detail in Chapter 5, are calculated using Eqs. (5.75) and (5.77), respectively.

The original experimental data are not given in Figs. 4.10-4.25 because we shall not dis-cuss the discrepancies between these data and the theoretical values of $P(x)$.

Crosses are used in Figs. 4.10-4.25 to represent the experimental values of the Hugoniot pressures P_{He} for some substances (there are as yet no dynamic data for the other substances referred to in Figs. 4.10-4.25). These experimental values are not used in calculations and are shown in order to provide a check against experimental data. Thus, we can see that, in the case of Sb (Fig. 4.10), we can draw a Hugoniot curve P_H^n (shown by a dashed curve) through the points represented by crosses. The relative positions of the shock Hugoniot curves P_H and

TABLE 4.2

Substance	$A \cdot 10^{-5}$, bar	b	$K \cdot 10^{-5}$, bar	δ
Noble metals				
Cu [21]	5.5115	9.6801	5.7322	—0.234
	5.2703	9.9448	5.4910	0.076
	5.1328	10.0894	5.3535	0.400
Ag [21]	2.8255	13.1208	3.0059	—0.408
	2.6887	13.5293	2.8691	—0.114
	2.5599	13.7990	2.7803	0.198
Au [21]	9.0762	7.8020	9.2975	1.116
	8.8721	7.9029	9.0935	1.447
	8.7741	7.9519	8.9954	1.781
Divalent metals				
Be [19]	2.0961	18.1506	2.3614	—2.604
	1.7918	19.9553	2.0571	—2.523
	1.5533	21.6696	1.8187	—2.425
Mg [20]	3.0239	5.6097	3.1044	—0.080
	2.8386	5.7902	2.9191	0.239
	2.7619	5.8636	2.8424	0.569
Zn [19]	1.9991	11.6065	2.1950	—0.274
	1.8403	12.1914	2.0362	0.008
	1.7412	12.5764	1.9371	0.315
Cd [9]	25.0957	2.5881	25.2243	1.738
	24.9938	2.5933	24.9224	2.072
	24.5797	2.5996	24.7083	2.394
Trivalent metals				
Al [21]	2.6081	10.5782	2.7828	—0.322
	2.4433	10.9916	2.6180	—0.024
	2.3466	11.2392	2.5213	0.293
In [8]	1.2994	11.1626	1.4040	—0.325
	1.2274	11.5108	1.3320	—0.010
	1.1806	11.7496	1.2852	0.315
Tetravalent metals				
Sn [21]	1.7798	11.3293	1.8744	—0.536
	1.6923	11.6741	1.7868	—0.235
	1.6379	11.8899	1.7324	0.082
Pb [8, 9]	1.8506	9.1318	1.9623	0.580
	1.7719	9.3515	1.8836	0.910
	1.7261	9.4798	1.8378	1.245
Pentavalent metals				
Sb [21]	1.1087	12.5654	1.1440	—1.190
	1.0437	13.0215	1.0790	—0.626
	0.9998	13.3424	4.0351	—1.334
Transition elements				
Ti [19, 21]	1.4824	22.8721	1.5702	—3.328
	1.3882	23.8180	1.4761	—3.130
	1.3124	24.6274	1.4003	—2.910
Zr [19]	4.0325	8.5256	4.0741	—1.275
	3.8994	8.6963	3.9410	—0.966
	0.8267	8.7832	3.8683	—0.645

Substance	$A \cdot 10^{-5}$, bar	b	$K \cdot 10^{-5}$, bar	δ
Nb [21]	20.1220	4.5890	20.2408	0.369
	19.8079	4.6213	19.9267	0.700
	19.7046	4.6321	19.8234	1.034
FeI [21]	10.2963	6.9790	10.4859	—0.102
	9.9743	7.0985	10.1639	0.222
	9.8336	7.1521	10.0232	0.553
Co [21]	90.8696	2.6417	91.1056	1.401
	90.2906	2.6438	90.5267	1.736
	89.7886	2.6480	90.0245	2.063
Ni [21]	4.9647	13.6027	5.1924	1.034
	4.7646	13.9474	4.9922	0.737
	4.6336	14.1700	4.8612	—0.424
Pd [21]	5.5911	12.1236	5.7776	—0.500
	5.4307	12.3419	5.6172	—0.188
	5.3285	12.4757	5.5150	0.13$_6$
Semiconductors				
Si [21]	4.5320	8.6631	4.5869	—1.332
	4.3669	8.8513	4.4217	—1.024
	4.2764	8.9499	4.3313	—0.704
Ge [21]	4.7402	6.9707	4.7822	—1.050
	4.5765	7.1077	4.6185	—0.736
	4.4973	7.1715	4.5394	—0.418
Ionic crystals				
CsI [8]	0.5866	8.7926	0.6319	—0.160
	0.5456	9.0538	0.5968	0.179
	0.5239	9.2207	0.5752	0.519
NaCl [8, 9]	1.1829	8.5644	1.2713	—0.567
	1.0849	8.9488	1.1733	—0.257
	1.0282	9.1413	1.1166	0.066
NaI [8]	0.4874	11.7059	0.5449	—1.100
	0.4397	12.3490	0.4972	—0.818
	0.4068	12.8345	0.4643	—0.518
NaBr [8]	0.7547	10.3534	0.8280	—0.883
	0.6910	10.8196	0.7644	—0.578
	0.6494	11.1449	0.7227	—0.262
CsCl [8]	0.6016	11.0558	0.6708	—0.608
	0.5567	11.4866	0.6259	—0.289
	0.5264	11.7941	0.5956	0.038
CsBr [8]	0.9510	7.2944	1.0105	0.045
	0.8952	7.5090	0.9547	0.380
	0.8545	7.6300	0.9240	0.718
Minerals and rocks				
Fe₃O₄ [21]	27.2828	3.9049	27.4227	0.248
	26.6830	3.9333	26.8730	0.581
	26.5189	3.9420	26.7088	0.915
MgO [21]	11.1815	5.5919	11.4238	—0.316
	10.7374	6.7287	10.9796	0.007
	10.5506	6.7817	10.7931	0.339
FeS₂ [21]	4.3637	12.3944	4.5287	1.233
	4.1727	12.7322	4.3377	—0.938
	4.0502	12.9413	4.2153	—0.626

TABLE 4.2 (continued)

Substance	$A \cdot 10^{-5}$, bar	b	$K \cdot 10^{-5}$, bar	δ	Substance	$A \cdot 10^{-5}$, bar	b	$K \cdot 10^{-5}$, bar	δ
SiO$_2$I [19, 21]	1.2832	11.6762	1.3826	−1.747	Ortho-clase [19] (Spain)	3.7961	6.4649	3.8622	−1.188
	1.1129	12.6690	1.2123	−1.546		3.4593	6.7857	3.5255	−0.902
	0.9991	13.4320	1.0986	−1.310		3.3025	6.9402	3.3687	−0.589
Olivine [19]	26.0949	3.4212	26.2837	0.187	Diopside [19]	2.4403	16.1651	2.5819	−2.484
	25.5397	3.4413	25.7285	0.522		2.2236	17.0914	2.3652	−2.282
	25.3518	3.4496	25.5406	0.855		2.0648	17.8262	2 2065	−2.050
Garnet [21]	30.0398	3.8039	30.3319	0.270	Labra-dorite [19]	10.6484	4.0746	10.7022	−0.792
	29.4744	3.8243	29.7665	0.609		10.0589	4.1678	10.1128	−0.476
	29.3142	3.8301	29.6062	0.944		9.5723	4.1989	9.9262	−0.146
Andrad-ite [21]	4.9551	11.4749	5.1574	−1.480	Grossul-arite [19]	2.9241	16.5697	3.1285	−2.462
	4.6282	11.9457	4.8305	−1.207		2.6384	17.6332	2.8428	−2.278
	4.4281	12.2422	4.6304	−0.908		2.4274	18.4982	2.6318	−2.063

TABLE 4.3

Substance	ρ_0 g/cm^3	γ_0	θ_0,°K	Substance	ρ_0 g/cm^3	γ_0	θ_0,°K	Substance	ρ_0 g/cm^3	γ_0	θ_0,°K
Cu	8.90	2.04	315	Ta	16.46	1.69	225	NaBr	3.16	1.56	243
Ag	10.50	2.47	215	Mo	10.20	1.58	380	CsCl	3.95	1.97	166
Au	19.24	3.05	170	W	19.17	1.55	310	CsI	4.51	2.01	100
Be	1.845	1.17	1000	FeI	7.84	1.68	420	CsBr	4.45	1.93	119
Mg	1.725	1.46	318	Co	8.82	1.99	385	Fe$_3$O$_4$	5.20	1.4	600
Zn	7.135	2.38	235	Rh	12.42	2.26	340	Al$_2$O$_3$	3.99	1.6	600
Cd	8.64	2.27	120	Ni	8.86	1.91	375	MgO	3.56	1.4	800
Al	2.785	2.13	390	Pd	11.95	2.18	275	FeS$_2$	5.02	1.5	600
In	7.27	2.24	129	Pt	21.37	2.63	230	Olivine	3.32	1.2	900
Sn	7.28	2.03	260	Si	2.34	0.74	625	Garnet	3.58	1.4	1100
Pb	11.34	2.78	88	Ge	5.40	0.72	360	Andradite	3.86	1.1	800
Sb	6.67	0.86	200	Th	11.68	1.12	100	Orthoclase	2.56	0.5	1000
Ti	4.51	1.18	380	U	18.9	1.83	160	Diopside	3.29	0.9	900
Zr	6.49	0.771	250	NaCl	2.165	1.55	299	Labradorite	2.70	0.4	900
V	6.10	1.29	273	NaI	3.64	1.59	140	Grossularite	3.53	1.0	900
Nb	8.60	1.68	280								

P_H^{π} of the type shown in Fig. 4.10 indicates a phase transition (see Chapter 5). This transition does indeed occur at a pressure of 85 kbar [7]. The agreement between the theoretical Hugoniot curves and the independent experimental points P_H is good for substances such as NaCl, NaI, and CsI. This means that the static and dynamic data for these substances are in agreement, and that the equations of state of these substances are reliable. On the other hand, the discrepancy between the theoretical Hugoniot curve P_H and the experimental points P_{He} for NaBr, CsCl, and CsBr is considerable. This shows that the static and dynamic data for these substances are not in agreement. There are two possibilities: either the static data are incorrect and then the calculated equation of state is wrong; or the dynamic data are insufficiently accurate. Further experimental work would be required to determine which of these possibilities applies.

If we use the parameters A, b, K, and δ from Table 4.2 for m = 0 and m = 2, we find that the P_H and P_p curves differ by a few percent from those shown in Figs. 4.10–4.25 and the discrepancies in P_p are smaller than in P_H.

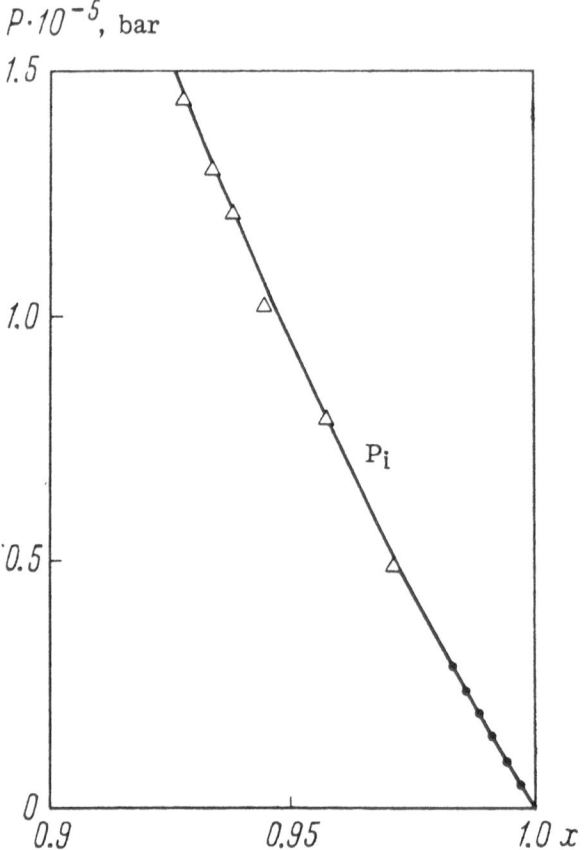

Fig. 4.9. Isotherm of iron: ● — experimental points P_{ie} taken from [21] and used to obtain the equation of state; △ — points deduced from x-ray diffraction data [24].

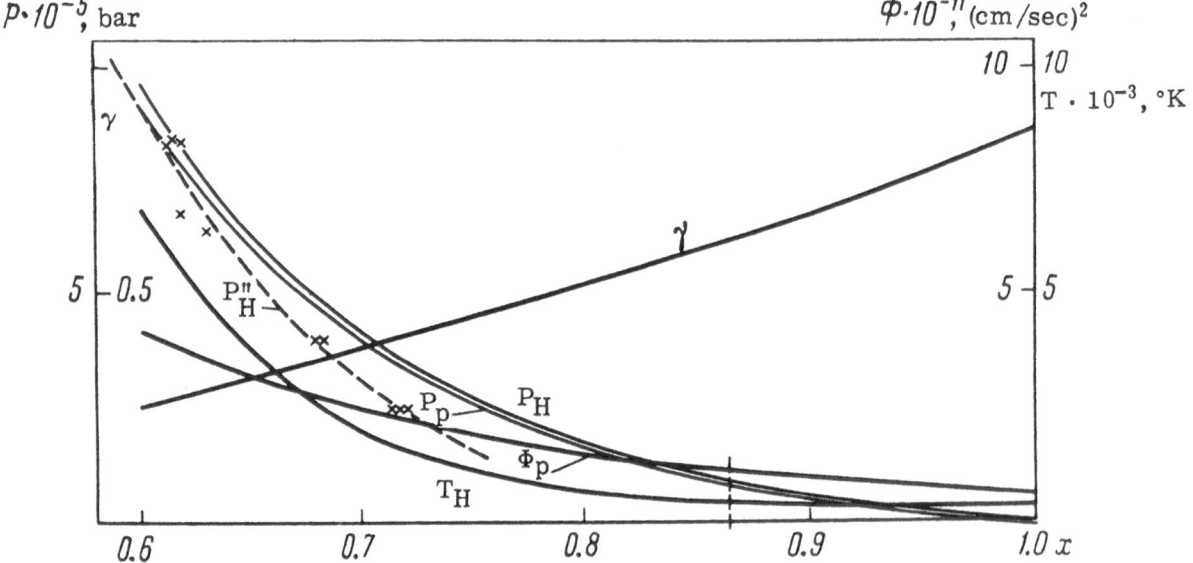

Fig. 4.10. Equation of state of antimony (× — points P_{He} taken from [8] in Chapter 5).

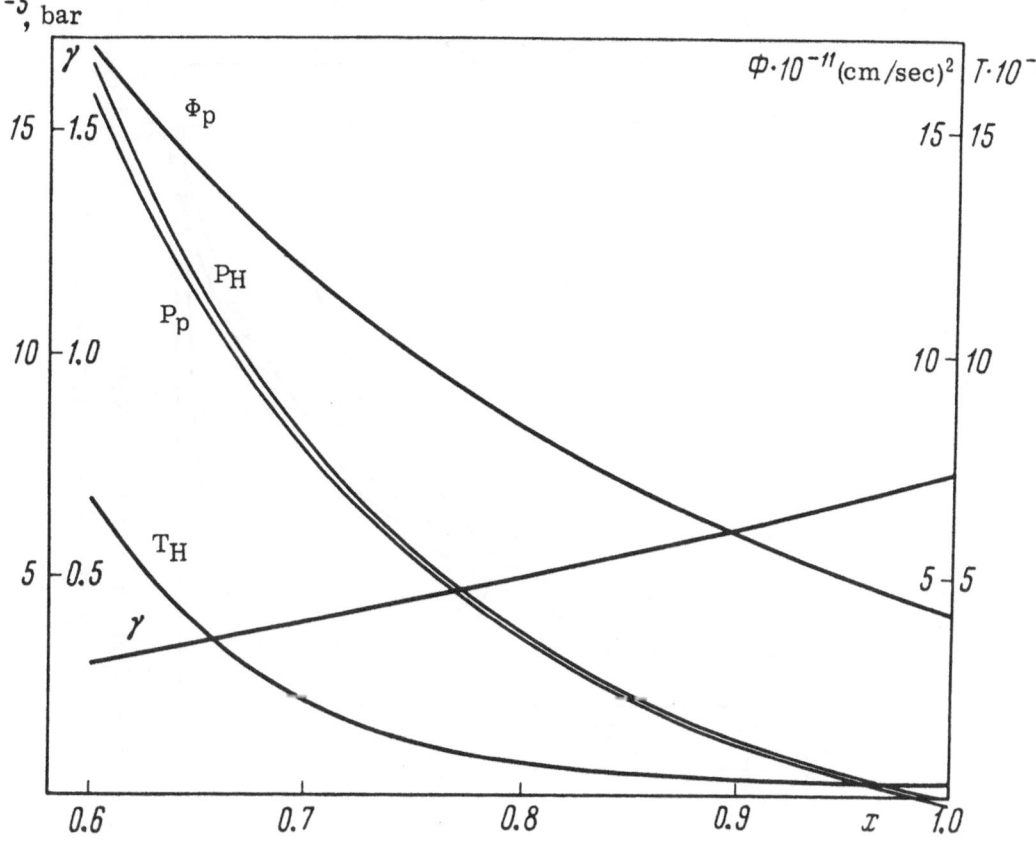

Fig. 4.11. Equation of state of silicon.

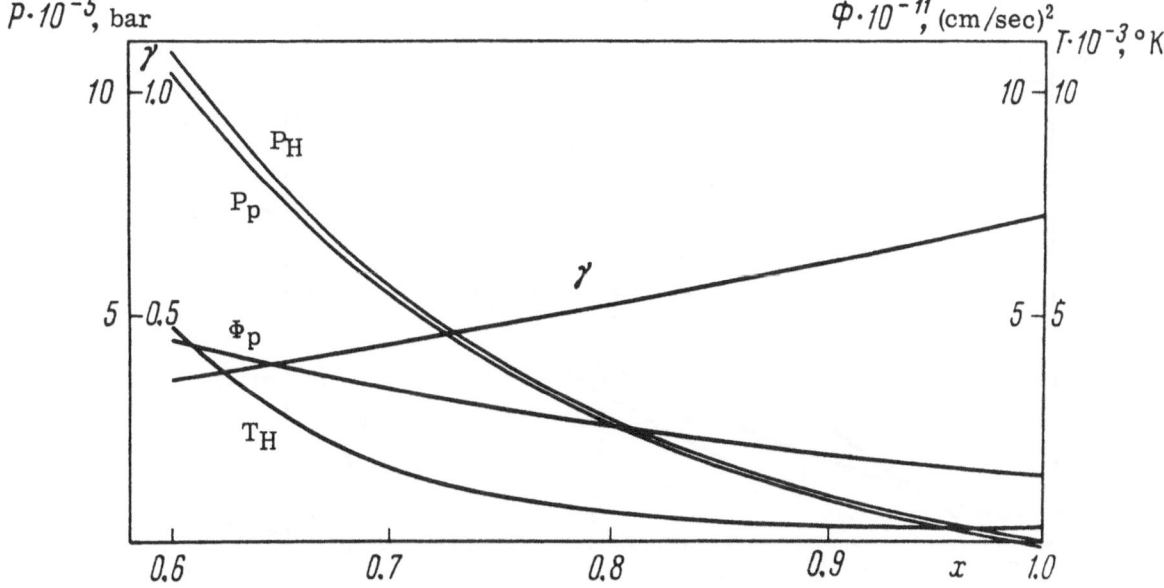

Fig. 4.12. Equation of state of germanium.

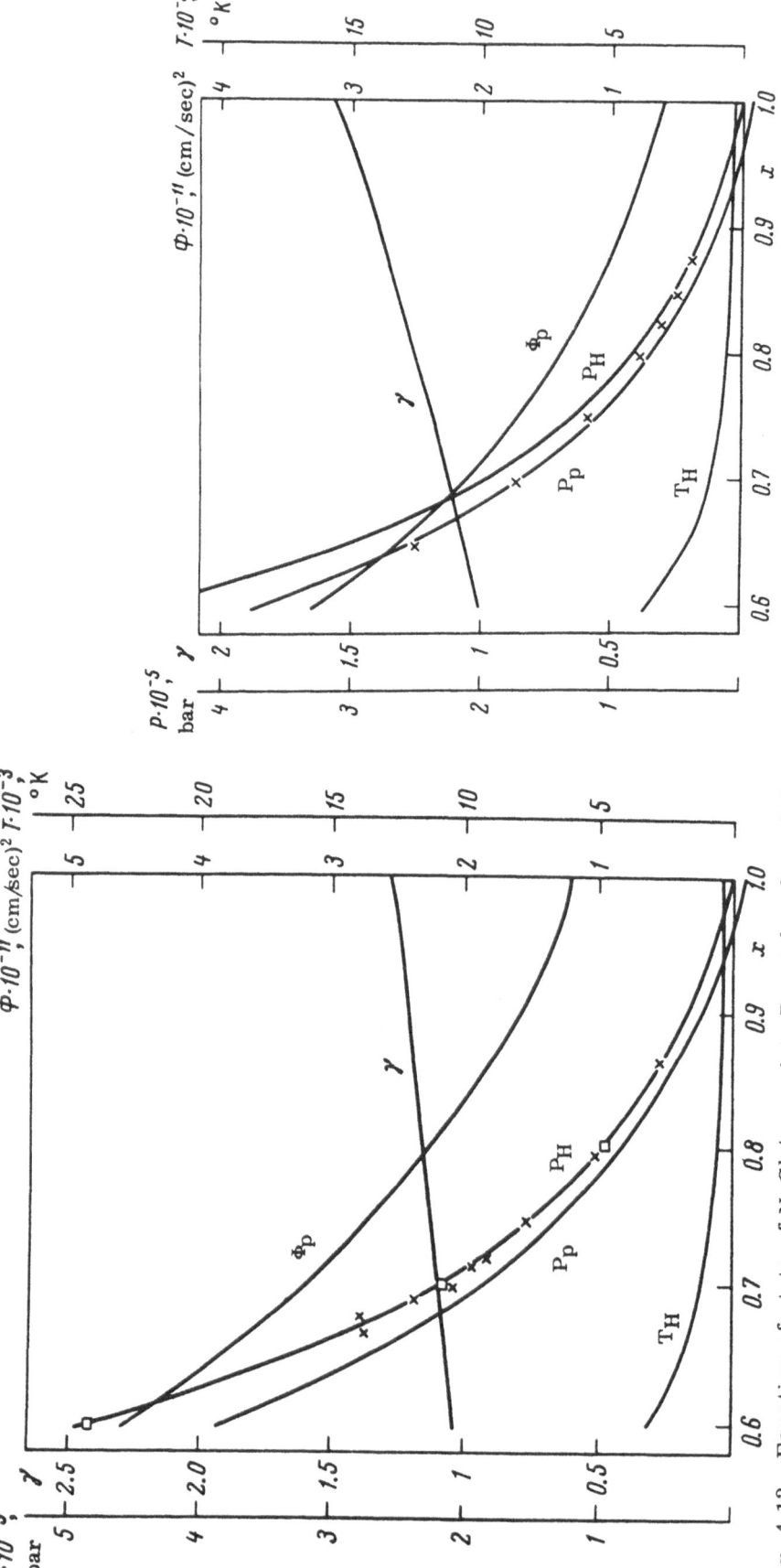

Fig. 4.14. Equation of state of NaBr (× − points P_H calculated on the basis of the data given in Table 4.2).

Fig. 4.13. Equation of state of NaCl (× − points P_{He} taken from [13, 39] in Chapter 5; □ − values calculated on the basis of the data given in Table 4.2).

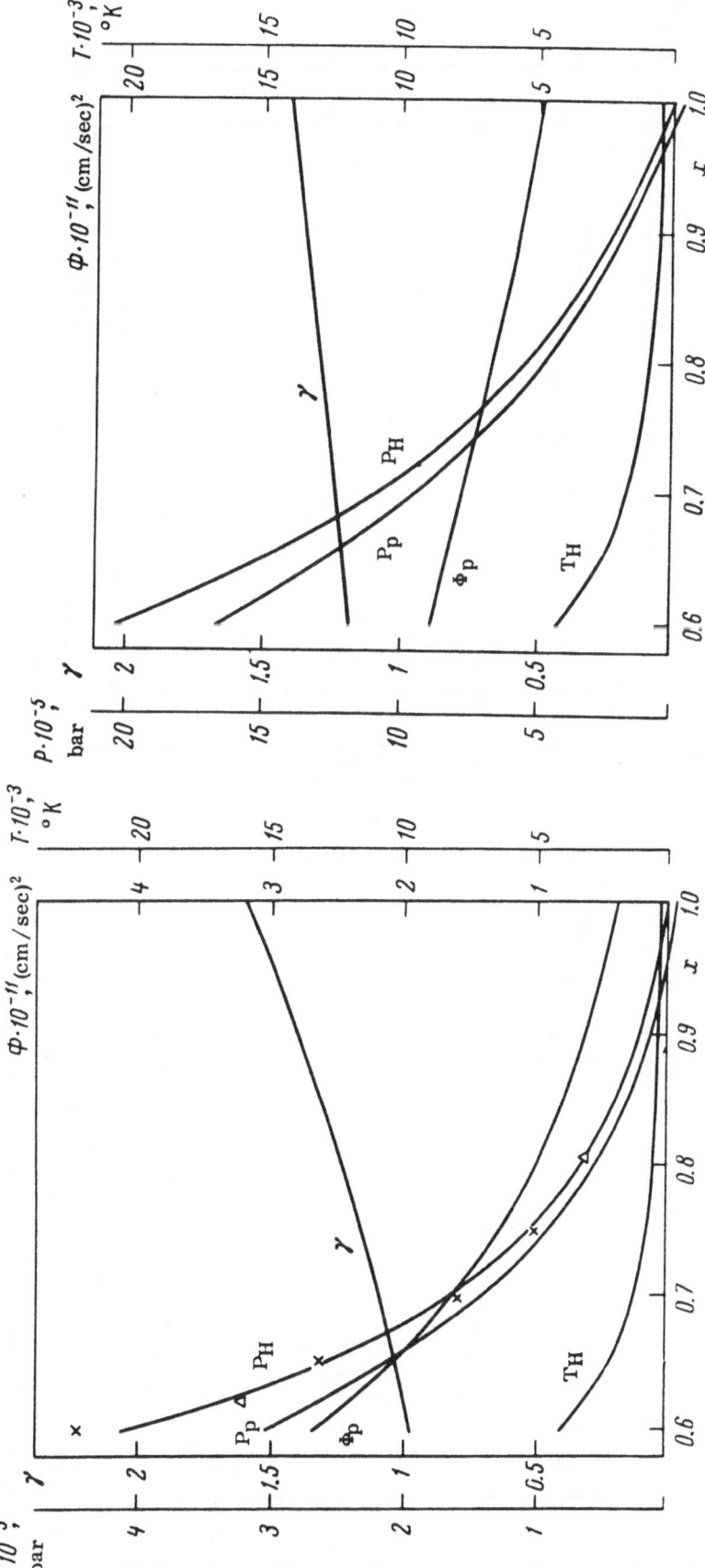

Fig. 4.16. Equation of state of garnet.

Fig. 4.15. Equation of state of NaI ($\triangle$ — points P_{He} taken from [15] in Chapter 5; $\times$ — represents the values calculated on the basis of the data given in Table 4.2).

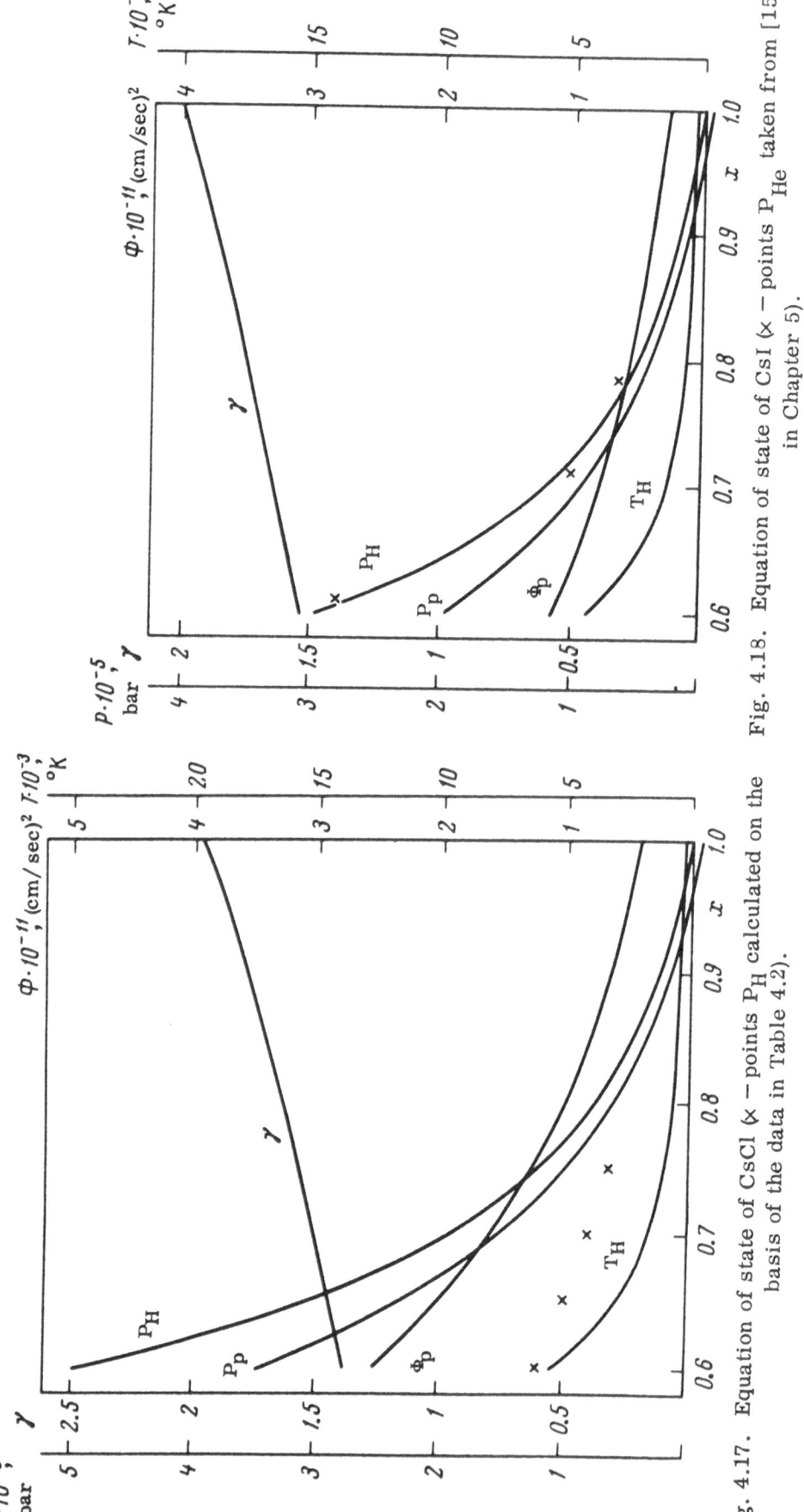

Fig. 4.18. Equation of state of CsI (x – points P_{He} taken from [15] in Chapter 5).

g. 4.17. Equation of state of CsCl (x – points P_H calculated on the basis of the data in Table 4.2).

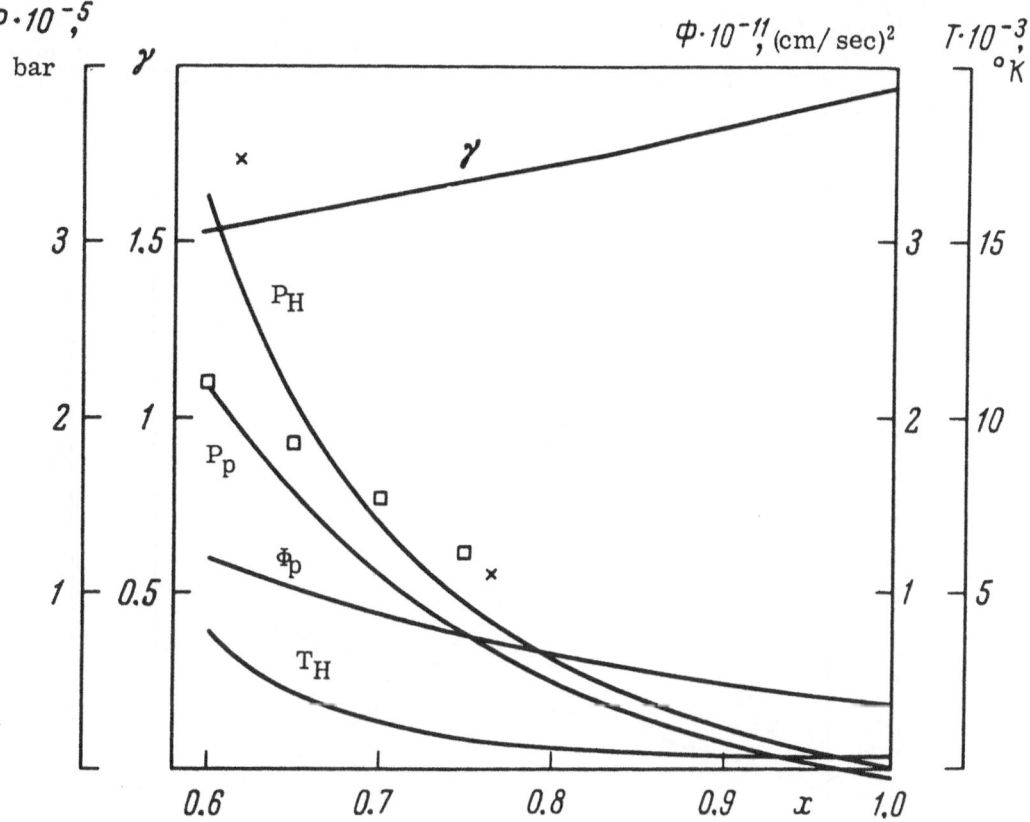

Fig. 4.19. Equation of state of CsBr ($\times$ — points P_{He} taken from [39] in Chapter 5; $\square$ — values calculated on the basis of the data given in Table 4.2).

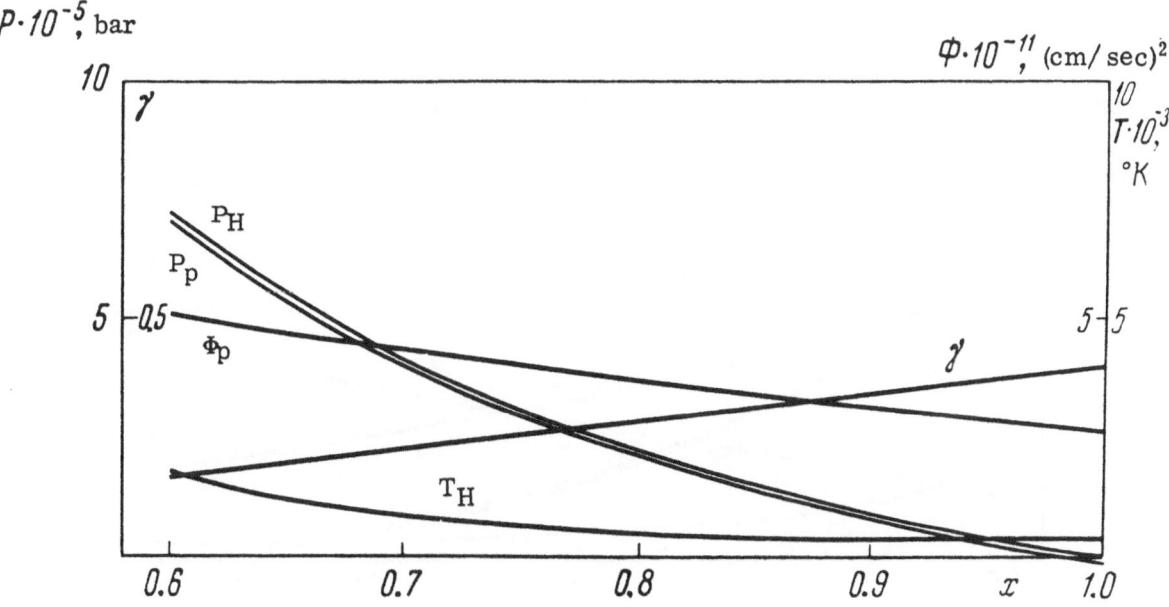

Fig. 4.20. Equation of state of labradorite.

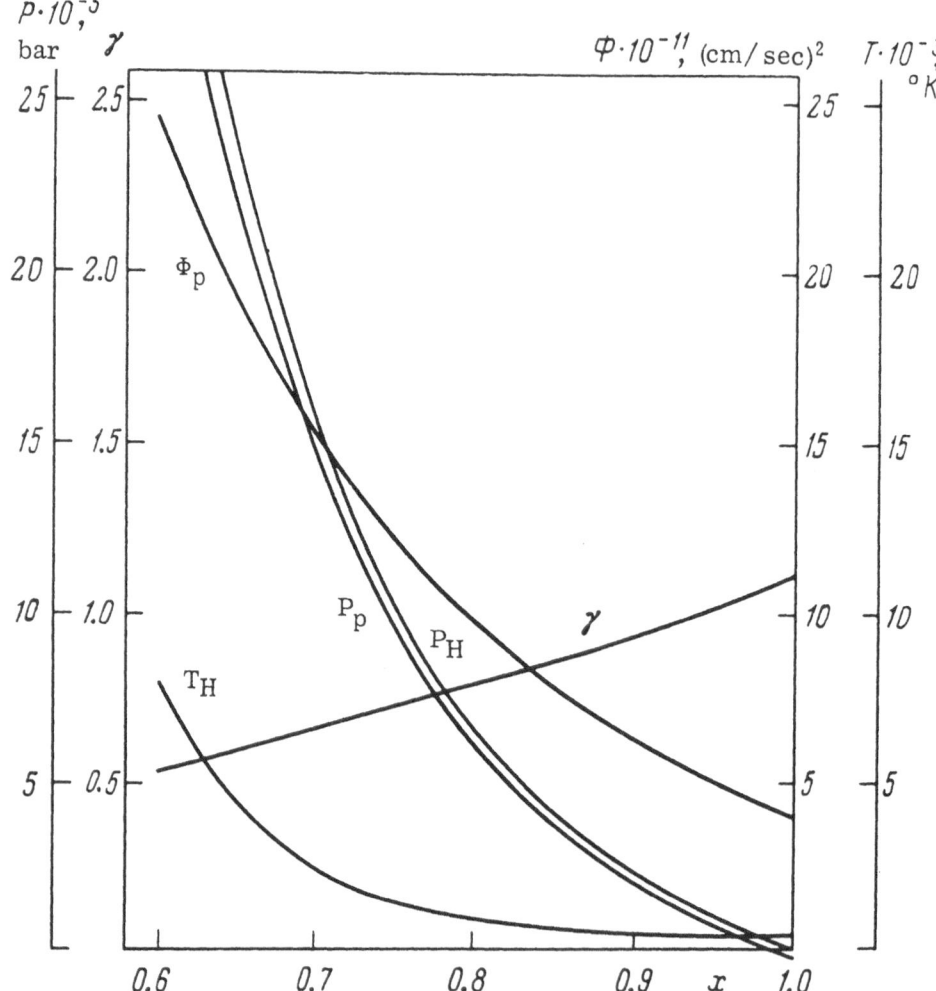

Fig. 4.21. Equation of state of andradite.

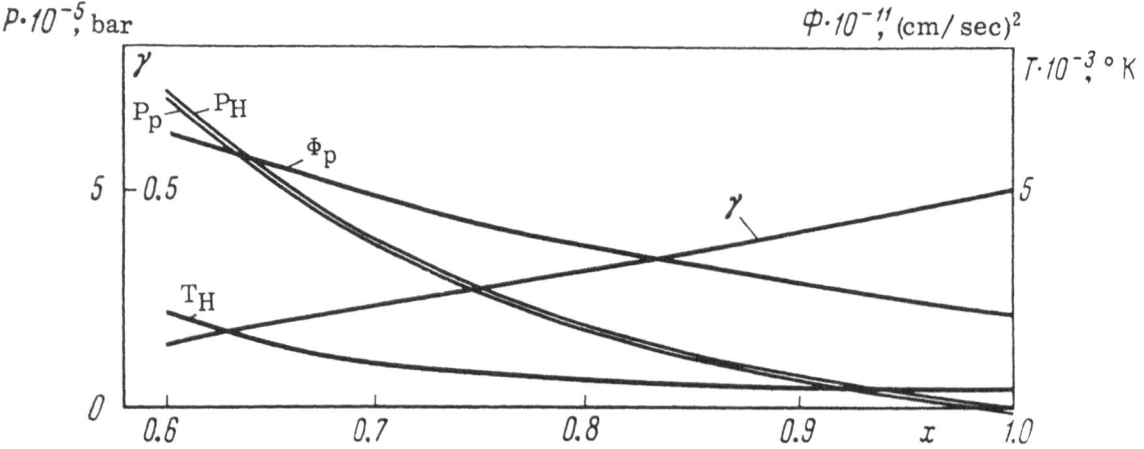

Fig. 4.22. Equation of state of Spanish orthoclase.

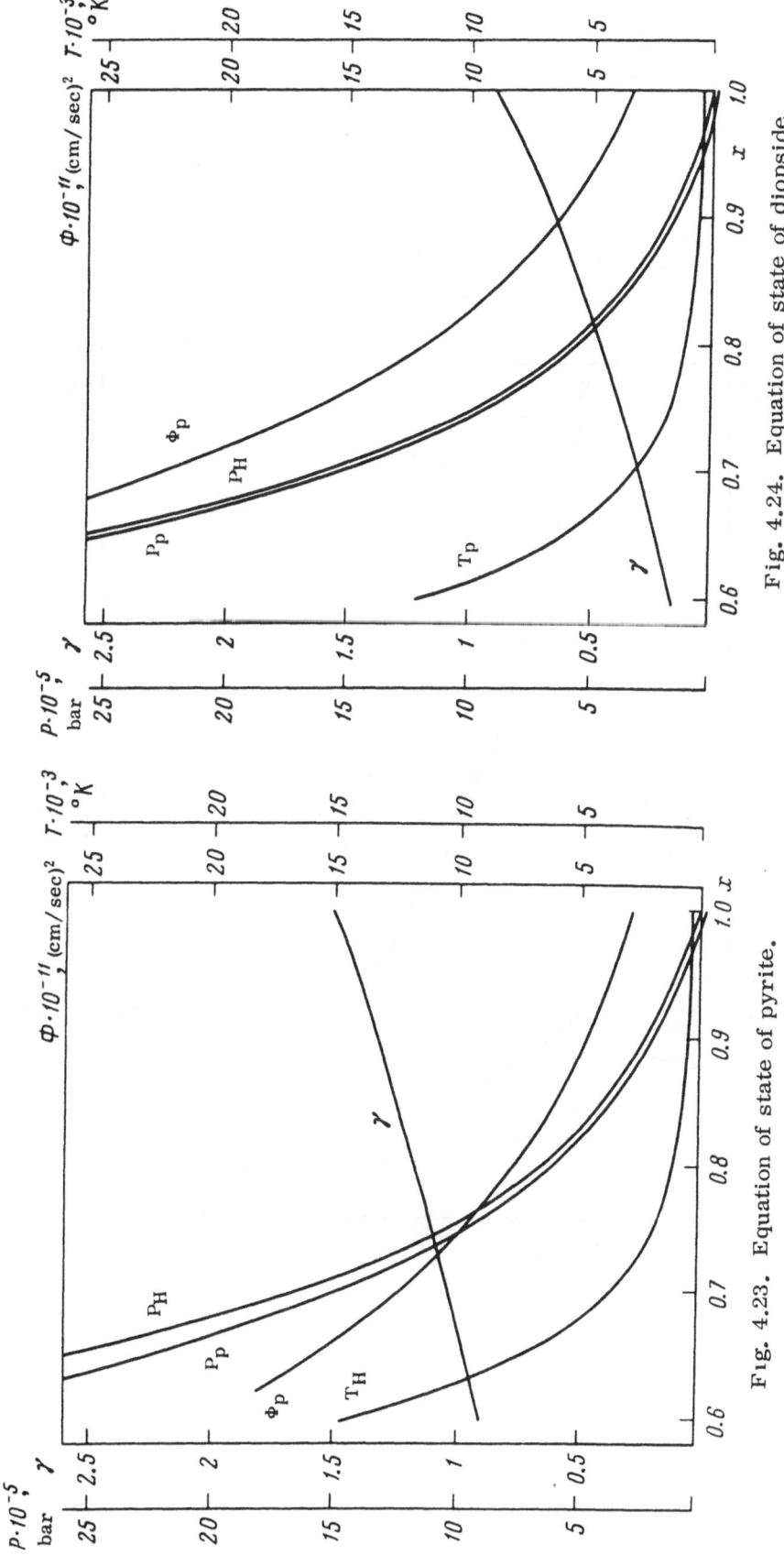

Fig. 4.24. Equation of state of diopside.

Fig. 4.23. Equation of state of pyrite.

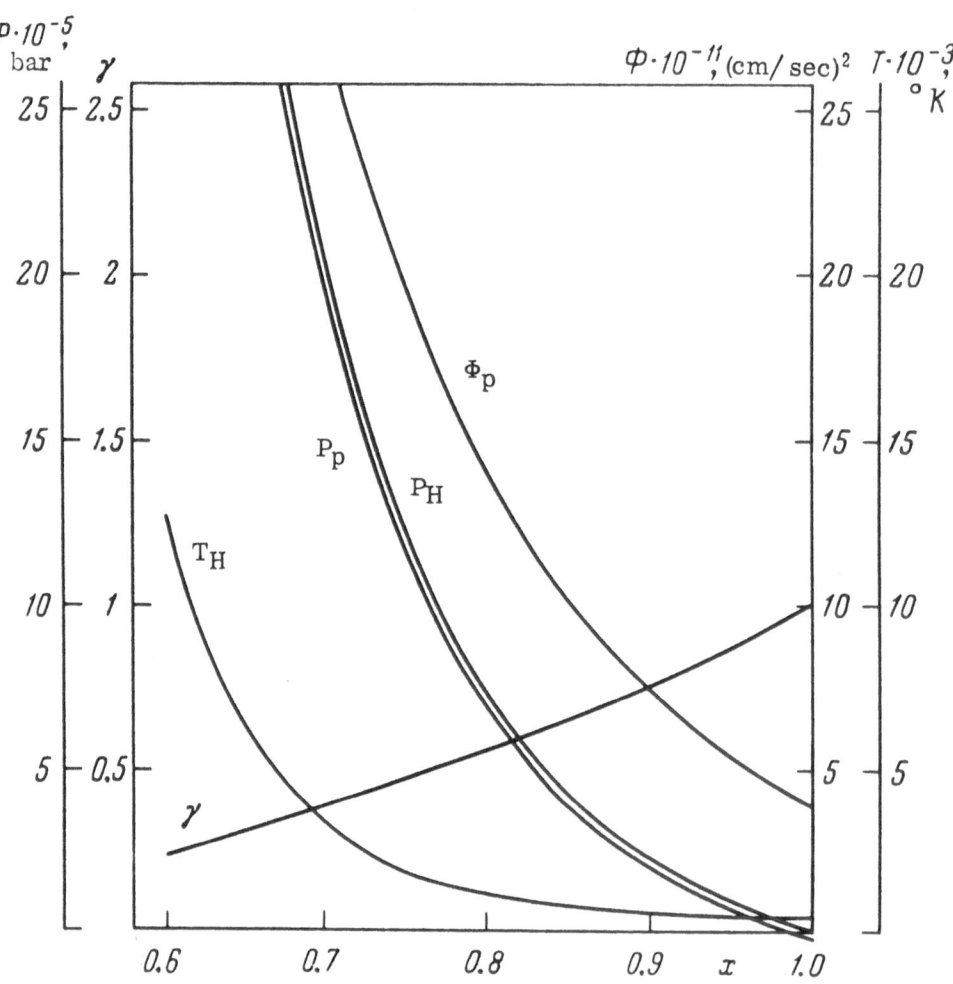

Fig. 4.25. Equation of state of grossularite.

Literature Cited

1. P. W. Bridgman, The Physics of High Pressure, Bell, London (1949).
2. P. W. Bridgman, Proc. Am. Acad. Arts Sci., 74:11 (1940).
3. P. W. Bridgman, "Recent work in the field of high pressures," Rev. Mod. Phys., 18:1 (1946).
4. K. Swenson, "Physics at high pressure," Solid State Phys., 11:41 (1960).
5. R. H. Wentorf, Jr. (ed.), Modern Very High Pressure Techniques, Butterworths, London (1962).
6. Yu. S. Genshaft, L. D. Livshits, and Yu. N. Ryabinin, Zh. Prikl. Mekh. Tekh. Fiz., No. 5, p. 107 (1962).
7. P. W. Bridgman, Proc. Am. Acad. Arts Sci., 74:425 (1942).
8. P. W. Bridgman, Proc. Am. Acad. Arts Sci., 76:1 (1945).
9. P. W. Bridgman, Proc. Am. Acad. Arts Sci., 76:9 (1945).
10. P. W. Bridgman, Proc. Am. Acad. Arts Sci., 76:55 (1948).
11. S. P. Clark, Jr. (ed.), Handbook of Physical Constants, rev. ed., Geological Society of America, New York (1966).
12. W. Paul and D. M. Wareschauer (eds.), Solids under Pressure, McGraw-Hill, New York (1963).
13. D. Lazarus, Phys. Rev., 76:545 (1949).
14. Ultrasound in Geophysics [Russian translation], Mir, Moscow (1964).
15. Electrical and Mechanical Properties of Rocks at High Pressures [in Russian], Nauka, Moscow (1966).
16. F. F. Voronov and V. A. Goncharova, Zh. Éksp. Teor. Fiz., 50:1173 (1966).
17. M. P. Volarovich and A. I. Levykin, Dokl. Akad. Nauk SSSR, 165:1287 (1965).
18. L. F. Vereshchagin, Appendix to C. A. Swenson, "Physics at high pressure," Solid State Phys., 11:41 (1960) [Russian translation], IL, Moscow (1963).
19. P. W. Bridgman, Proc. Am. Acad. Arts Sci., 76:71 (1948).
20. P. W. Bridgman, Proc. Am. Acad. Arts Sci., 76:89 (1948).
21. P. W. Bridgman, Proc. Am. Acad. Arts Sci., 77:187 (1949).
22. P. W. Bridgman, Proc. Am. Acad. Arts Sci., 83:1 (1954).
23. N. P. Klepikov and S. N. Sokolov, Analysis and Planning of Experiments using the Maximum Probability Method [in Russian], Nauka, Moscow (1964).
24. Ho-Kwang Mao, W. A. Bassett, and T. Takahashi, J. Appl. Phys., 38:272 (1967).

CHAPTER 5

COMPRESSION OF SOLIDS
BY STRONG SHOCK WAVES

5.1. Introduction

In this chapter we shall consider a wide range of topics associated with the determination of the equations of state from dynamic data obtained up to pressures of several million bars.

It is practically impossible to determine the equation of state of condensed matter in the range of pressures up to 10^8 bar at which electron shells are not yet "squashed." However, dynamic experimental methods for the compression of condensed matter by shock waves producing pressures of several million bars are already available. Using these methods, it has been possible to determine experimentally the Hugoniot curves or shock adiabats of many metals, ionic crystals, some liquids, and rocks. The Hugoniot curve $P_H(V)$, which represents the dependence of the shock pressure (the pressure behind a shock-wave front) on the specific volume of a compressed substance is insufficient for the determination of the equation of state of a solid or liquid. This is because the equations of state of condensed substances are represented at moderate temperatures by two independent functions of the volume: the cold-compression isotherm $P_p(V)$ and the Grüneisen parameter $\gamma(V)$. By moderate temperatures we understand the phonon range of temperatures in which only the thermal vibrations of atoms (phonons) are excited in a condensed substance. At high temperatures, the following phenomena may become important: thermally excited conduction electrons, interactions between phonons, anharmonicity, the excitation of defects, thermal ionization, etc. Naturally, each of these effects introduces some unknown function of the volume. However, the Hugoniot curve is characterized by the fact that, initially, it is very close to the corresponding isotherms and isentropes. When the pressure is increased, the temperature along the Hugoniot curve begins to increase rapidly and it goes over from the phonon region into the region where high-temperature corrections are necessary. Therefore, even in the simplest case, where only the Hugoniot curve for the phonon region is employed, we encounter the difficulty that, in general, it is impossible to determine two functions $\gamma(x)$ and $P_p(V)$ from a single known curve $P_H(V)$. This difficulty is usually avoided by assuming a definite relationship between $P_p(V)$ and $\gamma(V)$ and expressing $\gamma(V)$ in terms of $P_p(V)$.

In this connection, we meet the problem of the determination of two experimental functions of the volume in the phonon temperature range. One such function is $P_H(V)$. The other function may be obtained in several ways. For example, we may determine several Hugoniot curves $P_H(V)$ by varying the initial state (V or T) of the substance being investigated. Secondly, we may use the static isotherm $P_i(V)$. Finally, we can determine the temperature on the Hugoniot curve $T_H(V)$ or $T_H(P_H)$. However, we must mention that it is not easy to determine a second experimental function of the volume and, therefore, it is often necessary to determine the equation of state from the dynamic data assuming some relationship between $\gamma(V)$ and $P_p(V)$.

Even greater difficulties are encountered in the use of the high-temperature part of the Hugoniot curve in the determination of the equation of state. As already pointed out, the number of unknown functions in this region is greater. In view of this, equations of state valid to pressures up to several million atmospheres are based on the dynamic data, as well as on the static data, high-temperature data obtained under normal conditions, and (in the case of metals) the data on the specific heat and thermal expansion near absolute zero.

In deriving the equations of state, we shall describe $P_p(V)$, $\gamma(V)$, $A_a(V)$, and $A_e(V)$ by certain functions in which numerical coefficients will be determined from the experimental data. This approach to the determination of the equation of state will be called the method of potentials. It originates in the classical investigations of Born (Chapter 1), and a description can be found in [1-5] of the form in which it will be used in the present chapter. There are other methods for the determination of the equation of state from the dynamic data [6-20].

Although frequently the results obtained by different methods are quite similar, we shall concentrate on the method of potentials. This is because we shall assume that if the potential is selected satisfactorily, the equation of state can be extrapolated to a wider range of values of V and T. Secondly, if the potential is selected properly, the quantities obtained by differentiation of the equation of state are relatively accurate. Finally, this method yields the equation of state in the form of a compact formula rather than a complex table.

When the dynamic experimental data are used to determine equations of state, it is customary to make the following two assumptions: 1) the measured values of P, V, and E refer to states of thermodynamic equilibrium; 2) compression at a given pressure is equal to the compression which would be produced by a hydrostatic pressure of the same magnitude.

The first condition is satisfied if thermodynamic equilibrium is established in the short time of the passage of a shock wave (which is of the order of 10^{-7} sec). We may expect some polymorphic transitions to require longer times. In such cases, the experimental shock data represent a quasiequilibrium state immediately behind a shock-wave front.

The second condition is not satisfied exactly because the compression in dynamic experiments is one-dimensional. However, the range of pressures investigated by means of powerful shock waves is well beyond the yield stress of solids; consequently, the effects associated with the rigidity of solids can be ignored and the compressed substance can be regarded as an ideal liquid.

We shall postulate that these assumptions are satisfied in all cases. Some justification for these assumptions is provided by the agreement between the many results obtained under dynamic and static conditions. In particular, the dynamic compressibility data can be joined smoothly to the static data of Bridgman's obtained at lower pressures. This chapter will give the hydrodynamic relationships, illustrated by examples for specific substances, which will be required subsequently. This will be followed by descriptions of the experimental methods used to obtain the Hugoniot curves and to determine the equations of state from the dynamic data using the method of potentials.

5.2. Hydrodynamic Relationships [12-22]

We shall be interested only in one-dimensional hydrodynamic flow in which all the quantities depend on the coordinate ξ and on time t.

Equations of Motion

A moving fluid can be described by functions representing the distribution of velocity in the fluid $u = u(\xi, t)$ (we shall call this the mass velocity) and any two thermodynamic quantities, for example, the pressure $P(\xi, t)$ and the density $\rho(\xi, t)$. Thermodynamic relationships and equations of state can be used to transform the selected thermodynamic quantities to different pairs of thermodynamic variables (Appendix A). Thus, if we specify three functions $u(\xi, t)$, $P(\xi, t)$, and $\rho(\xi, t)$, we describe completely the state of a moving fluid. These functions can be regarded as fields in the (ξ, t) plane. The first equation of hydrodynamics

$$\frac{\partial \rho}{\partial t} + \frac{\partial}{\partial \xi}(\rho u) = 0 \qquad (5.1)$$

is the equation of continuity. It expresses the fact that the rate of increase of mass $(\partial \rho / \partial t)d\xi$, in an element of volume, which is $d\xi$ long and has a unit cross section, should be equal to the total flow of mass into that volume element. The quantity

$$j = \rho u \qquad (5.2)$$

is known as the density of fluid flow.

If there are no volume forces (such as the force of gravity), the only hydrodynamic force F_P acting in a fluid is equal to the pressure gradient with its sign reversed: $F_P = -\partial P/\partial\xi$. This is deduced from the balance of forces acting on the volume element $d\xi$: $P - [P + (\partial P/\partial\xi)d\xi]$. Equating the force to the product of the element of mass $\rho d\xi$ and the acceleration given by

$$\frac{du}{dt} = \frac{\partial u}{\partial t} + \frac{\partial \xi}{\partial t}\frac{\partial u}{\partial \xi},$$

we obtain the equation of motion of a fluid, known as Euler's equation

$$\rho u_t + \rho u u_\xi = -P_\xi. \qquad (5.3)$$

The subscripts in this equation indicate partial differentiation with respect to the arguments ξ, t.

We shall consider fluid flow in which we can neglect heat conduction processes, in which the energy can be transferred from one element of mass to another. In such flow, known as adiabatic flow, the entropy of each element of mass remains constant

$$dS / dt = 0,$$

or, in terms of the total derivative with respect to time,

$$S_t + uS_\xi = 0. \qquad (5.4)$$

We shall consider the flow in which the entropy is the same throughout the fluid at some initial moment.

According to the adiabatic equation (5.4), this value of the entropy is conserved during subsequent motion:

$$S = \text{const.} \qquad (5.5)$$

The motion of a fluid in which the entropy remains constant at all points in the fluid, is known as the isentropic flow. The thermodynamic description of the isentropic flow is quite simple. Taking the density (or the specific volume $v = 1/\rho$) and the entropy as the independent thermodynamic variables, we can write the equation of state for the isentrope in the form

$$P = f_1(\rho, S_0), \qquad (5.6)$$

where S_0 represents the constant value of the entropy.

The specific internal energy [given by Eq. (A.1) in Appendix A] also depends on the specific volume

$$dE = -Pdv. \tag{5.7}$$

Consequently, all quantities are functions only of the density. Thus, for example, the velocity of sound

$$c(\rho) \equiv \left(\frac{\partial P}{\partial \rho}\right)_{s}^{1/2} \tag{5.8}$$

in the isentropic flow depends only on the density of the fluid.

The energy flux Π_e and the momentum flux Π_p in a medium are given by the relationships [21]

$$\Pi_e = \rho u \left(\frac{u^2}{2} + w\right), \tag{5.9}$$

where w is the specific enthalpy, and

$$\Pi_p = P + \rho u^2. \tag{5.10}$$

The energy per unit volume of the fluid is

$$\rho \frac{u^2}{2} + \rho E,$$

where the first term is the kinetic energy and the second term is the internal specific energy.

The law of conservation of energy [21]

$$\frac{\partial}{\partial t}\left(\frac{\rho u^2}{2} + \rho E\right) + \frac{\partial}{\partial \xi}\left[\rho u \left(\frac{u^2}{2} + w\right)\right] = 0 \tag{5.11}$$

shows that the change in the energy per unit volume of the fluid per unit time is equal to the efflux of energy from this volume per unit time. Similarly, the law of conservation of momentum [21]

$$\frac{\partial}{\partial t} \rho u + \frac{\partial}{\partial \xi} (P + \rho u^2) = 0 \tag{5.12}$$

shows that the change in the momentum per unit volume of the fluid per unit time is equal to the efflux of momentum per unit time from this volume. Using the equation of continuity, we can easily show that Eq. (5.12) is equivalent to Euler's equation (5.3).

In the presence of dissipative processes in the fluid, such as viscosity and heat conduction, the expressions for the energy and momentum fluxes acquire additional terms [21]:

$$\Pi_{e'} = \rho u \left(\frac{u^2}{2} + w\right) - \left(\frac{4}{3}\eta + \zeta\right) u \frac{\partial u}{\partial \xi} - \varkappa \frac{\partial T}{\partial \xi}, \tag{5.9a}$$

$$\Pi_{p'} = P + \rho u^2 - \left(\frac{4}{3}\eta + \zeta\right)\frac{\partial u}{\partial \xi}, \tag{5.10a}$$

where η is the usual Newtonian viscosity; $\varkappa$ is the thermal conductivity; ζ is the second viscosity. The second viscosity gives rise to energy dissipation because of the inhomogeneity of the deformation of elements of volume in the fluid. The dissipative terms in the energy and momentum fluxes of Eqs. (5.9a) and (5.10a) are important only in discussions of the width of a shock-wave front.

Flow in a Simple Rarefaction Wave.
Characteristics. Riemann Invariants

We shall now write the equation of continuity and Euler's equation in the form

$$\rho_t + \rho_\xi u + \rho u_\xi = 0,$$

$$\frac{\rho}{c}(u_t + u u_\xi) + \rho_\xi c = 0.$$

We have replaced P_ξ in Euler's equation with $\rho_\xi c^2$, and we have divided the whole expression by the velocity of sound given by Eq. (5.8). Forming a sum and a difference of these equations, we obtain

$$\rho_t + (u \pm c)\rho_\xi \pm \frac{\rho}{c}[u_t + (u \pm c)u_\xi] = 0. \qquad (5.13)$$

For an observer moving with the fluid at a velocity u = $d\xi$/dt, the total derivative with respect to time is

$$\frac{d}{dt} = \frac{\partial}{\partial t} + u\frac{\partial}{\partial \xi}.$$

Consequently, the first expression in Eq. (5.13) yields the relationships

$$\frac{d\rho}{dt} + \frac{\rho}{c}\frac{du}{dt} = 0, \qquad \text{or} \qquad du + \frac{c}{\rho}d\rho = 0 \qquad (5.14)$$

for $\dfrac{d\xi}{dt} = u + c.$

The second expression in Eq. (5.13) yields

$$\frac{d\rho}{dt} - \frac{\rho}{c}\frac{du}{dt} = 0, \qquad \text{or} \qquad du - \frac{c}{\rho}d\rho = 0 \qquad (5.15)$$

for $\dfrac{d\xi}{dt} = u - c.$

According to the equation of state (5.6) and the definition of the velocity of sound (5.8), we obtain $d\rho = (1/c^2)dP$ for the isentropic flow, and this makes it possible to rewrite Eqs. (5.14) and (5.15) in the form

$$du + \frac{dP}{c\rho} = 0 \quad \text{along} \quad \frac{d\xi}{dt} = u + c \qquad (C_+), \qquad (5.14a)$$

$$du - \frac{dP}{c\rho} = 0 \quad \text{along} \quad \frac{d\xi}{dt} = u - c \qquad (C_-). \qquad (5.15a)$$

The two pairs of relationships (5.14) and (5.15) [or (5.14a) and (5.15a)] form a set of four "characteristic" equations which describe the isentropic flow. The quantities

$$\left.\begin{aligned} J_+ &= u + \int\frac{dP}{\rho c} = u + \int\frac{c}{\rho}d\rho, \\ J_- &= u - \int\frac{dP}{\rho c} = u - \int\frac{c}{\rho}d\rho \end{aligned}\right\} \qquad (5.16)$$

are known as Riemann invariants. Equations (5.14) and (5.15) show that these invariants remain constant along the corresponding characteristic directions. Since $(c + u)$ and $(c - u)$ represent the local velocities of propagation of acoustic perturbations along the $+\xi$ and $-\xi$ directions, the characteristic directions in the (ξ, t) plane are lines along which acoustic perturbations are propagated. The family of curves, corresponding to the propagation of perturbations along the $+\xi$ direction, is known as the C_+ characteristics. Similarly, the family of curves corresponding to the propagation of acoustic perturbations along the $-\xi$ direction is known as the C_- characteristics. Equations (5.14) and (5.15) show that the Riemann invariants J_+ or J_- remain constant along the C_+ or C_- characteristics, respectively.

The one-dimensional isentropic flow is fully determined by specifying two functions, for example, $u(\xi, t)$ and $\rho(\xi, t)$. The formulas in Eq. (5.16) show that a complete description of such flow is given also by specifying J_+ and J_- rather than u and ρ. The expressions in Eq. (5.16) establish a relationship between the (u, ρ) and (J_+, J_-) planes.

Let us now consider the motion of a gas under the action of a moving piston. Let us assume that the gas occupies semi-infinite space $0 \leq \xi \leq \infty$, and that this space is bounded by the piston at the origin of the coordinate $(\xi = 0)$. As long as the piston is at rest, the gas represents a uniform mass of constant density ρ_0 and entropy S_0 so that the pressure P_0 in the gas and the velocity of sound c_0 are also constant. The system of characteristics of the gas at rest $d\xi/dt = c_0$ (C_+) and $d\xi/dt = -c_0$ (C_-) is shown in Fig. 5.1a. The characteristics C_+ and C_- are families of parallel straight lines inclined at the same angle to the 0ξ axis. The characteristics of the system C_+ originate on the 0ξ axis and go to infinity. The uppermost C_+ characteristic, shown in Fig. 5.1a by a thick line, represents the path, in the (ξ, t) plane, of acoustic perturbations which originate at the piston surface and travel into the gas at rest. All the other C_+ characteristics lie below the characteristic starting from the origin of the coordinates. The C_- characteristics are bounded intercepts starting on the 0ξ axis and ending at the surface of the piston, which is represented by the $0t$ axis. The "paths" of the particles in a gas at rest are represented by the dashed lines in the (ξ, t) plane; these lines begin on the 0ξ axis and go to infinity parallel to the $0t$ axis.

Let us consider how this pattern changes when the piston moves (Fig. 5.1b). We shall assume that the acceleration of the piston along the $-\xi$ direction ends at a moment t_1. Beginning from this moment, the velocity of the piston will be taken to be constant. Head waves, which are generated at the beginning of the motion of the piston, are represented by the head C_+ characteristic in Fig. 5.1b. This characteristic describes the propagation of acoustic perturbations in a gas at rest, in the same way as in Fig. 5.1a. Consequently, the whole pattern in the (ξ, t) plane, below the head C_+ characteristic remains unchanged. In the (u, ρ) plane, this characteristic corresponds to a gas at rest: $u = 0$; $\rho = \rho_0$. We thus see that, as in the case of a gas at rest, all the C_- characteristics which originate from the 0ξ axis at the moment $t = 0$ intersect the head C_+ characteristic. This means that the boundary conditions $u = 0$, $\rho = \rho_0$ are the same for all the C_- characteristics and, consequently, the integration of Eq. (5.15) makes it possible to determine the mass velocity as a function of the density for the whole flow

$$u = \int_{\rho_0}^{\rho} \frac{c d\rho}{\rho} \tag{5.17}$$

(we recall that in the isentropic flow the velocity of sound, like all the other thermodynamic quantities, depends only on the density). Equation (5.17) can be regarded as a curve $u = u(\rho_0, \rho)$ in the (u, ρ) plane, which represents the results of the transformation of all the C_- characteristics. Waves in such a flow are known as simple waves. A rarefaction wave considered here is such a simple wave.

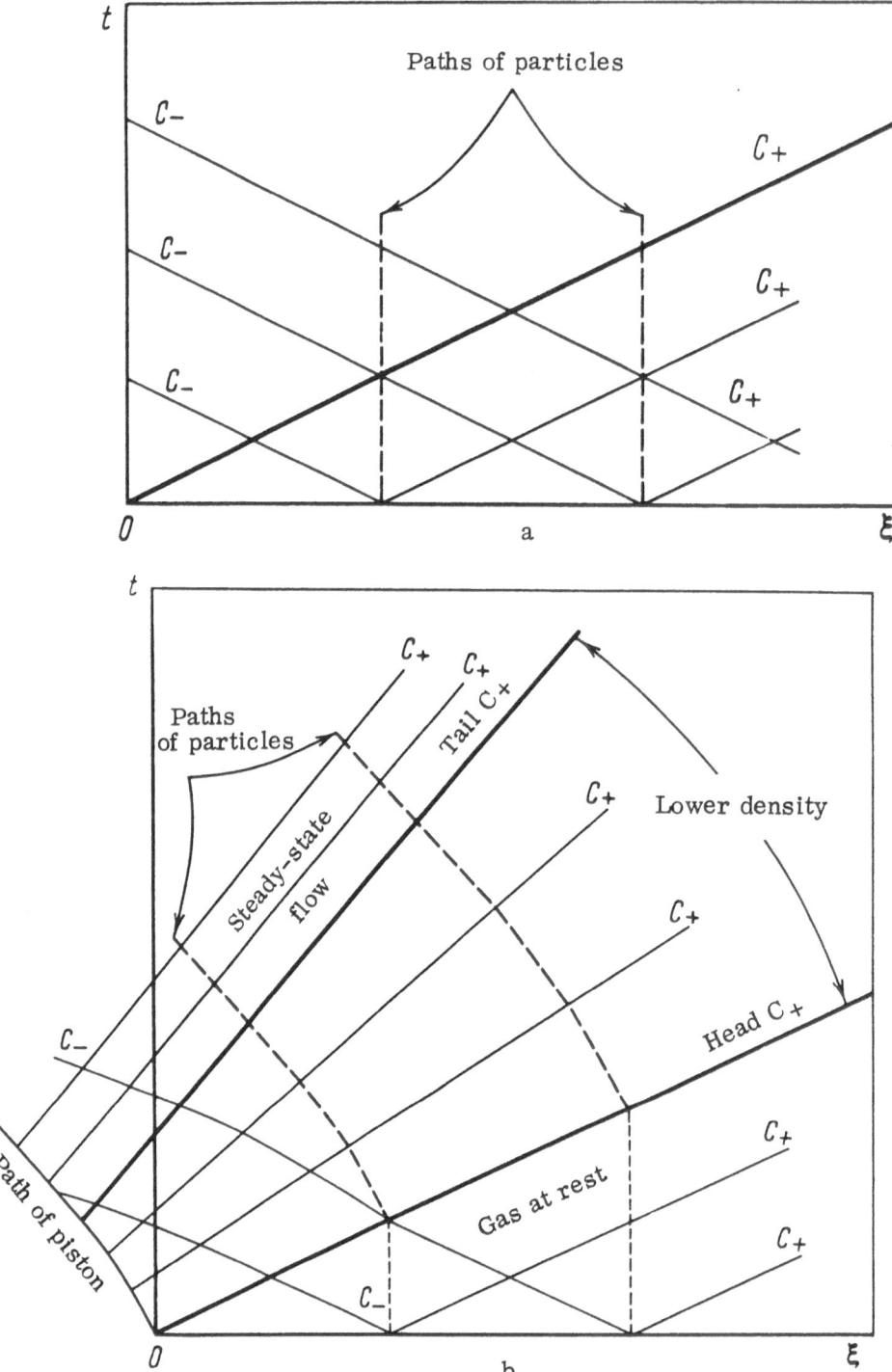

Fig. 5.1. Characteristics in a gas at rest (a) and in a moving gas (b).

We shall now consider a region in the (ξ, t) plane which is bounded by the head and tail C_+ characteristics beginning at the initial and final moments of the accelerated motion of the piston. This region represents a gradual decrease in the density (or pressure) and the C_+ characteristics in this region are rectilinear. In this case, the flow can be described by a "field" of the Riemann invariants J_- and J_+. We have already mentioned that the invariant J_- is constant over the whole region of flow and J_+ is constant, by definition, for each of the C_+

characteristics. Consequently, both $u(J_+, J_-)$ and $c(J_+, J_-)$ are constant along each of the C_+ characteristics and the equation $d\xi/dt = u + c$ describes a straight line. When we cross the C_+ characteristic, we reach the region of a steady-state flow, where J_- and J_+ are constant for the whole flow. The paths of the particles in Fig. 5.1b are shown by the dashed lines. They are parallel to the $0t$ axis until they intersect the head C_+ characteristic; then they bend in the negative direction of the 0ξ axis, but after intersecting the tail C_+ characteristic, they again become linear.

Using Eqs. (5.17) and (5.8) and denoting the steady-state values of the density and pressure by ρ_1 and P_1, we obtain an equation for the tail C_+ characteristic in the form

$$\frac{d\xi}{dt} = \int_{\rho_0}^{\rho_1} \frac{c\,d\rho}{\rho} + \left(\frac{\partial P}{\partial \rho}\right)^{1/2}_{S,\,\rho=\rho_1}. \tag{5.18}$$

The value of the mass velocity in the steady-state flow region is

$$u = \int_{\rho_0}^{\rho_1} \frac{c\,d\rho}{\rho} = \int_{P_0}^{P_1} \left(-\frac{\partial v}{\partial P}\right)^{1/2}_{s} dP, \tag{5.19}$$

where $v = 1/\rho$ is the specific volume.

In order to gain a better insight into the meaning of the general formulas describing a rarefaction wave, we shall apply them to some simple cases.

Rarefaction Wave in an Ideal Gas with Constant Specific Heats

The equations for Poisson's adiabat (the isentropes)

$$\rho = \rho_0 \left(\frac{T}{T_0}\right)^{1/(\gamma-1)}, \qquad P = P_0 \left(\frac{\rho}{\rho_0}\right)^{\gamma}, \qquad \gamma = \frac{c_P}{c_V} \tag{5.20}$$

and the expression for the velocity of sound

$$c^2 = \gamma \frac{P}{\rho} = \gamma \frac{RT}{\mu} \tag{5.21}$$

in an ideal gas with constant specific heats are derived in Appendix C. Using Eq. (5.21), we can express the velocity of sound (5.8) in terms of the density

$$c = \left(\frac{\gamma P_0}{\rho_0^{\gamma}}\right) \rho^{(\gamma-1)/2} = c_0 \left(\frac{\rho}{\rho_0}\right)^{(\gamma-1)/2}. \tag{5.22}$$

The quantities ρ_0, P_0, and c_0 represent the initial state of a gas, corresponding to a region in Fig. 5.1b bounded by the 0ξ axis and the head C_+ characteristic. Integrating the expressions in Eq. (5.16) and assuming the constants of integration to be zero, we obtain the Riemann invariants

$$\left. \begin{aligned} J_+ &= u + \frac{2}{\gamma-1}c, \\ J_- &= u - \frac{2}{\gamma-1}c = -\frac{2}{\gamma-1}c_0. \end{aligned} \right\} \tag{5.16a}$$

It is evident from Eq. (5.16a) that the J_- invariants are governed by the velocity of sound c_0 in the unperturbed region and are constant for the whole flow.

We shall describe the flow by the field of the Riemann invariants J_- and J_+ and we shall express all other quantities in terms of these invariants. This can be done easily using Eqs. (5.20)-(5.22) and (5.16a):

$$
\left.\begin{aligned}
c &= \frac{\gamma-1}{4}(J_+ - J_-), \qquad u = \frac{1}{2}(J_+ + J_-), \\
T &= T_0 \left[\frac{\gamma-1}{4c_0}(J_+ - J_-) \right]^2, \\
P &= P_0 \left[\frac{\gamma-1}{4c_0}(J_+ - J_-) \right]^{2\gamma/(\gamma-1)} \\
\rho &= \rho_0 \left[\frac{\gamma-1}{4c_0}(J_+ - J_-) \right]^{2/(\gamma-1)}.
\end{aligned}\right\}
\tag{5.23}
$$

The velocity of flow u and the velocity of sound c are related by

$$
u = \frac{2}{\gamma-1}(c - c_0), \qquad c = c_0 + \frac{\gamma-1}{2}u.
\tag{5.24}
$$

We shall now describe the flow in each of the three regions in Fig. 5.1b. In the region at rest (u = 0), the equations for the C_+ and C_- characteristics are $(d\xi/dt)_+ = c_0$, $(d\xi/dt)_- = -c_0$, and the two Riemann invariants are given by $J_\pm = \pm(2/\gamma - 1)c_0$. In the region of accelerated motion of the piston, bounded by the head C_+ and the tail C_+ characteristics in Fig. 5.1b, the velocity of the piston U(t) is a function of time and increases from zero to a constant value U_1. Since the flow is continuous, the velocity of gas next to the piston is equal to the velocity of the piston itself. Consequently,

$$
J_+ = U(t) + \frac{2}{\gamma-1}c = \frac{2}{\gamma-1}c - |U(t)|.
$$

All other quantities can be obtained using the expressions given in Eq. (5.23). In particular, it follows from these expressions that the characteristics $(d\xi/dt)_+ = u + c$ (C_+) are linear.

The region of nonsteady flow ends at the tail C_+ characteristic (Fig. 5.1b). Beyond this characteristic we have a region of steady-state flow with constant values of the Riemann invariants

$$
J_+ = \frac{2}{\gamma-1}c_1 - |U_1|,
$$

$$
u_1 = -|U_1|, \qquad c_1 = c_0 - \frac{\gamma-1}{2}|U_1|.
$$

All other quantities can be obtained using the general formulas of Eq. (5.23). In the region of steady-state flow, the characteristics C_+ and C_- are linear and they form two families of parallel lines.

Finally, in the case of an ideal gas with constant specific heats the mass velocity (5.17) and the equation for the tail C_+ characteristic (5.18) can be written in the form of explicit functions of the density:

$$
u = \frac{2c_0}{\gamma-1}\left[\left(\frac{\rho}{\rho_0}\right)^{(\gamma-1)/2} - 1\right], \qquad c_0^2 = \gamma\frac{P_0}{\rho_0},
\tag{5.17a}
$$

$$
\left(\frac{d\xi}{dt}\right)_+ = \frac{2c_0}{\gamma-1}\left[\left(\frac{\rho_1}{\rho_0}\right)^{(\gamma-1)/2} - 1\right] + c_0\left(\frac{\rho_1}{\rho_0}\right)^{(\gamma-1)/2}.
\tag{5.18a}
$$

Rarefaction Wave in a Degenerate Fermi Gas

All the thermodynamic formulas for a degenerate Fermi gas are derived in Appendix C. The equations for the isentropes are of the form

$$P = P_0 \left(\frac{\rho}{\rho_0} \right)^{5/3}, \qquad \rho = \rho_0 \left(\frac{T}{T_0} \right)^{3/2}, \left. \begin{array}{c} \\ \\ \\ \end{array} \right\}$$
$$P = P_0 \left(\frac{T}{T_0} \right)^{5/2}. \qquad\qquad\qquad\qquad (5.25)$$

The velocity of sound is now

$$c^2 = \frac{5}{3} \frac{P_0}{\rho_0} \left(\frac{\rho}{\rho_0} \right)^{2/3} = c_0^2 x^{-2/3}. \qquad (5.26)$$

Here, ρ_0, P_0, and c_0 denote the initial state of the gas, corresponding to the region in Fig. 5.1b bounded by the 0ξ axis and the head C_+ characteristic.

There is no need to derive again all the formulas for a rarefaction wave in a degenerate Fermi gas. This is because all the relationships for an ideal gas with constant specific heats are derived using Eqs. (5.20) and (5.22). We can easily see that the corresponding relationships (5.25) and (5.26) for a degenerate Fermi gas are special cases of Eqs. (5.20) and (5.22), in which we substitute $\gamma = 5/3$. This means that the isentropic flow of a degenerate Fermi gas is analogous to the flow of a monatomic ideal gas with constant specific heats. It must be pointed out that, in the case of a Fermi gas, the exponent of the adiabat $\gamma = 5/3$ does not represent the ratio of the specific heats at constant pressure and at constant volume, i.e., $\gamma \neq c_P/c_V$. Substituting $\gamma = 5/3$ in the formulas given in the preceding subsection, we obtain all the necessary relationships:

$$J_+ = u + 3c, \qquad J_- = u - 3c = -3c_0; \qquad (5.16b)$$

$$c = \frac{1}{6}(J_+ - J_-), \qquad u = \frac{1}{2}(J_+ + J_-), \left. \begin{array}{c} \\ \\ \\ \\ \end{array} \right\}$$
$$T = T_0 \left[\frac{1}{6c_0}(J_+ - J_-) \right]^2, \qquad P = P_0 \left[\frac{1}{6c_0}(J_+ - J_-) \right]^5, \qquad (5.23a)$$
$$\rho = \rho_0 \left[\frac{1}{6c_0}(J_+ - J_-) \right]^3;$$

$$u = 3(c - c_0), \qquad c = c_0 + \frac{1}{3}u. \qquad (5.24a)$$

In the steady-state flow region beyond the tail C_+ characteristic in Fig. 5.1b, we have

$$J_+ = 3c_1 - |U_1|, \qquad J_- = u_1 - 3c_1 = -3c_0,$$
$$u_1 = -|U_1|, \qquad c_1 = c_0 - \frac{1}{3}|U_1|.$$

Finally, the mass velocity (5.17) and the equation for the tail C_+ characteristic (5.18) can be given as explicit functions of the density

$$u = 3c_0 \left[\left(\frac{\rho}{\rho_0} \right)^{1/3} - 1 \right], \qquad c_0^2 = \frac{5}{3} \frac{P_0}{\rho_0}, \qquad (5.17b)$$

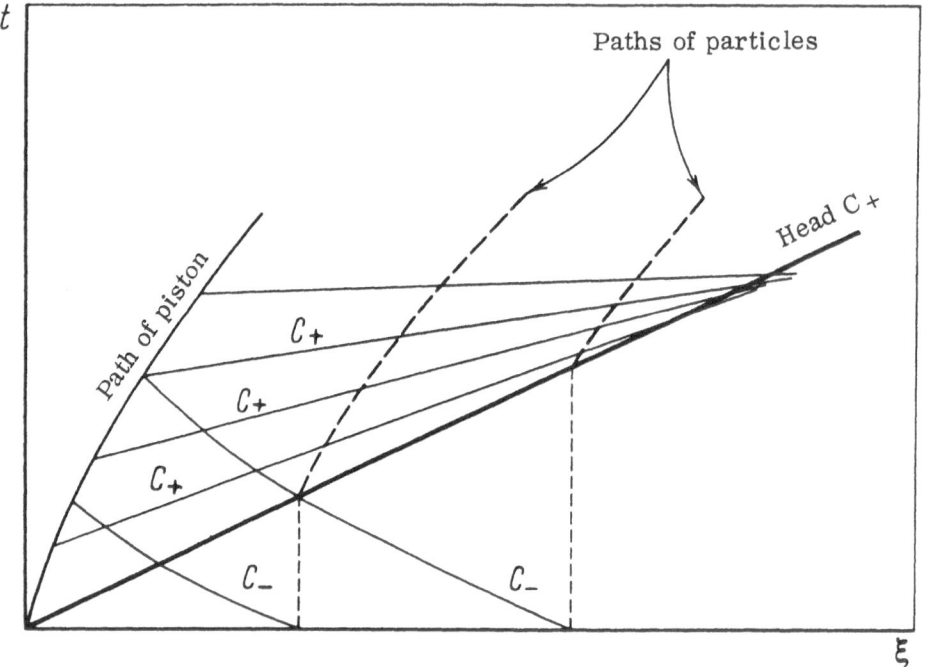

Fig. 5.2. Intersection of characteristics in a gas compressed by a piston.

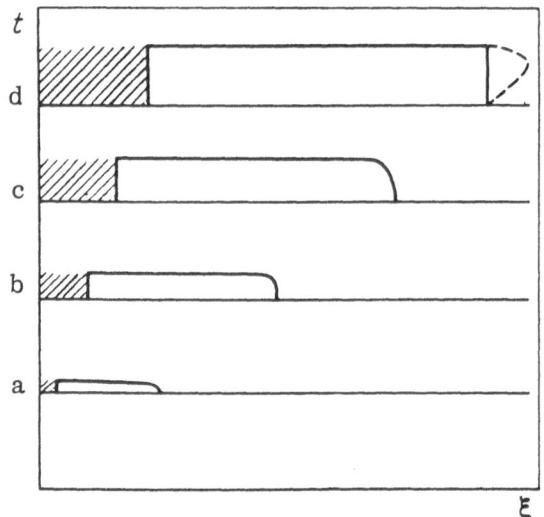

Fig. 5.3. Formation of a shock wave.

$$\left(\frac{\partial \xi}{dt}\right)_+ = c_0\left[4\left(\frac{\rho_1}{\rho_0}\right)^{1/3} - 3\right]. \qquad (5.18b)$$

Shock Waves

We shall now consider a compression wave in a fluid. Such a wave is generated by the motion of a piston in a gas or liquid: the velocities of the particles next to the piston are equal to the velocity of the piston and they increase with time. Before the beginning of the motion of the piston, at a time t < 0, the system of characteristics of a gas at rest is that shown in Fig. 5.1a. As the piston is pushed into the gas, the pattern shown in that figure changes to that given in Fig. 5.2. If the flow is assumed to be continuous, much of the analysis given in the preceding subsections still applies. Thus, the derivation of Eq. (5.17) and the proof that the C_+ characteristics are lines along which u and c are constant, remain unaltered. Since c increases with increasing density of the gas being compressed and u increases as per definition, the slopes of the C_+ characteristics $(d\xi /dt)_+ = u + c$ increase along the path of the piston (Fig. 5.2). Consequently, the C_+ characteristics intersect. In the case of continuous flow, the intersection of characteristics has no physical meaning, since a given point in the (ξ, t) plane would correspond to several values of ρ, u, c, etc. This difficulty is only apparent because a compression wave becomes unstable and discontinuous.

We shall now consider how such a discontinuity is generated. As soon as we start the motion of the piston in a gas at rest, we start generating disturbances which travel away from

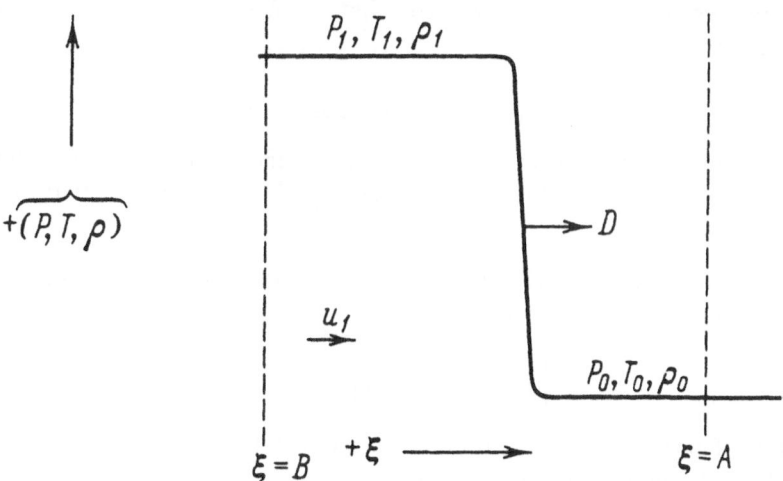

Fig. 5.4. Pressure, density, and temperature profiles in a shock wave.
The surface $\xi = B$ moves with the fluid.

the piston. Let us divide all disturbances generated by the piston into a large number of elementary disturbances, each of which can be regarded as an acoustic wave. Since the velocity of sound in a liquid or a gas is increased by compression, each subsequent elementary disturbance is propagated at a velocity higher than the preceding disturbance. This means that at some moment a given elementary disturbance can catch up with and overtake the preceding elementary disturbances (Fig. 5.3). However, in continuous flow, it is physically impossible to overtake a preceding disturbance because this would give rise to indeterminacy. This means that the velocity profile u in a compression wave, shown dashed in Fig. 5.3d, is fictitious. In fact, an increasingly steeper step of the velocity profile (a velocity discontinuity) appears at the head of a wave which is penetrating the gas at rest. This velocity discontinuity is accompanied by discontinuities of other thermodynamic variables (pressure, density, temperature, velocity of sound, etc.). In a narrow region of space in which this discontinuity is located, there are strong gradients of thermodynamic variables (pressure, density, temperature, etc.). In this region, the flow cannot be regarded as isentropic and, because of these strong gradients, the processes of viscosity and heat conduction become important and they increase the entropy.

This discontinuity of the thermodynamic variables and of the mass velocity is known as a shock front. The thickness of the shock front is governed by dissipative processes and, in general, it is small. In most cases, a shock front can conveniently be considered as a geometrical discontinuity (Fig. 5.4). Usually the term "shock wave" is intended to denote both the shock-wave front as well as the region of high values of P, T, and ρ behind the front. A shock wave may be propagated through a gas at rest or through a gas which is in motion. An important property of steady-state shock waves is the fact that the motion ahead of and behind the shock front is isentropic and the entropy changes only in the shock front itself. Another characteristic of shock waves is that all their properties can be established using the laws of conservation and the equations of state of a given substance. These properties are independent of the thickness of the shock-wave front.

Laws of Conservation for a Shock Wave

We shall now formulate the laws of conservation which should be obeyed by a shock wave passing through a gas. We shall select a system of coordinates which is moving with the shock front and we shall take the normal to the shock front as our 0ξ axis.

The law of conservation of mass postulates that the flow of a gas ρu should be the same ahead of and behind a shock-wave front:

$$[\rho u_\xi] = \rho_1 u_{1\xi} - \rho_0 u_{0\xi} = 0, \tag{5.27}$$

where the subscripts 0 and 1 refer to the state of matter ahead of and behind the front; the square brackets indicate that we are taking a difference of any given quantity on both sides of the front.

Using the definition of the flow of energy (5.9), the law of conservation of energy for a steady-state shock front can be written in the form

$$\left[\rho u_\xi \left(\frac{u_\xi^2}{2} + w \right) \right] = 0. \tag{5.28}$$

Finally, the law of conservation of momentum postulates that the momentum flux (5.10) on both sides of a shock front should be the same:

$$[P + \rho u_\xi^2] = 0. \tag{5.29}$$

Let us now return to the system of coordinates at rest and use u_S for the velocity of a shock front in the system of coordinates at rest. Then,

$$u_{\xi i} = u_i - u_s,$$

and the laws of conservation (5.27)-(5.29) in the laboratory system of coordinates assume the form

$$\rho_0(u_s - u_0) = \rho_1(u_s - u_1), \tag{5.27a}$$

$$\frac{(u_s - u_0)^2}{2} + w_0 = \frac{(u_s - u_1)^2}{2} + w_1, \tag{5.28a}$$

$$P_0 + \rho_0(u_s - u_0)^2 = P_1 + \rho_1(u_s - u_1)^2. \tag{5.29a}$$

The most important practical case is the propagation of a shock wave in a gas at rest. Substituting $u_0 = 0$ into Eqs. (5.27a)-(5.29a), we obtain the following formulas:

$$\rho_0 u_s = \rho_1(u_s - u_1), \tag{5.27b}$$

$$w_0 = w_1 + \frac{u_1^2}{2} - u_s u_1, \tag{5.28b}$$

$$P_0 + \rho_0 u_s u_1 = P_1. \tag{5.29b}$$

It follows from these formulas that if the velocity of a shock front u_S and the mass velocity u_1 are determined experimentally and the state in a gas or liquid at rest ahead of a shock-wave front is assumed to be known, Eqs. (5.27b)-(5.29b) provide a full description of the state behind the shock-wave front, given by the quantities P_1, ρ_1, and w_1.

Hugoniot Curve

We shall now draw some conclusions from the general laws of conservation at a discontinuity, given by Eqs. (5.27)-(5.29). We shall use specific volumes $V_0 = 1/\rho_0$, $V_1 = 1/\rho_1$ and introduce an invariant density of mass flow (5.2), given by $j = \rho_0 u_{0\xi} = \rho_1 u_{1\xi}$. We then obtain from Eq. (5.27):

$$u_{0\xi} = jV_0, \qquad u_{1\xi} = jV_1. \tag{5.30}$$

Substituting the expressions in Eq. (5.30) into Eq. (5.29), we have

$$P_0 + j^2 V_0 = P_1 + j^2 V_1, \tag{5.31}$$

or

$$j^2 = \frac{P_1 - P_0}{V_0 - V_1}. \tag{5.32}$$

Since $j^2 > 0$, we should have either $P_1 > P_0$, $V_0 > V_1$, or $P_1 < P_0$, $V_0 < V_1$. In fact, only the first case is possible [21]. The difference between the mass velocities $u_{0\xi} - u_{1\xi} = j(V_1 - V_2)$ can be transformed, using Eq. (5.32), to the following expression:

$$u_{0\xi} - u_{1\xi} = \sqrt{(P_1 - P_0)(V_0 - V_1)}. \tag{5.33}$$

The positive value of the square root is used since $u_{0\xi} > u_{1\xi}$. We shall now introduce j in Eq. (5.28):

$$w_0 + \frac{j^2 V_0^2}{2} = w_1 + \frac{j^2 V_1^2}{2}. \tag{5.34}$$

Using Eq. (5.32) to eliminate j from Eq. (5.34), we obtain the required equation

$$w_0 - w_1 + \frac{1}{2}(V_0 - V_1)(P_1 - P_0) = 0, \tag{5.35}$$

which relates the thermodynamic variables on both sides of a shock front. An equivalent form of Eq. (5.35) can be obtained by expressing the thermal function (enthalpy) in terms of the energy:

$$w = E + PV.$$

Then,

$$E_0 - E_1 + \tfrac{1}{2}(V_1 - V_0)(P_0 + P_1) = 0. \tag{5.36}$$

For a given initial state of a gas or a liquid (P_0, V_0), Eq. (5.35) or Eq. (5.36) gives the relationship between P_1 and V_1 (it is assumed that the dependence of the internal energy, or of the thermal function, on the volume and pressure is known). This dependence

$$P_H \equiv P_1 = P_H(V_1, P_0, V_0) \tag{5.37}$$

is known as the Hugoniot curve or shock adiabat (derived by W. J. M. Rankine in 1870 and by H. Hugoniot in 1889). The difference between the Hugoniot curve of Eq. (5.37) and the isentrope or Poisson's adiabat of Eq. (5.6) is this: the Hugoniot curve is a two-parameter family of curves (the parameters are P_0 and V_0) and the isentrope is a single-parameter family of curves (the parameter is S_0). We shall assume that V_1 and the Hugoniot pressure $P_H = P_1$ have been determined experimentally; then, all the other quantities can be expressed in terms of these two parameters and in terms of the initial parameters P_0, V_0, and u_0. Thus, it follows from Eq. (5.33) that

$$u_1 = u_0 + \sqrt{(P_1 - P_0)(V_0 - V_1)}. \tag{5.33a}$$

If $u_0 = 0$, we have

$$u_1 = \sqrt{(P_1 - P_0)(V_0 - V_1)}. \tag{5.33b}$$

It follows from Eqs. (5.29a), (5.27a), and (5.33a) that

$$u_s = u_0 + V_0 \sqrt{\frac{P_1 - P_0}{V_0 - V_1}}. \tag{5.38}$$

When a shock wave is propagated in a gas initially at rest ($u_0 = 0$), we obtain

$$u_s = V_0 \sqrt{\frac{P_1 - P_0}{V_0 - V_1}}. \tag{5.38a}$$

The Hugoniot curve can be considered not only in the (P_1, V_1) plane but also in any other plane. Let us consider this curve in the (P_1, u_1) plane. Then, V_1 and u_s are easily found in terms of P_1, u_1, and in terms of the initial values P_0, V_0, u_0 by means of Eqs. (5.33a) and (5.38):

$$V_1 = V_0 - \frac{(u_1 - u_0)^2}{P_1 - P_0}, \qquad u_s = u_0 + V_0 \frac{P_1 - P_0}{u_1 - u_0}, \qquad (P_1, u_1). \tag{5.39}$$

Similarly, we can obtain the formulas for the Hugoniot curve in terms of other variables:

$$V_1 = V_0 \left[1 - \frac{V_0(P_1 - P_0)}{(u_s - u_0)^2} \right], \quad u_1 = u_0 + \frac{V_0(P_1 - P_0)}{u_s - u_0}, \quad (P_1, u_s), \tag{5.40}$$

$$P_1 = P_0 + \frac{(u_1 - u_0)^2}{V_0 - V_1}, \qquad u_s = u_0 + V_0 \frac{u_1 - u_0}{V_0 - V_1}, \qquad (V_1, u_1), \tag{5.41}$$

$$P_1 = P_0 + \frac{V_0 - V_1}{V_0^2}(u_s - u_0)^2, \qquad u_1 = u_0 + \frac{V_0 - V_1}{V_0}(u_s - u_0), \qquad (V_1, u_s), \tag{5.42}$$

$$V_1 = V_0 \left(\frac{u_s - u_1}{u_s - u_0} \right), \quad P_1 = P_0 + \frac{1}{V_0}(u_s - u_0)(u_1 - u_0), \quad (u_s, u_1). \tag{5.43}$$

In the case of strong shock waves in solids,* the important Hugoniot curve is that expressed in terms of (u_s, u_1), since these quantities can be determined experimentally. It is also convenient to employ the Hugoniot curve expressed in terms of (P_1, u_1).

Weak Shock Waves

The discontinuities of all the thermodynamic variables in weak shock waves are naturally small. The required formula (5.45), which relates the entropy discontinuity ($S_1 - S_0$) to the pressure discontinuity ($P_1 - P_0$), is obtained by expanding Eq. (5.35) in terms of these quantities:

$$w_1 - w_0 = \left(\frac{\partial w}{\partial S_0}\right)_P (S_1 - S_0) + \left(\frac{\partial w}{\partial P_1}\right)_S (P_1 - P_0) + \frac{1}{2}\left(\frac{\partial^2 w}{\partial P_1^2}\right)_S (P_1 - P_0)^2 + \frac{1}{6}\left(\frac{\partial^3 w}{\partial P_1^3}\right)_S (P_1 - P_0)^3.$$

This expansion is carried out up to terms of the third order in ($P_1 - P_0$). In the case of ($S_1 - S_0$), we need expand only to terms of the first order, since this quantity is itself of the order of ($P_1 - P_0$)^3. Substituting the values of the derivatives using Eq. (A.6) from Appendix A,

$$T = \left(\frac{\partial w}{\partial S}\right)_P \qquad V = \left(\frac{\partial w}{\partial P}\right)_S,$$

we obtain

$$w_1 - w_0 = T_0(S_1 - S_0) + V_0(P_1 - P_0) + \frac{1}{2}\left(\frac{\partial V}{\partial P_0}\right)_S (P_1 - P_0)^2 + \frac{1}{6}\left(\frac{\partial^2 V}{\partial P_0^2}\right)(P_1 - P_0)^3.$$

---

*Under the action of a strong shock wave, a solid becomes plastic and, to a good approximation, can be regarded as a liquid. We shall consider this point in more detail later.

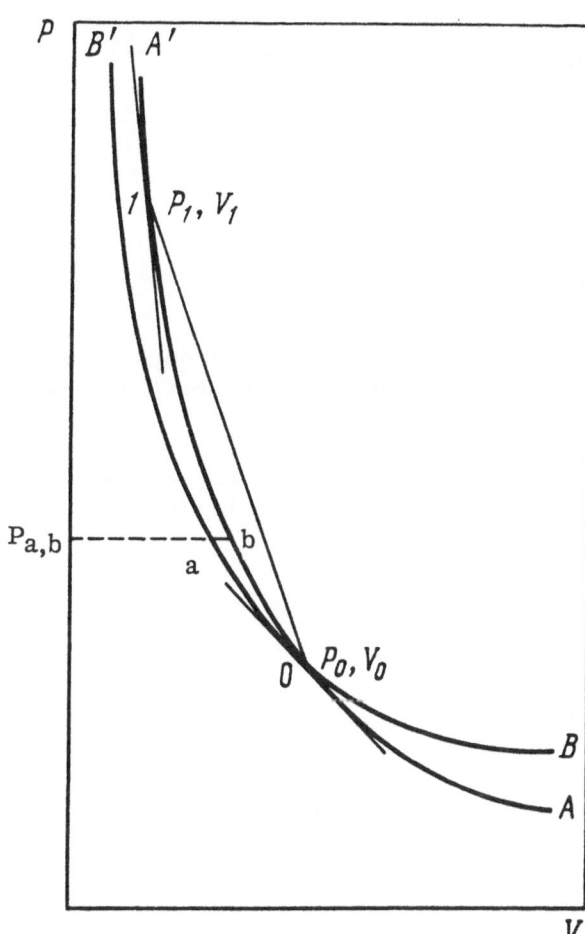

Fig. 5.5. Relative positions of the Hugoniot curve (A — A') and of the isentrope (B — B').

We shall expand the volume V_1 up to terms of the second order since the volume in Eq. (5.35) is multiplied by the small quantity $(P_1 - P_0)$:

$$V_1 - V_0 = \left(\frac{\partial V}{\partial P_0}\right)_S (P_1 - P_0) + $$
$$+ \frac{1}{2}\left(\frac{\partial^2 V}{\partial P_0^2}\right)_S (P_1 - P_0)^2. \qquad (5.44)$$

Substituting these expansions in Eq. (5.35), we obtain the required relationship

$$S_1 - S_0 = \frac{1}{12 T_0}\left(\frac{\partial^2 V}{\partial P_0^2}\right)_S (P_1 - P_0)^3. \qquad (5.45)$$

We shall now draw some conclusions from this relationship using the inequality

$$\left(\frac{\partial^2 V}{\partial P^2}\right)_S > 0 \qquad (5.46)$$

which applies to the great majority of solids, liquids, and gases. We shall draw through the point 0 (P_0, V_0) in the (P, V) plane a Hugoniot curve as well as an isentrope whose equation is

$$S_1 - S_0 = 0. \qquad (5.47)$$

We shall now expand the volume in terms of $(S_1 - S_0)$ and $(P_1 - P_0)$ and we shall use the equation for the isentrope (5.47) and the equation for the Hugoniot curve (5.45). If we restrict ourselves to terms of the second order in $(P_1 - P_0)$, we obtain the same expression (5.44) for the two curves. It follows that the two curves touch at this point and that this is a contact point of the second order.

We shall now consider the positions of the Hugoniot curve and the isentrope near the point P_0, V_0. According to Eqs. (5.45) and (5.46), $S_1 > S_0$ on the Hugoniot curve if $P_1 > P_0$. Moreover, it follows from the thermodynamic formula (A.27)

$$\left(\frac{\partial V}{\partial S}\right)_P = \frac{T}{c_P}\left(\frac{\partial V}{\partial T}\right)_P$$

that the entropy increases with increasing volume at constant pressure in the case of those substances which expand during heating: $(\partial V / \partial T)_P > 0$. We shall draw an isobar $P = P_1$, which intersects the Hugoniot curve and the isentrope at points a and b in Fig. 5.5. At a fixed value of the pressure P_1 the state on the Hugoniot curve corresponds to a higher entropy than the state on the isentrope. Consequently, the specific volume on the Hugoniot curve is higher than the specific volume on the isentrope (for a fixed value of P_1). This means that point a lies on the isentrope and point b lies on the Hugoniot curve. We can similarly show that the reverse is true below the point P_0, V_0 and that the relative positions of the two curves are as shown in Fig. 5.5.

In the case of weak discontinuities and sufficiently low values of $(P_1 - P_0)$ and $(V_1 - V_0)$, the relationship given by Eq. (5.32) can be written, in the first approximation, in the form

$$j^2 = -\left(\frac{\partial P}{\partial V}\right)_s$$

In this approximation, the velocities $u_{0\xi}$ and $u_{1\xi}$ are equal and given by

$$u_\xi = jV = \sqrt{\left(\frac{\partial P}{\partial \rho}\right)_s} = c. \tag{5.48}$$

The velocity of weak shock waves is equal to the velocity of sound. Since the entropy always increases, it follows that $S_1 > S_0$ for a shock wave and, therefore, $P_1 > P_0$ and point 1 (P_1, V_1) lies above point 0 (P_0, V_0).

Let us now consider the relative positions of tangents to the Hugoniot curve at points 0 and 1 as well as the chords joining these points (Fig. 5.5). Since the chord 01 is steeper than the tangent at point 0, it follows that

$$j^2 > -\left(\frac{\partial P}{\partial V_0}\right)_{S_0}.$$

Multiplying the above inequality by V_0^2, we obtain

$$j^2 V_0{}^2 = u_{0\xi}^2 > c_0{}^2, \quad u_{0\xi} > c_0.$$

Similarly, since the chord 01 is less steep than the tangent to the Hugoniot curve at point 1, it follows that

$$u_{1\xi} < c_1.$$

Summarizing these results, we can say that the increase in the entropy in a shock wave, $S_1 > S_0$, has the following consequences:

$$P_1 > P_0, \qquad u_{0\xi} > c, \qquad u_{1\xi} < c_1. \tag{5.49}$$

The expressions in Eq. (5.49), together with Eqs. (5.32) and (5.30), yield

$$V_0 > V_1, \qquad u_{0\xi} > u_{1\xi}. \tag{5.50}$$

It is shown in [21] that the above relationships are of universal validity and are valid for shock waves of any intensity.

Going over to the laboratory system of coordinates and considering the case of a shock wave in a gas or liquid at rest, we find that the expressions in Eq. (5.49) yield the conclusion that the shock-wave velocity u_s is higher than the velocity of sound c_0:

$$u_s > c_0. \tag{5.51}$$

Consequently, no disturbances originating in a shock front can overtake it. It is also easy to show that

$$u_s < c_1, \tag{5.52}$$

i.e., the shock-wave velocity is lower than the velocity of sound behind the shock front.

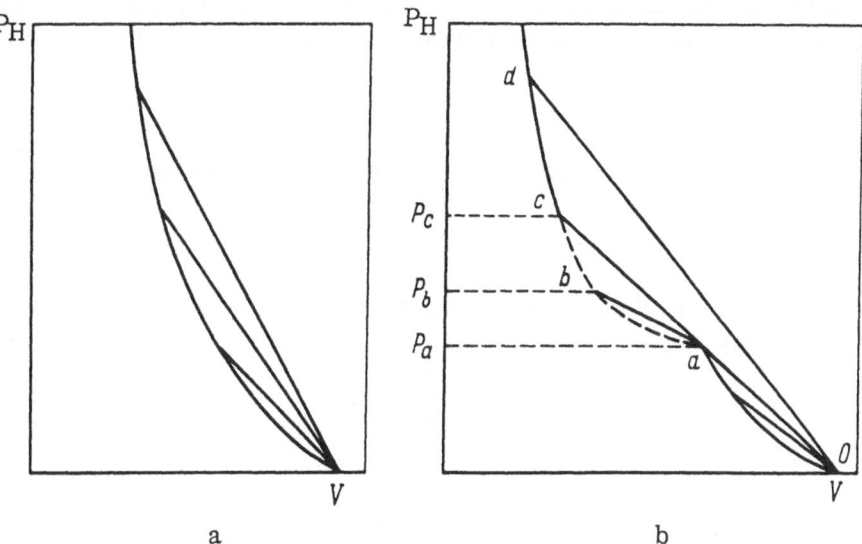

Fig. 5.6. Normal shape of the Hugoniot curve (a) and a kink in the curve (b).

5.3. Stability of Shock Waves.

Two-Wave Structures

A shock wave is stable when an increase in the shock pressure corresponds to an increase in the shock-front velocity. Using the example of a shock wave in a gas (or liquid) at rest, we can see from Eq. (5.38a) that the steepness of the Hugoniot curve of a stable shock wave should increase (Fig. 5.6a). Consequently, the following relationships are satisfied on the Hugoniot curve:

$$\frac{dP}{dV} < 0, \qquad \frac{d^2P}{dV^2} > 0. \tag{5.53}$$

In practice, there are two important cases which do not obey the expressions in Eq. (5.53). They are elastoplastic wave structures, resulting from the finite value of the shear modulus of solids, and two-wave structures associated with phase transitions that are accompanied by a decrease in volume. In both cases, the Hugoniot curve has a kink at some pressure P_a (Fig. 5.6b), above which the curve is deflected toward the pressure axis.

In order to understand the physical situation, we must turn back to Fig. 5.3. We can see that when the pressure is increasing, later disturbances catch up with earlier ones, since the velocity is a monotonically increasing function of the pressure. This continues until the Hugoniot curve passes through states in the interval $0a$ (Fig. 5.6b). The situation changes in the region abc. The velocity of disturbances at pressures $P > P_a$ is lower than the velocity of the total perturbation which is the shock wave P_1 joining the states 0 and a. Consequently, later disturbances cannot catch up and enhance the shock wave P_1, and they are propagated in a state established by the wave P_1. At amplitudes $(P - P_a) > 0$, these disturbances give rise to a second shock wave which joins the state (P_a, V_a) with some state in the region between a and c (Fig. 5.6b). The second wave is propagated at a lower velocity than the first and, consequently, the two-wave configuration produced in this way is stable. As the amplitude of the second wave P_2 increases, a moment is reached when the velocities of both waves become equal (point c in Fig. 5.6b). At high amplitudes of the second wave, the two-wave structure becomes unstable because then the second wave catches up with and enhances the first wave so that we again have a single shock wave at pressures $P > P_c$.

This can be expressed easily in the analytic form. The first wave, which joins the states (P_0, V_0) and (P_a, V_a), is propagated — according to Eqs. (5.30) and (5.32) — at a velocity

$$u_{s1} - u_a = V_a \sqrt{\frac{P_a - P_0}{V_0 - V_a}} \tag{5.54}$$

relative to the state (P_a, V_a). The second wave moves at a velocity given by Eq. (5.38a)

$$u_{s2} = V_a \sqrt{\frac{P_2 - P_a}{V_a - V_2}} \tag{5.55}$$

relative to the same state (P_a, V_a). Thus, the shock wave becomes unstable and a two-wave structure arises when $u_{s1} - u_a > u_{s2}$, i.e., when

$$\frac{P_a - P_0}{V_0 - V_a} > \frac{P_2 - P_a}{V_a - V_2}. \tag{5.56}$$

Two-Wave Structures due to the Rigidity of a Substance

At low stresses, Hooke's law applies and the stress tensor σ_{ik} is related linearly to the strain (deformation) tensor u_{ik}. In the case of isotropic bodies, this relationship is governed by two coefficients — the bulk modulus K and the shear modulus μ:

$$\sigma_{ik} = K \operatorname{div} \mathbf{u} \cdot \delta_{ik} + 2\mu \left(u_{ik} - \frac{1}{3} \delta_{ik} \operatorname{div} \mathbf{u} \right), \tag{5.57}$$

where $\mathbf{u}$ is the displacement vector and δ_{ik} is Kronecker's delta (i, k = x, y, z, which are the Cartesian coordinates). When the stresses are hydrostatic:

$$P = \sigma_{xx} = \sigma_{yy} = \sigma_{zz} = K \frac{\Delta v}{v}, \qquad \sigma_{ik} = 0 \quad \text{when} \quad i \neq k,$$

the diagonal components of the stress tensor are equal to the applied pressure and are governed only by the bulk modulus K.

In the case of a one-dimensional shock stress (pressure) along the x axis, we have $u_{xx} = \Delta v / v$ and all the other components vanish ($u_{ik} = 0$). It follows from Eq. (5.57) that, in this case,

$$\sigma_{xx} = \left(K + \frac{4}{3}\mu \right) \frac{\Delta v}{v}, \qquad \sigma_{yy} = \sigma_{zz} = \left(K - \frac{2}{3}\mu \right) \frac{\Delta V}{V},$$

$$\sigma_{ik} = 0 \quad \text{when} \quad i \neq k.$$

It is known [21] that when the stress tensor is of the diagonal form the maximum tangential stress τ is equal to the largest difference between the principal stresses, divided by 2,

$$\tau = \frac{\sigma_{xx} - \sigma_{yy}}{2} = \mu \frac{\Delta V}{V}. \tag{5.58}$$

The maximum tangential stresses in crystalline solids are limited by the yield stress Y

$$\tau \leqslant Y, \tag{5.59}$$

and when $\tau > Y$ solids "flow" and behave as liquids in the hydrodynamic sense.

It is very important to have a correct understanding of the physical nature of the phenomena discussed here, especially as the physics of shock waves in solids is rarely discussed in a satisfactory manner. The problem is that the yield stress for real solids is $Y \sim 10^3$ bar and the theoretical value of Y_i for ideal single crystals should be of the order of the shear

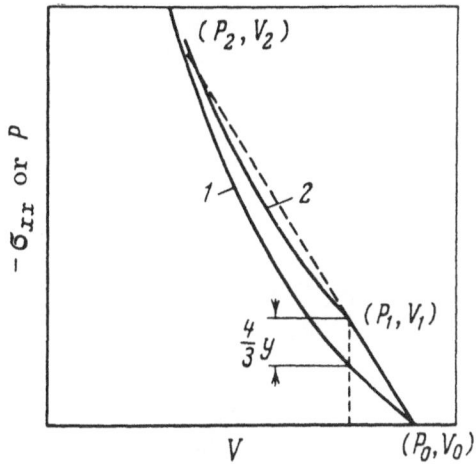

Fig. 5.7. Formation of an elastoplastic two-wave structure. (P_1, V_1) is the Hugoniot elastic limit; (P_2, V_2) is the limit of stability of a shock wave. 1) Hydrostatic curve; 2) $-\sigma_{xx}$ for one-dimensional compression.

modulus: $Y_i \sim 0.1\,\mu$. Consequently, real yield stresses are two or three orders of magnitude lower than the ideal stresses Y_i. This discrepancy was first pointed out by the eminent Soviet physicist Ya. I. Frenkel' and explained fully in [23, 24].

The discrepancy is due to the fact that real crystals are not perfect but have some defects (recently, dislocation-free whiskers have been produced whose yield stress is, as expected, equal to the theoretical value Y_i). Defects are classified as one-dimensional or point (which are associated with diffusion, Chapter 3), linear (screw and edge dislocations which are active in plastic flow), and two-dimensional (boundaries between crystallites). The most important defects in plastic flow are dislocations. Each dislocation is an element or "quantum" of plastic shear. Dislocations are created in plastic flow, so that a body which has been subjected to considerable plastic deformation contains a large number of dislocations. An important property of dislocations is that they are localized in definite crystallographic planes known as slip planes. In order to make a solid flow plastically not along one plane but along various directions like a liquid, it is necessary to set in motion dislocations belonging to several different systems of slip planes. The minimum number of different slip planes necessary for viscous flow depends on the crystal structure. The low values of the yield stresses of some solids are related to the presence of dislocations, which give rise to plastic flow at considerably lower stresses than $\tau \sim Y_i$.

We shall now consider a quantitative analysis given in [7]. Let us assume that the stress — strain relationship is linear at stresses lower than the yield point Y (5.59), corresponding to a constant value of the shear modulus μ (Hooke's law). We shall also assume that the tangential stress τ of Eq. (5.58) is constant for deformations exceeding those which occur at the yield point; it follows then from Eq. (5.58) and (5.59) that the shear modulus $\mu = Y(V/\Delta V)$ decreases with increasing deformation. Consequently, the dependence $\sigma_{xx} = f(\Delta V/V)$ should have a kink at the point corresponding to the yield stress (Fig. 5.7). The pressure corresponding to this kink, P_1, is

$$P_1 = \frac{K}{\mu} Y + \frac{4}{3} Y \qquad (5.60)$$

and is known as the Hugoniot elastic limit. Figure 5.7 shows also the hydrostatic curve. The pressure on the hydrostatic curve at the value of $\Delta V/V$ corresponding to P_1 is $P_{01} = KY/\mu$ and the difference between these two curves is $4Y/3$. The kink in the $\sigma_{xx} = f(\Delta V/V)$ curve gives rise to a two-wave structure at $P > P_1$. The first wave, joining the states (P_0, V_0) and (P_1, V_1), is known as the elastic wave. Its velocity, measured relative to the undisturbed matter ahead of it, is

$$u_s = V_0 \sqrt{\frac{P_1 - P_0}{V_0 - V_1}} \qquad (5.61)$$

and it is equal to the velocity $c_l = \sqrt{(K + {}^4/_3\mu)/\rho}$ of a plane longitudinal acoustic wave. The second wave, which follows the elastic wave, is called the plastic wave. The steepness of the

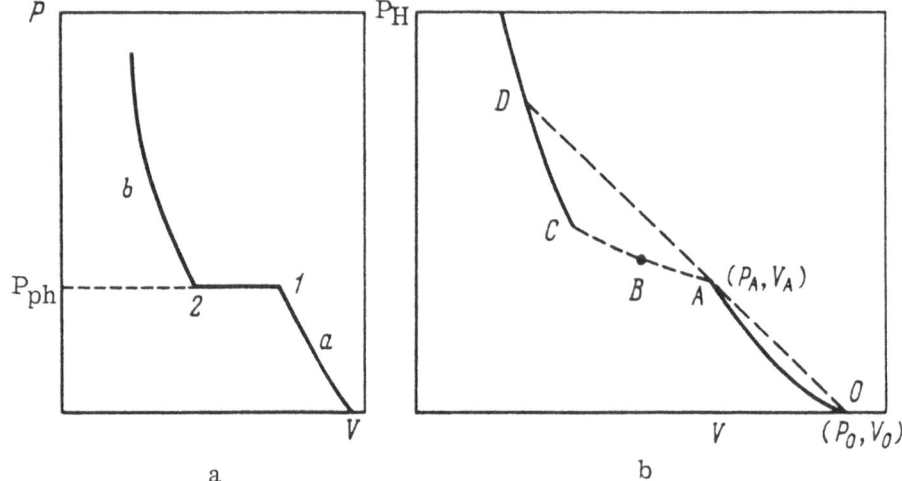

Fig. 5.8. Phase transition on an isotherm (a) and on a Hugoniot curve (b).

second region of the $\sigma_{xx} = f(\Delta V / V)$ curve increases with increasing compression and at the point (P_2, V_2) the curve intersects the continuation of the straight line joining (P_0, V_0) with (P_1, V_1).

The pressure P_2 is known as the limit of stability of the shock wave. At pressures $P > P_2$, the plastic and elastic waves merge and there is only one shock wave. It is evident from Fig. 5.7 that the departure of the stress σ_{xx} from the hydrostatic kind decreases with increasing pressure.

Two-Wave Structures due to Phase Transitions

Phase transitions that are accompanied by a decrease in volume at high pressures are of the greatest interest. The Hugoniot curve has a kink at the phase transition point and this gives rise to a two-wave structure for the same reasons as those given in the preceding subsection. Phase transitions take place at relatively low shock pressures when a substance is not yet strongly heated in a state which is on the Hugoniot curve. Consequently, a phase transition can be described by the following isothermal process (Fig. 5.8a). The pressure increases along the isotherm of phase a until it intersects the phase equilibrium line $P_{ph}(T)$. A finite reduction in volume, $V_1 - V_2 > 0$, occurs on the phase equilibrium line; this reduction in volume is accompanied by a change in the enthalpy $Q = w_2 - w_1$, which represents the latent heat of 1 g of the substance which has undergone a phase transition. The pressure and temperature are constant in the range of coexistence of phases a and b. Higher pressures are reached along the isotherm of phase b. The latent heat Q is related to the slope of the phase equilibrium line by the Clapeyron — Clausius equation

$$\frac{dP_{ph}}{dT} = -\frac{Q}{T(V_a - V_b)}. \tag{5.62}$$

In the case of a shock wave (Fig. 5.8b), a phase transition is "smeared out" over a range of pressures and temperatures. The shock wave is stable in the region 0A, which covers the range of existence of the first phase and ends at the phase equilibrium line. A stronger shock wave, capable of reaching D along ACD is unstable because the condition (5.56) is then satisfied. A two-wave structure appears in this region: one wave joins the state (P_0, V_0) with the state (P_A, V_A), while the other wave joins the state (P_A, V_A) with some state on the Hugoniot curve in the region ACD. At the point D the two waves merge and at higher pressures only one shock wave is stable. We have ignored here the elastoplastic two-wave structures, which are observed at much lower pressures.

The quantity Q in Eq. (5.62) may be positive (heat is absorbed in an endothermic transition) or negative (heat is evolved in an exothermic transition). We thus have

$$\frac{dP_{ph}}{dT} < 0, \qquad Q > 0 \tag{5.63a}$$

and

$$\frac{dP_{ph}}{dT} > 0, \qquad Q < 0; \tag{5.63b}$$

moreover, some differences will be observed between the static (Fig. 5.8a) and dynamic (Fig. 5.8b) cases. The temperature increases somewhat in the 0A region of the Hugoniot curve (the increase is approximately identical for the isentrope). Consequently, if Q > 0, a shock wave reaches the phase equilibrium line at pressures lower than P_{ph} (Fig. 5.8a). Conversely, if Q < 0, we have $P_a > P_{ph}$. This difference becomes more pronounced when we record a series of Hugoniot curves starting from different initial temperatures.

The two-wave region ACD can be used to obtain quantitative information on a phase transition [25]. Let us consider the point B on the Hugoniot curve in the region of coexistence of two phases (Fig. 5.8b). We shall assume that the state B is thermodynamically stable. We shall use subscripts 1 and 2 to denote the quantities referring to the first and second phases, respectively. We shall employ λ_B for the mass fraction (relative concentration) of phase 2 at the point B. Then,

$$w_B = (1 - \lambda_B) w_1(T_B, P_B) + \lambda_B w_2(T_B, P_B) = w_1 + \lambda_B \Delta w_B,$$
$$V_B = (1 - \lambda_B) V_1(T_B, P_B) + \lambda_B V_2(T_B, P_B) = V_1 + \lambda_B \Delta V_B,$$

where

$$\Delta w_B = w_2(T_B, P_B) - w_1(T_B, P_B),$$
$$\Delta V = V_2(T_B, P_B) - V_1(T_B, P_B).$$

Equations for the Hugoniot curves of the first wave at the point A (beginning of the phase transition) and for the second wave at the point B are of the form

$$\left. \begin{array}{l} w_A = w_0 + \dfrac{1}{2} P_A (V_0 + V_A), \\[2ex] w_B = w_A + \dfrac{1}{2} (P_B - P_A)(V_A + V_B). \end{array} \right\} \tag{5.64}$$

We shall now expand $w_1(T_B, P_B)$ and $V_1(T_B, P_B)$ in terms of $\Delta T = T_B - T_A$ and $\Delta P = P_B - P_A$ near A:

$$V_1 = V_A (1 - \beta \Delta P + \alpha \Delta T)_A,$$

$$w_1 = w_A + V_A (1 - \alpha T)_A \Delta P + c_{PA} \Delta T.$$

We have used here some of the formulas given in Appendix A.

We shall now calculate the quantity

$$\left(\frac{\Delta w}{\Delta V}\right)_B = \frac{w_B - w_{1B}}{V_B - V_{1B}} = \frac{\frac{1}{2}(V_A + V_B) - V_A(1 - \alpha T)_A - c_{P_A}\frac{\Delta T}{\Delta P}}{\frac{V_B - V_A}{\Delta P} + \beta_A V_A - \alpha_A V_A \frac{\Delta T}{\Delta P}}.$$

Our next step is to make the point B approach the point A. Then

$$\left.\begin{aligned}\lim_{B \to A} \frac{w_B - w_{1B}}{V_B - V_{1B}} &= \frac{\lambda_B \Delta w_B}{\lambda_B \Delta V_B}\bigg|_{B \to A} = \frac{\Delta w_A}{\Delta V_A} = T_A\left(\frac{dP}{dT}\right)_{\text{phA}}, \\[2mm] \beta_{AB} &= -\frac{1}{V}\lim_{B \to A}\left[\frac{V_B - V_A}{P_B - P_A}\right], \\[2mm] T_A\left(\frac{dP}{dT}\right)_{\text{phA}} &= \frac{[\alpha V T - c_P(\Delta T/\Delta P)]_A}{[V(\beta - \beta_{AB}) - \alpha V(\Delta T/\Delta P)]_A},\end{aligned}\right\} \qquad (5.65)$$

where

$$\frac{\Delta T}{\Delta P} = \lim_{B \to A}\frac{T_B - T_A}{P_B - P_A}.$$

In the range of coexistence of the two phases, the substance is in thermodynamic equilibrium and, therefore, P and T in this region lie on the phase equilibrium curve

$$\left(\frac{\Delta T}{\Delta P}\right)_{B \to A} = \left(\frac{dT}{dP}\right)_{\text{ph}},$$

which gives the required expression

$$\left(\frac{dP}{dT}\right)_{\text{ph}}^2 + \frac{2\alpha}{\beta_{AB} - \beta}\frac{dP}{dT} - \frac{c_P}{TV(\beta_{AB} - \beta)} = 0. \qquad (5.66)$$

Equation (5.66) can be used to determine as important a quantity as the slope of the phase equilibrium curve at the point A using the known values of the thermal expansion coefficient α, the isothermal compressibility β, and the specific heat c_P of the first phase at the point A and the compressibility β_{AB} along the Hugoniot curve.

Usually, the thermodynamic properties of the first phase are known quite accurately and the Hugoniot curve in the two-phase region can be determined experimentally. This makes it possible to express $(dP/dT)_{\text{ph}}$ in terms of known quantities [7]. Let us now express the difference of the enthalpies for the first and second waves in Eq. (5.64) in terms of the enthalpy of the first phase Q, and λ_B

$$w_B - w_A = \left[\int_{P_A}^{P_B} dw\right] + \lambda Q(P_B). \qquad (5.67)$$

Using Eq. (5.62), the thermodynamic identity

$$dw = c_v\,dT + T\left(\frac{\partial p}{\partial T}\right)_V dV + V\,dP,$$

and the definition of λ_B, we obtain the required formula

$$\left(\frac{dP}{dT}\right)_{\text{ph}} = -\frac{(P_B - P_A)(V_{1A} - V_B)}{2T_B(V_{1B} - V_B)} + \left[\int_{T_A}^{T_B} c_v \, dT + \int_{V_A}^{V_{1B}} T\left(\frac{\partial P}{\partial T}\right)_v dV + \int_{P_A}^{P_B} V \, dP\right]_1, \qquad (5.68)$$

where the subscript 1 outside the square brackets indicates that integration is carried out with respect to parameters of the first phase along the phase equilibrium curve. Knowing the initial parameters (T_A, P_A) and integrating Eq. (5.68), we can determine the whole phase equilibrium curve in the two-phase region. Conversely, knowing the equilibrium curve and the parameters of the first phase, we can calculate theoretically the Hugoniot curve in the two-wave region. This method is applicable only in the two-phase region. A method for the determination of the equation of state of the second phase from the Hugoniot curve at high pressures $(P > P_D$ in Fig. 5.8b) is developed in [3] and will be described below. This method makes it possible to plot the isotherm of the high-pressure phase, which is the boundary of the two-phase region on the second phase side.

In dynamic investigations, the pressures are applied for time intervals of the order of a microsecond or less. Because of this short interval, the nature of the mechanisms of phase transitions induced by shock waves is still controversial. It seems to us that the phase-transition mechanism is associated with dislocations. The plastic flow of crystalline solids subjected to shock waves produces a large number of dislocations in different slip planes. The simplest change in the structure of a crystal lattice caused by the motion of dislocations is twinning, which is well known and has been investigated quite thoroughly [24]. In particular, the twinning mechanism is invoked in [26] in connection with the phase transition in iron that was discovered by Bancroft, Peterson, and Minshall [27]. The processes of dislocation growth of crystals are also well known. Consequently, it is quite natural to assume that the changes in the lattice structure caused by shock waves are due to dislocations. It is difficult to imagine any other mechanism which could alter the crystal structure in a time of the order of a microsecond (we are not dealing here with electron transitions, which are practically instantaneous). The dislocation hypothesis of phase transitions in shock waves explains qualitatively why phase transitions in valence (covalent) crystals require higher amplitudes and longer times than transitions in ionic crystals or in metals [28]. This is because exceptionally high shear stresses are required for the motion and creation of dislocations in the slip planes of valence (covalent) crystals. Therefore, transitions in complex crystals which have covalent bonds can occur only when a shock wave is strongly nonhydrostatic, which is the case in substances having a low yield stress.

5.4. Hugoniot Curve and Method of Potentials

Hugoniot Curve in the Phonon Region

The equation of state in the phonon region was considered in detail in Chapter 2. It is governed by two functions of volume: the potential $E_p(x)$ and the Grüneisen parameter $\gamma(x)$, where $x = V/V_0$ is dimensionless volume. The Debye approximation can be used quite generally in the phonon region. In this case, the free energy F, the internal energy E, and the pressure P are of the form

$$F = E_p(x) + \frac{9}{8}\frac{Rv\theta}{\mu} + \frac{RvT}{\mu}\left[3\ln(1 - e^{-\theta/T}) - D\left(\frac{\theta}{T}\right)\right], \qquad (5.69)$$

$$E = E_p(x) + \frac{9}{8}\frac{Rv\theta}{\mu} + \frac{3RvT}{\mu}D(\theta/T), \qquad (5.70)$$

$$P = P_p(x) + P_{zv}(x) + P_T(x, T),$$ (5.71)

$$P_p(x) = -\rho_0 \frac{\partial E_p}{\partial x}, \qquad P_{zv} = \frac{9}{8} \frac{Rv\theta}{\mu} \frac{\rho_0\gamma}{x},$$

$$P_T = \frac{3RvT}{\mu} \frac{\rho_0\gamma}{x} D\left(\frac{\theta}{T}\right),$$

where ν is the number of atoms in one molecule; μ is the molecular weight; R is the gas constant; θ is the Debye temperature; P_{zv} is the pressure corresponding to the zero-point vibrations. In the classical limit $T > \theta$, we have (Appendix B)

$$F = E_p(x) - \frac{RvT}{\mu}\left[1 - 3\ln\frac{\theta}{T} - \frac{3}{40}\frac{\theta^2}{T^2}\right],$$ (5.69a)

$$E = E_p(x) + \frac{3RvT}{\mu}\left[1 + \frac{1}{20}\frac{\theta^2}{T^2}\right],$$ (5.70a)

$$P = -\rho_0\frac{\partial E_p}{\partial x} + \frac{3RvT}{\mu}\frac{\rho_0\gamma}{x}\left[1 + \frac{1}{20}\frac{\theta^2}{T^2}\right].$$ (5.71a)

Quantum corrections for temperatures $T > \theta$ are proportional to θ^2/T^2 and in practical calculations these corrections to F, E, and P can be neglected. The potential $E_p(x)$ can be represented by any of the functions discussed in Chapter 1. We shall use mainly the Born—Mayer potential

$$E_p(x) = \frac{3}{\rho_0}\left[\frac{A}{b}\exp(b(1 - x^{1/3})) - Kx^{-1/3}\right].$$ (5.72)

When the substance ahead of a shock wave is at rest, the Hugoniot equation (5.36) is of the form

$$E_H = E_0 + \frac{P_H}{2\rho_0}(1 - x).$$ (5.73)

This equation relates the Hugoniot pressure P_H in the shock wave and the energy E_H of a body subjected to this pressure. Here, E_0 is the energy per unit mass of a crystal ahead of a shock wave. Using Eq. (5.70), we obtain

$$E_0 = E_p(1) + \frac{Rv}{\mu}\left[\frac{9}{8}\theta_0 + 3T_0 D\left(\frac{\theta_0}{T_0}\right)\right].$$ (5.74)

Eliminating E_H from Eqs. (5.73), (5.70), and (5.71), we obtain the required equation

$$P_H = \frac{P_p(x) + \frac{\rho_0\gamma}{x}[E_0 - E_p(x)]}{1 - \frac{\gamma(1 - x)}{2x}},$$ (5.75)

which relates the experimentally determined quantity P_H to two unknown functions: $E_p(x)$, $\gamma(x)$ or $P_p(x)$, $\gamma(x)$. As described in Chapter 2 (Sec. 2.7), $\gamma(x)$ is usually expressed in terms of $P_p(x)$. Thus, we now have a problem which can be solved completely: we can find the unknown function $P_p(x)$ from the experimental dependence $P_H(x)$. The cold-compression isotherm $P_p(x)$, corresponding to the potential $E_p(x)$ of Eq. (5.72), is of the form

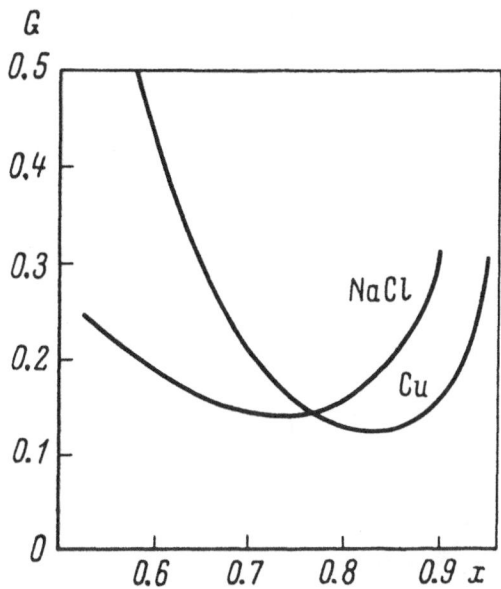

Fig. 5.9. Ratio $G = \mathscr{P}_\gamma / \Gamma_1$ as a function of the relative volume x for sodium chloride and copper.

$$P_p(x) = Ax^{-2/3} \exp[b(1 - x^{1/3})] - Kx^{-4/3}. \qquad (5.76)$$

Thus, we must determine three constants A, b, and K from the experimentally found Hugoniot curve P_H. The actual method of determination of A, b, and K from P_H will be described in a special section, whereas here we shall concentrate on some general aspects of the problem, assuming that A, b, K, and consequently $P_p(x)$ and $\gamma(x)$, are known.

First we can calculate easily the temperature on the Hugoniot curve T_H. A good approximation for this temperature is given by the formula for the classical limit (5.70a):

$$T_H = \frac{E_H - E_p(x)}{3R} \frac{\mu}{\nu}, \qquad T_H > \theta, \qquad (5.77)$$

where E_H is calculated from the theoretical Hugoniot curve (smoothed curve).

The formulas given in Chapter 2 for the Grüneisen parameter normalize γ to the thermodynamic value γ_0 under normal conditions

$$\gamma_0 = \frac{axK_T}{\rho_0 c_v} = \frac{axK_S}{\rho_0 c_P} \bigg|_{x=1}. \qquad (5.78)$$

Consequently, we start from the known value of the Grüneisen parameter for x = 1. As the pressure increases and the relative volume x decreases, the Grüneisen parameter γ decreases in accordance with the semiempirical relationship assumed in calculations of $\gamma(x)$. The question now arises of the influence of small errors in γ on the determination of the cold-compression isotherm $P_p(x)$ by means of Eq. (5.75). Let us represent γ in the form

$$\gamma = \gamma^0 (1 + \Gamma_1), \qquad (5.79)$$

where Γ_1 is some indeterminacy in γ (we shall assume that $\Gamma_1 \lesssim 0.1$). We shall write P_H of Eq. (5.75) in the form

$$P_H = P_H^0 (1 + \mathscr{P}_\gamma), \qquad (5.80)$$

where P_H^0 is the theoretical Hugoniot curve and $\mathscr{P}_\gamma$ is the error in the determination of the Hugoniot curve, which is assumed to be due to the inaccuracy in $\gamma(x)$. We shall now find the relationship between $\mathscr{P}_\gamma$ with Γ_1. Expanding P_H of Eq. (5.75) in terms of Γ_1, we obtain the required relationship

$$\mathscr{P}_\gamma = e_1 \left(1 - \frac{P_p}{P_H^0}\right) \Gamma_1, \qquad (5.81)$$

$$e_1 = \frac{1}{1 - \dfrac{\gamma^0 (1 - x)}{2x}}.$$

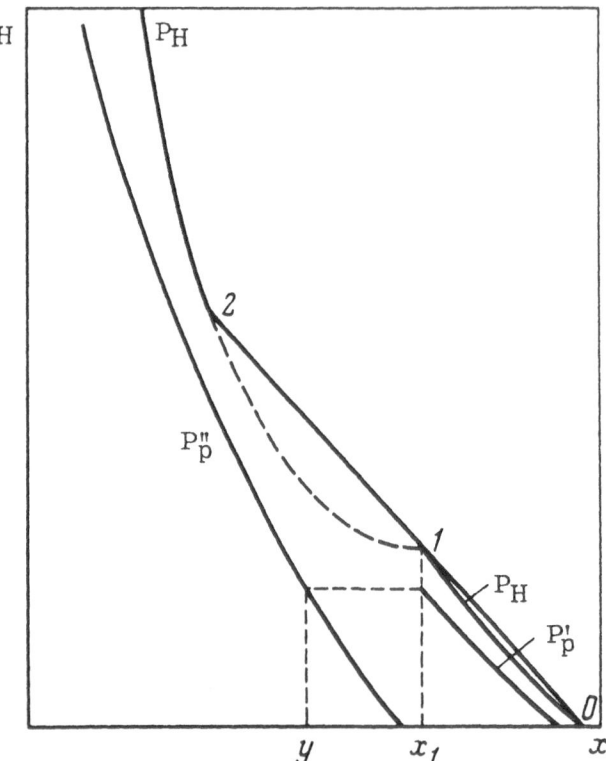

Fig. 5.10. Calculation of the cold-compression isotherm
of the high-pressure phase from dynamic data.

The ratio $G = \mathscr{P}_\gamma / \Gamma_1$ is shown in Fig. 5.9 as a function of x for sodium chloride and for copper. It follows from this figure that the Hugoniot curve is relatively insensitive to inaccuracies in the Grüneisen parameter when $x \geqslant 0.8$–0.7. Bearing this in mind, we can determine the cold-compression isotherm from the low-temperature part of the Hugoniot curve in the range $x > 0.8$–0.7.

We shall now consider the problem of the determination of the cold-compression isotherm of the high-pressure phase of substances which undergo a phase transition [3, 29, 30]. Let us now turn to Fig. 5.10, which shows the Hugoniot curve as well as the cold-compression isotherms of some substance (for example, iron or the mineral called gabbro) for the low- and high-pressure phases. We shall be interested only in modifications of the method for the determination of the parameters A, b, and K of the high-pressure phase due to the occurrence of a phase transition. The Hugoniot curve consists of three regions: a single-wave region $0 - 1$, a two-wave region $1 - 2$, and a single-wave region $P_H > P_2$, where the shock wave is stable and is propagated in the high-pressure phase. In the third region, the equation for the Hugoniot curve has its ordinary form (5.73) and Eq. (5.75) describes the high-pressure phase. The only quantity in Eq. (5.75) which refers to the low-pressure phase is E_0. We shall use double primes for the quantities representing the high-pressure phase and single primes for the low-pressure phase. Our problem is to determine the constants A″, b″, and K″. The internal energy of the second phase E_p'' should be correctly normalized with respect to the internal energy of the first phase E_p'. This can be done easily using the fact that the difference between the energies on the cold-compression isotherms of the first and second phases, taken at some fixed pressure P on the assumption that the Debye temperatures are equal ($\theta'' = \theta'$), is given by

$$\frac{P}{\rho_0}[x'(P) - x''(P)].$$

(5.82)

The error in the deviation of Eq. (5.82) is equal to the difference between the energies at zero temperature of the two phases

$$\sim \frac{9}{8} \frac{Rv}{\mu} (\theta' - \theta''),$$ (5.83)

which may be assumed to be small compared with Eq. (5.82). This condition is satisfied both in the case of Fe [3] and in the case of gabbro [30]. Let us assume that $x' = x_1$ is the transition point; we shall denote the corresponding point for the high-pressure phase by $x'' = y$. The point y can be found easily from the intersection of the cold-compression isotherm of the high-pressure phase with the straight line $P_p'(x_1) = P_1$ in the (P, x) diagram. Consequently, out of the three parameters A'', b'', and K'', the parameter K'' can be expressed in terms of A'', b'', y, and the known characteristics of the first phase

$$K'' = y^{1/3} \frac{A''}{b''} e^{b''(1-y^{1/3})} - \frac{\rho_0}{3} y^{1/3} \left[E_p'(x_1) + \frac{P_1}{\rho_0} (x_1 - y) \right].$$ (5.84)

The other two parameters, A'' and b'', should then be determined by the method of least squares using the experimental Hugoniot curve and the theoretical formula for this curve

$$P_H = \frac{P_p''(x) + \frac{\gamma'' \rho_0}{x}[E_0 - E_p''(x)]}{1 - \frac{\gamma''(1-x)}{2x}}.$$ (5.75a)

Equation (5.75a) includes the Grüneisen parameter of the high-pressure phase γ'', whose thermodynamic value γ_0'' under normal conditions [Eq. (5.78)] is not usually known. Consequently, in actual calculations, we must use unnormalized formulas of Slater and Landau (2.35) or of Dugdale and MacDonald (2.42); or we must normalize γ'' under normal conditions using the values of γ_0 [Eq. (5.78)] for similar substances.

We shall now consider the problem of the unloading of a shock-compressed substance. The process of unloading can be regarded as an isentropic flow with the initial coordinates $P_1 = P_{H1}$, $T_1 = T_{H1}$, $x = x_1$, etc. Consequently, during the unloading the substance passes through all the states on the isentrope which originates from the point (P_{H1}, x_1). We can easily obtain a formula for the unloading isentrope in terms of the variables (P, x). For simplicity, we shall consider the classical limiting case $T > \theta$, which will be denoted by the subscript "c." Using Eq. (5.69a), we shall determine the entropy

$$S = -\left(\frac{\partial F}{\partial T}\right)_v = c_{vc}\left(\frac{4}{3} - \ln\frac{\theta}{T}\right), \qquad c_{vc} = \frac{3Rv}{\mu}.$$ (5.85)

The dependence of the Debye temperature on the volume is expressed in terms of the Grüneisen parameter

$$\theta(x) = \theta_0 \exp \int_x^1 \frac{\gamma}{x} dx; \qquad \theta_0 = \theta(1).$$ (5.86)

Equation (5.86) for a constant value of γ is very simple:

$$\theta(x) = \theta_0 x^{-\gamma}, \qquad \gamma = \text{const.}$$ (5.86a)

We shall use $S_{H1} = S(T_{H1}, x_1)$ to denote the entropy at that point on the Hugoniot curve from which the unloading isentrope begins. Then, the condition

$$S(T, x) = S_{H1}$$

defines the temperature on the unloading isentrope

$$T = \theta \exp\left(\frac{S_{H1}}{c_{vc}} - \frac{4}{3}\right) = \theta_0 \exp\left[\left(\frac{S_{H1}}{c_{vc}} - \frac{4}{3}\right) + \int_x^1 \gamma(x)\frac{dx}{x}\right]. \tag{5.87}$$

Eliminating the temperature from the equation for the pressure (5.71a), we obtain the formula for the required isentrope

$$P_{s1} = P_p(x) + \frac{3Rv}{\mu}\frac{\rho_0\gamma(x)}{x}\theta_0\exp\left[\left(\frac{S_{H1}}{c_{vc}} - \frac{4}{3}\right) + \int_x^1 \gamma(x)\frac{dx}{x}\right]. \tag{5.88}$$

We shall consider now a point at which the Hugoniot curve differs considerably from the isotherm. This can be seen easily from the basic equation (5.75). We find that when

$$\bar{\gamma} = \frac{2\bar{x}}{1 - \bar{x}}, \qquad \bar{x} = \frac{\bar{\gamma}}{2 + \bar{\gamma}}, \qquad \bar{\gamma} = \gamma(\bar{x}), \tag{5.89}$$

the pressure on the Hugoniot curve $P_H(\bar{x})$ becomes infinite and the volume $\bar{x}$ is called the limiting compression. This result is obtained from the equation of the Hugoniot curve in the phonon region but, as in the case of gases [19], it is of general application. The explanation of this effect is well known. As the shock pressure increases, the temperature on the Hugoniot curve increases exponentially. Consequently, when $x \to \bar{x}$, the rise of P_H is solely due to the thermal motion.

Our discussion has been limited to the determination of the equation of state from the Hugoniot curve. It follows, of course, that if we know the equation of state, we can use Eq. (5.75) to determine the Hugoniot curve and we can employ Eqs. (5.33)-(5.43) to find all the other parameters of a shock wave.

Geometrical Interpretation.
Compression of Porous Substances

Graphical methods are very helpful in the investigation of complex processes occurring in shock waves [19, 31]. In particular, the Hugoniot curve itself can be given a simple geometrical interpretation (Fig. 5.11). The total work done by a shock wave is $P(1 - x)/\rho_0$ per 1 g of a given substance. This work is divided equally between the increase in the internal energy (5.73)

$$\Delta E = E_H - E_0 = \frac{P_H}{2\rho_0}(1 - x)$$

and the kinetic energy (5.39)

$$\frac{u_1^2}{2} = \frac{P_H}{2\rho_0}(1 - x)$$

(it is assumed that the substance ahead of a shock wave is at rest). Figure 5.11 shows, in addition to the Hugoniot curve P_H, the cold-compression isotherm, P_p, which is drawn (for simplicity) from the point $x = 1$, i.e., it is assumed that the substance ahead of the shock front is at $T = 0°K$. Usually, the initial conditions are the normal temperature ($T \sim 300°K$) and pressure ($P = 1$ bar). In this case, the potential pressure $P_p(1) < 0$ and its absolute value is of the order of several kilobars; we thus have $P_p(1) + P_T(1,300°K) = 0$. The behavior of $P_p(x)$ near $x = 1$ is unimportant in the pressure range $\sim 10^2$–10^3 kbar. The area of the triangle OCA in

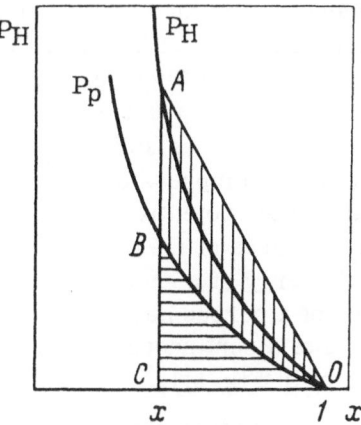

Fig. 5.11. Geometrical interpretation of the Hugoniot curve.

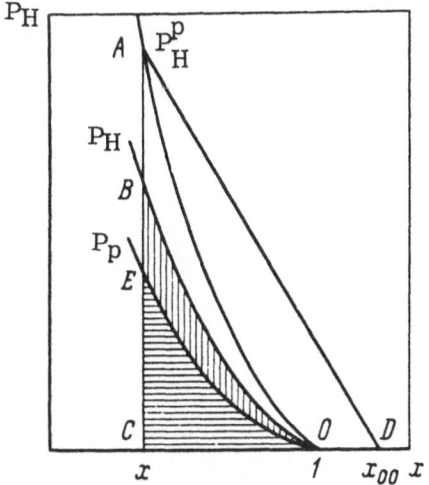

Fig. 5.12. Geometrical interpretation of the Hugoniot curve for a porous medium.

Fig. 5.11 is exactly equal to the total increase in the internal energy ΔE. The internal energy consists of the potential energy of the isothermal compression along the cold-compression isotherm $P_p(x)$, which is numerically equal to the area of the triangle OCB, one of whose sides is a curve, and the thermal energy E_T, which is represented by the area of the triangle OAB, which has two curvilinear sides. The transition through a shock front is accompanied by a sudden increase in the entropy. Consequently, the temperature on the Hugoniot curve increases appreciably. According to Eq. (5.45), the initial part of the Hugoniot curve is similar to the usual isentrope and, consequently, the temperature rise along both curves is similar. As the compression increases, the Hugoniot curve rises above the isentrope because of stronger heating.

The cold-compression isotherm $P_p(x)$ lies below the Hugoniot curve and divides the ordinates P_H into the potential $P_p(x)$ and thermal $P_T = P_H - P_p(x)$ pressures.

We shall now consider the problem of compression of porous bodies by shock waves. Zel'-dovich [32] suggested the use of the Hugoniot curve of porous samples for obtaining additional information on the properties of a solid. The first experiments on the dynamic compression of porous iron were carried out by Al'tshuler et al. [9]. Later, extensive experimental investigations on tungsten were carried out by Krupnikov et al. [33], and on nickel, aluminum, copper, and lead, by Kormer et al. [18]. The results of these investigations were reviewed in detail by Al'tshuler [31].

The Hugoniot curve for the compression of porous samples is shown schematically in Fig. 5.12.

We can easily obtain the formulas used to draw the curves in this figure. We shall use $x_{00} = v_{00}/v_0$ for the relative volume of a porous sample ($x_{00} > 1$). We shall call the quantity x_{00} the porosity. In this case, the equation of the Hugoniot curve (5.36) or (5.73) assumes the form (the superscript "p" denotes porous substances)

$$E_H^p = E_0 + P_H^p \frac{x_{00} - x}{2\rho_0}, \qquad (5.90)$$

and Eq. (5.75) transforms into

$$P_H^p = \frac{P_p(x) + \frac{\rho_0 \gamma}{x} [E_0 - E_p(x)]}{1 - \frac{\gamma}{2x}(x_{00} - x)}. \qquad (5.91)$$

Intuitive reasoning shows that large pressures are not necessary for the collapse of pores since under static conditions such collapse occurs at pressures of the order of kilobars. Consequently, during the initial stage of compression from $x = x_{00}$ to $x = 1$, the Hugoniot curve should lie along the volume axis. In fact, it follows from Eq. (5.91) that $P_H^p = 0$ at $x = 1$. Thus, the Hugoniot curve for a porous substance, P_H^p, like the curve for a nonporous solid, P_H, starts at the point $x = 1$. The increase in the internal energy under dynamic compression, $E_H^p - E_0$ of Eq. (5.90), is equal to the area of the triangle DAC. The increase in the potential energy is independent of the porosity x_{00} and, as before, it is represented by the area of the curvilinear triangle OEC. Thus, porosity increases the fraction of the thermal energy of a shock-compressed sample and, consequently, the temperature on the Hugoniot curve of a porous substance (P_H^p) rises more rapidly than the temperature along the Hugoniot curve of a continuous solid (P_H), i.e., $P_H^p > P_H$, and the mutual positions of the curves are as shown in Fig. 5.12.

The dynamic compression of porous samples rapidly raises their temperature, and the state of the substance behind a shock front quickly goes over from the phonon region to the region of high-temperature effects such as anharmonicity, thermal excitation of conduction electrons, thermal ionization, etc. Consequently, the equation for the Hugoniot curve is no longer as simple as that given by Eq. (5.91). However, the majority of the basic features of the compression of porous samples can be deduced from Eq. (5.91). This equation describes a family of the Hugoniot curves for a substance of initial porosity x_{00} and all the curves of this family start at the point $x = 1$. The limiting compression is found from the condition

$$1 - \frac{\overline{\gamma}}{2\overline{x}}(x_{00} - \overline{x}) = 0$$

or

$$\overline{x} = \frac{\overline{\gamma}x_{00}}{2 + \overline{\gamma}}, \qquad \overline{\gamma} = \gamma(\overline{x}).$$

Consequently, the limiting volume of a porous substance increases by a factor of x_{00}.

When $\overline{x} < 1$, $x_{00} < (2 + \overline{\gamma})/\gamma$, the volume is reduced by shock compression and the higher the value of x_{00}, the smaller is the reduction in the volume. When $\overline{x} = 1$, $x_{00} = (2 + \gamma)/\overline{\gamma}$, the Hugoniot curve rises vertically from the point $x = 1$. When $\overline{x} > 1$, $x_{00} > (2 + \overline{\gamma})/\overline{\gamma}$, the volume is increased by shock compression and all the Hugoniot curves are deflected to the right. All these remarkable properties of porous substances were found in the experimental studies reported in [33, 18].

Grüneisen Parameter and Hugoniot Curve

The key problem in the dynamics of high pressures is the determination of the functional dependence of the Grüneisen parameter on the volume. The inaccuracies in the semiempirical formulas for $\gamma(x)$ affect the cold-compression isotherm and influence even more strongly the high-temperature corrections to the equation of state at high pressures. It is not yet possible to determine the function $\gamma(x)$ by purely theoretical reasoning. Therefore, it would be very desirable to determine the dependence $\gamma(x)$ experimentally. Such a determination is difficult. At moderate temperatures, corresponding to the phonon region, the thermal pressure is relatively small compared with the potential pressure on the Hugoniot curve. At first sight, it seems natural to use the Hugoniot curve of porous samples. However, the rising temperature rapidly shifts the Hugoniot curve to the region where high-temperature corrections are necessary. Consequently, although Hugoniot curves of porous substances are available, they have not provided much information on the function $\gamma(x)$.

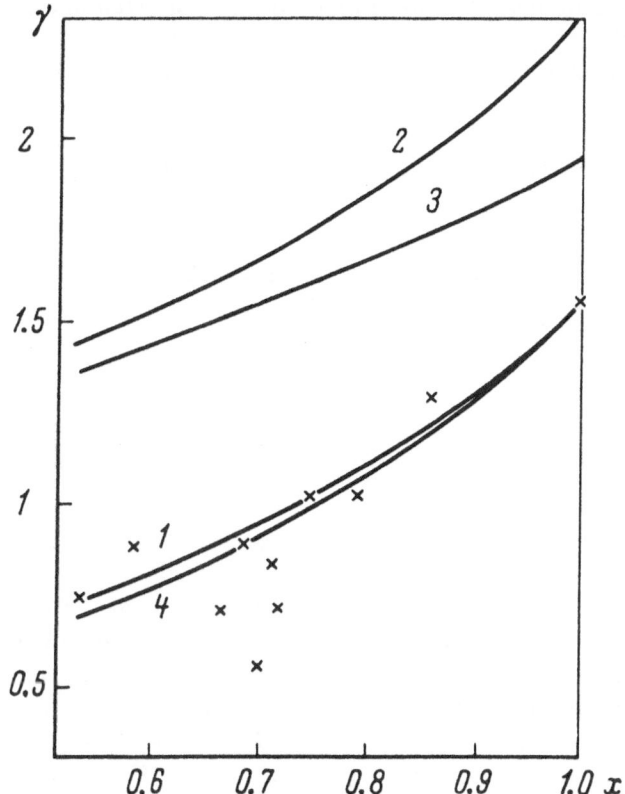

Fig. 5.13. Dependence of the Gruneisen parameter on
the relative volume x of sodium chloride.

The following guidelines should be followed in the determination of $\gamma(x)$ from the experimental Hugoniot curves: a) the initial conditions should be selected so as to reduce the temperature rise along the Hugoniot curve; b) substances with high melting points (for example W or MgO) should be used because their phonon range of temperatures is considerably wider than that of substances with low and moderately high melting points.

Bearing in mind these guidelines, we shall consider the method for the experimental determination of $\gamma(x)$, used first by Zharkov and Kalinin [4]. The Grüneisen parameter can be determined by deducing $P_H(x)$ and $P_p(x)$ from independent measurements. Then, rewriting Eq. (5.75) in such a way as to obtain γ, we have

$$\gamma = \frac{P_H - P_p(x)}{\dfrac{\rho_0}{\tau}[E_0 - E_p(x)] + \dfrac{P_H(1-x)}{2x}} \tag{5.92}$$

The cold-compression isotherm $P_p(x)$ is found by the method of potentials from the static data. The dependence $\gamma(x)$ was determined in [4] by applying Eq. (5.92) to NaCl and calculating $P_p(x)$ from the static data due to Bridgman [1]. The same procedure can be carried out even more satisfactorily using more modern data. High-pressure x-ray diffraction cameras allow us to follow changes of the lattice constant with pressure and thus to determine the isotherm $P(x, T_0)$, up to $\sim 5 \cdot 10^5$ bar.

Figure 5.13 shows the values of γ calculated using Eq. (5.92) and the experimental data for P_H of NaCl [13]. The considerable scatter of the points is due to the experimental errors. The scatter is increased by the fact that the thermal pressure $P_T \sim \gamma$ is only 10–16% of P_H.

Applying the least-squares method, we obtain the Grüneisen parameter dependence represented by curve 1. The parameters γ_1 [see the Landau—Slater formula (2.35)] and γ_2 [see the Dugdale—MacDonald formula (2.42)], calculated using the potential given by Eq. (5.72) and the constants of [1],

$$A = 0.965 \cdot 10^5 \text{ bar}, \quad b = 9.69, \quad K = 1.053 \cdot 10^5 \text{ bar}, \quad \left.\begin{array}{c} \\ \\ \end{array}\right\} \tag{5.93}$$
$$\rho_0 = 2.167$$

are represented by curves 2 and 3 in Fig. 5.13. It is evident from this figure that parallel translation of curve 2 makes it practically coincide with curve 1 and, therefore, γ can be represented by the expression

$$\gamma = \gamma_1 + \delta_1, \tag{5.94}$$

where the constant δ_1 is found by assuming that γ of Eq. (5.94) is equal to γ_0 of Eq. (5.78) at P = 0, i.e.,

$$\delta_1 = \gamma_0 - \gamma_1(1). \tag{5.95}$$

The values of γ normalized in this way are represented by curve 4 in Fig. 5.13. Comparing curves 1 and 4, we find that δ_1 decreases somewhat with decreasing x but this effect is slight and it can be ignored. The theoretical Hugoniot curves of Eq. (5.75), calculated using normalized values of $\gamma_1(x)$ and normalized or unnormalized values of $\gamma_2(x)$, lie considerably higher than the experimental curves [13]. Normalizing the Landau—Slater and the Dugdale—MacDonald formulas to the thermodynamic value of γ_0 of Eq. (5.78), we obtain the following expressions for γ:

$$\gamma_{1n} = -\frac{x}{2}\frac{\partial^2 P_p/\partial x^2}{\partial P_p/\partial x} + \frac{1}{2}\left[\frac{\partial^2 P_p/\partial x^2}{\partial P_p/\partial x}\right]_{x=1} + \gamma_0, \tag{5.96}$$

$$\gamma_{2n} = -\frac{x}{2}\frac{\partial^2 (P_p x^{2/3})/\partial x^2}{\partial (P_p x^{2/3})/\partial x} + \frac{1}{2}\left[\frac{\partial^2 (P_p x^{2/3})/\partial x^2}{\partial (P_p x^{2/3})/\partial x}\right]_{x=1} + \gamma_0. \tag{5.97}$$

It follows from Fig. 5.13 that Eq. (5.96), with the constants given by Eq. (5.93), describes satisfactorily the pressure dependence of the Grüneisen parameter of NaCl up to $\sim 7 \cdot 10^5$ bar. If the pressure dependence of γ given by Eq. (5.96) were universal (which is not the case), it would be sufficient to know only one experimentally determined Hugoniot curve in order to determine the equation of state. Then, instead of two unknown functions, we would need to determine only three parameters A, b, and K using Eq. (5.75). In the case of NaCl, this procedure yields the values

$$A = 0.800 \cdot 10^5 \text{ bar}, \qquad b = 10.666, \qquad K = 0.889 \cdot 10^5 \text{ bar}, \tag{5.93a}$$

which differ from those given in Eq. (5.93) but the agreement with the Bridgman data is still satisfactory.

The method described in [4] can now be improved somewhat. We can use the relationship between the shock-front velocity u_s and the mass velocity of a shock-compressed substance u_1, which was discovered experimentally and found to be universal. This relationship is linear:

$$u_s = C + S u_1, \tag{5.98}$$

where the numerical constants C and S are found experimentally. Equation (5.98) together with Eq. (5.33b) allows us to write P_H in the form of an explicit function of the volume:

$$P_H = \frac{\rho_0 C^2 (1 - x)}{[1 - S(1 - x)]^2} .$$

(5.99)

Equation (5.99) gives P_H in the form of a smooth experimental curve. Substituting Eq. (5.99) into Eq. (5.92), we obtain

$$\gamma = 2x \frac{\rho_0 C^2 (1 - x) - [1 - S(1 - x)]^2 P_p(x)}{2\rho_0 [1 - S(1 - x)]^2 [E_0 - E_p(x)] + \rho_0 C^2 (1 - x)^2}$$

(5.100)

which is a smooth experimental curve for γ. Comparing this experimental curve with the curves calculated using the formulas for $\gamma(x)$ given in Chapter 2, we can select that formula which fits best the experimental data.

Hugoniot Curve at High Temperatures

As the temperature rises, the high-temperature effects begin to make an increasing contribution to the energy and pressure. A detailed description of these effects and their theory are given in Chapter 3. The most important of the high-temperature effects are the anharmonicity and the thermal excitation of conduction electrons in metals. At very high temperatures, of the order of a few electron-volts (1 eV = $1.60 \cdot 10^{-12}$ erg = $1.16 \cdot 10^4$ °K), the processes of thermal ionization become important. The influence of the anharmonicity and of the thermally excited conduction electrons on the Hugoniot curve was included theoretically for the first time by Zharkov and Kalinin [3]. In contrast to Zharkov and Kalinin [3], Kormer et al. [11, 17] ignored the anharmonic effects and assumed that the electronic analog of the Grüneisen parameter γ_e is $\frac{1}{2}$. This is questionable both on theoretical grounds and in view of the experimental data on the specific heat and thermal expansion of metals near absolute zero.

It is shown in Chapter 3 that at high temperatures the energy E and the pressure P can be given by the following expansions (up to terms quadratic in temperature):

$$E = E_p(x) + \frac{3R}{\mu} T + \frac{A_a + A_e}{\rho_0} T^2,$$

(5.101)

$$P = P_p(x) + \frac{3RT}{\mu} \frac{\rho_0 \gamma}{x} + \frac{1}{x} (\gamma_a A_a + \gamma_e A_e) T^2,$$

(5.102)

where, in addition to the ordinary Grüneisen parameter γ, we find the Grüneisen parameter γ_a due to the anharmonicity, and the electron analog of the Grüneisen parameter γ_e

$$\left.\begin{array}{l} \gamma_a = \left(\dfrac{\partial \ln A_a}{\partial \ln x} \right)_T, \\[2mm] \gamma_e = \left(\dfrac{\partial \ln A_e}{\partial \ln x} \right)_T. \end{array}\right\}$$

(5.103)

The theoretical formula for the Hugoniot curve at high temperatures is obtained by eliminating the temperature from Eqs. (5.101), (5.102), and (5.73):

$$P_H = \frac{n_2{}^0 + n_2{}^1 - n_2{}^0 \sqrt{1 + \mu_1}}{2n_1{}^1},$$

(5.104)

where

$$n_1{}^1 = (A_a + A_e) \left[1 - \frac{1 - x}{2x} \Gamma \right]^2, \qquad \Gamma = \frac{A_a \gamma_a + A_e \gamma_e}{A_a + A_e},$$

$$n_2{}^0 = \frac{\rho_0{}^2 c_{VC}{}^2}{x}(\Gamma - \gamma)\left[1 - \frac{1-x}{2x}\gamma\right], \qquad n_2{}^1 = 2n_1{}^1 P_H{}^{\gamma \to \Gamma},$$

$$P_{H0}{}^{\gamma \to \Gamma} = \frac{1}{1 - \dfrac{\Gamma(1-x)}{2x}}\left\{P_p(x) + \frac{\rho_0 \Gamma}{x}[E_0 - E_p(x)]\right\},$$

$$\mu_1 = 4\frac{n_1{}^1}{n_2{}^0}(P_H{}^{\gamma \to \Gamma} - P_{H0}), \qquad P_{H0} = \frac{P_p(x) + \dfrac{\rho_0 \gamma}{x}[E_0 - E_p(x)]}{1 - \dfrac{\gamma(1-x)}{2x}}.$$

(5.105)

When $\mu_1 \ll 1$, Eq. (5.104) can be expanded in terms of μ_1, which gives

$$P_H = P_{H0} + \frac{1}{16}\frac{n_2{}^0}{n_1{}^1}\mu_1{}^2 = P_{H0}\left\{1 + \frac{\left(1 - \dfrac{1-x}{2x}\Gamma\right)^2}{\Gamma\left[1 - \dfrac{1-x}{2x}\gamma\right]\left(\dfrac{\Gamma}{\gamma} - 1\right)} \cdot \frac{(P_{H0}{}^{\gamma \to \Gamma} - P_{H0})^2}{P_{H0}P_d}\frac{E_2}{E_d}\right\}, \qquad (5.106)$$

where

$$E_d = c_{VC}T, \qquad P_d = \frac{c_{VC}\gamma\rho_0}{x}T, \qquad E_2 = \frac{A_a + A_e}{\rho_0}T^2. \qquad (5.107)$$

It is interesting to consider also the case when Γ is close to γ:

$$\Gamma = \gamma(1 + \Delta_1), \qquad \Delta_1 \ll 1. \qquad (5.108)$$

In this case, Eq. (5.106) becomes

$$P_H = P_{H0}\left\{1 + \left(1 - \frac{1-x}{2x}\gamma\right)\gamma^2\frac{\mathscr{P}^2}{P_{H0}P_d}\frac{E_2}{E_d}\Delta_1\right\}, \qquad (5.109)$$

where

$$\mathscr{P} \equiv \frac{\partial P_{H0}}{\partial \gamma} = \frac{\dfrac{1-x}{2x}P_{H0} + \dfrac{\rho_0}{x}[E_0 - E_p(x)]}{1 - \dfrac{1-x}{2x}\gamma}. \qquad (5.110)$$

Using Eq. (5.109), we can determine the relative positions of the extrapolated theoretical Hugoniot curve P_{H0} and the experimental curve P_H at high temperatures and pressures:

$$\begin{aligned}&\text{when} \quad \Gamma > \gamma, \quad P_H > P_{H0} \\ &\text{when} \quad \Gamma < \gamma, \quad P_H < P_{H0}\end{aligned}\Big\} \quad E_2 > 0 \qquad (5.111)$$

and the inequalities are reversed when $E_2 < 0$.

We note that when E_2 is dominated by the electron term ($A_e > A_a$), we find that $E_2 > 0$. When the anharmonic term is the dominant one ($A_a > A_e$), the value of E_2, like that of A_a, may be positive or negative. The sign of A_a can be determined only by analyzing the experimental data. In the case of metals, the values of A_e and γ_e can be determined from the specific heat and the thermal expansion coefficient near absolute zero. The values of A_e and γ_e found in this way are given for some metals in Table 5.1 [34, 35]. We can see that in the case of copper,

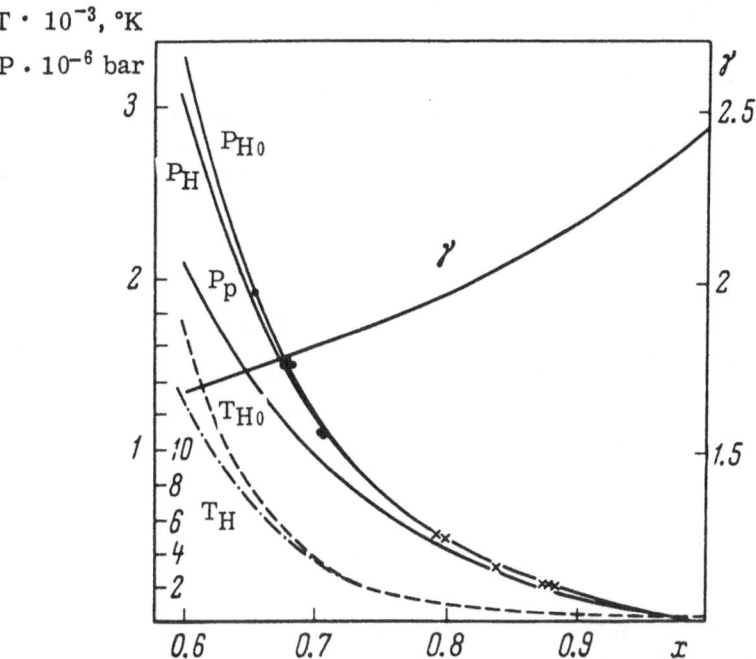

Fig. 5.14. Influence of the thermally excited conduction electrons on the temperature along the Hugoniot curve of silver: $\times$ — experimental points used to determine the potential component of the equation of state; $\bullet$ — experimental points which are not used in the calculations.

TABLE 5.1*

Metal	γ_e	$A_e \cdot 10^{-5}J \cdot cm^{-3} \cdot deg^{-2}$	Metal	γ_e	$A_e \cdot 10^{-5}J \cdot cm^{-3} \cdot deg^{-2}$	Metal	γ_e	$A_e \cdot 10^{-5}J \cdot cm^{-3} \cdot deg^{-2}$
Al	1.8 ± 0.1	7.54	Pd	2.1 ± 0.1	60.0	Co	1.9 ± 0.1	37.4
Cu	0.7 ± 0.2	5.04	Pt	2.4 ± 0.2	37.4	Mg (polycr.)	1.4 ± 0.2	(4.8)
Fe	2.1 ± 0.2	32.7	Ta	1.3 ± 0.1	24.7	Mg(single cr.)	1.4 ± 0.2	(4.8)
Mo	1.5 ± 0.3	12.7	V	1.65 ± 0.1	53.0	Re	3.5 ± 0.4	—
Nb	1.5 ± 0.2	39.4	W	0.2 ± 0.2	7.75	Ag	—	3 21
Ni	2.0 ± 0.1	55.8	Cr	-8.5 ± 1.0	10.6	Au	—	3.61
Pb	2.0 ± 0.1	8.2	Cd	0.5 ± 1.0	2.7			

* The values enclosed by parentheses are unreliable.

$\gamma_e \approx 0.7$, which is close to the value $2/3$ for an ideal electron gas. In the case of transition metals, the value of γ_e is considerably higher than $2/3$ and may reach 2 or 3.

The formulas given in this subsection, together with the discussion of the high-temperature effects in Chapter 3, show how the equation of state can be studied at high temperatures. We can see that, in order to determine A_a, A_e, and γ_a, γ_e, we have to use the low-temperature data for metals (for A_e and γ_e) and the high-temperature data for all types of solid (to obtain $A_a + A_e$ and Γ) at atmospheric pressure. The high-temperature part of the Hugoniot curve can be used to determine Γ and $A_a + A_e$. The first determination of the values of γ_e and A_e for iron from the high-temperature part of the Hugoniot curve was carried out by Zharkov and Kalinin [3]. However, the numerical values of γ_e and A_e reported in [3] are now only of methodological interest because the experimental data [11] used in that determination were somewhat inaccurate. In fact, the inequalities given by Eq. (5.111) can be masked in inaccuracies in the determination of the Grüneisen parameter γ. It is evident from Eq. (5.81) that if γ^0 is too low ($\Gamma_1 > 0$), we find that $P_H > P_{H0}$; and, conversely, if γ^0 is too high ($\Gamma_1 < 0$), we find that $P_H < P_{H0}$.

Figure 5.14 shows the Hugoniot curve for silver calculated using the theoretical value of the electronic parameter $\gamma_e = {}^2/_3$ (A_e was taken from Table 5.1). The temperature on the Hugoniot curve of Eq. (5.104) is given by the formula

$$T_H = \frac{E_H - E_p(x) - \dfrac{x}{\rho_0 \Gamma}[P_H - P_p(x)]}{c_{vc}\left(1 - \dfrac{\gamma}{\Gamma}\right)}, \tag{5.112}$$

which can be easily derived from the relationships given in this subsection. Only the smooth theoretical functions should be substituted into Eq. (5.112). The temperature T_H, calculated from Eq. (5.112), and the temperature T_{H0}, calculated from Eq. (5.77), are also given in Fig. 5.14. It follows from that figure that at $P_H \approx 3 \cdot 10^6$ bar the Hugoniot temperature of silver is $T_H \approx 1.4 \cdot 10^4$ °K, but without the electronic corrections (i.e., using the formulas valid in the phonon region), the temperature would be higher by about 30% ($T_{H0} \approx 1.8 \cdot 10^4$ °K). Even greater errors in the determination of the Hugoniot temperature would be committed if the thermal electronic terms were ignored in calculations referring to transition metals.

We shall now consider the problem of unloading of a shock-compressed substance taking into account the high-temperature corrections. The free energy, which includes terms up to the second power of temperature, is of the form (see Chapter 3)

$$F = E_p(x) - \frac{RT}{\mu}\left[1 - 3\ln\frac{\theta}{T}\right] - \frac{A_a + A_e}{\rho_0}T^2,$$

and the entropy is

$$S = -\left(\frac{\partial F}{\partial T}\right)_v = c_{vc}\left[\frac{4}{3} - \ln\frac{\theta}{T} + \frac{2(A_a + A_e)}{\rho_0 c_{vc}}\right], \tag{5.113}$$

where

$$c_{vc} = \frac{3R}{\mu}.$$

We shall use $S_{H1} = S(T_{H1}, x_1)$ to denote the entropy at that point on the Hugoniot curve which is the origin of the unloading isentrope. Then the condition

$$S(T, x) = S_{H1} \tag{5.114}$$

determines the temperature on the unloading isentrope. Having found the temperature as a function of the volume using Eq. (5.114) and having substituted it into the formula for the pressure (5.102), we obtain an expression for the required isentrope. This isentrope can be expressed in an implicit form using Eqs. (5.113), (5.114), and (5.102).

5.5. Experimental Methods for the
Determination of Shock-Wave Parameters

The advances made in the dynamics of high pressures since the Second World War is mainly due to two factors. First, considerable progress has been made in the preparation of large explosive charges. Consequently, blocks of explosive materials of various geometries and sizes (up to a few tens of centimeters) are now available and the dimensions of charges can be specified down to several microns. Secondly, quite accurate experimental (electrical and optical) methods have been developed for the recording of fast processes whose characteristic times are of the order of 0.1-1 μsec.

We are interested in the use of the experimental shock-wave data in the determination of the parameters in semiempirical equations of state of various types of solid. Therefore, we shall review only briefly the experimental methods used to obtain such data. More comprehensive information can be found in the relevant reviews and books [7, 19, 31, 36–38].

In principle, the Hugoniot curve can be expressed in terms of any pair of variables discussed in Sec. 5.2. In actual experiments, the Hugoniot curve is always found in the (u_s, u_1) plane. The velocity of a shock front (or the wave velocity) u_s can nowadays be found directly: the distance traversed by a wave is divided by the time taken to traverse this distance. It is much more difficult to determine directly the mass velocity u_1 behind a shock-wave front. Consequently, three indirect methods are usually employed which give not only u_s and u_1 but also all the other parameters [see Eq. (5.43)]. These methods were developed independently in the Soviet Union [31, 19] and in the USA [7, 6].

Free Surface Method*

The free surface method is used at relatively low pressures of the order of a few hundreds of kilobars. Walsh and Christian [6] showed that under these conditions the mass velocity u_1 and the velocity of a free surface u_{fs} are related by

$$u_1 \approx u_{fs}/2. \tag{5.115}$$

The free-surface velocity u_{fs}, like the wave velocity u_s, can be measured either by the method of electrical contacts or by an optical method. The approximation represented by Eq. (5.115) is called the free-surface velocity approximation [7], since it presupposes that a shock wave is reflected from the free surface and the reflection is not masked by secondary effects.

Let us assume that a shock wave emerges at a free surface of a solid, which is parallel to the shock-wave front. After reflection from this surface, it returns into the solid as a rarefaction wave, which reduces the pressure behind the front P_1 to zero. The free-surface velocity is then the sum of the mass velocity of the shock wave u_1 (5.33b) and the mass velocity of the rarefaction wave u_r (5.19):

$$u_{fs} = u_1 + u_r = \sqrt{(P_1 - P_0)(V_0 - V_1)} + \int_0^{P_1} \left(-\frac{\partial V}{\partial P}\right)_s^{1/2} dP. \tag{5.116}$$

---

*The authors call it the "spalling" method, which refers to the spalling (scabbing) which sometimes occurs when a shock wave emerges at the free surface of a solid. This method makes it possible to determine the velocity of the free surface from the velocity of spalled fragments. For details, the reader is referred to [39].

The approximate relationship

$$u_r / u_1 \approx 1.0, \tag{5.117}$$

which is valid for the reflection of weak shock waves, yields Eq. (5.115) and makes it possible to determine the point (P_1, x_1) on the Hugoniot curve if we use Eqs. (5.27b), (5.29b), and the initial density ρ_0.

In a rigorous approach, the measurements should be carried out by the method of successive approximations. In the zeroth approximation, the Hugoniot curve is found from the experimental data using the free-surface velocity approximation of Eq. (5.115). Then, the Hugoniot curve and the thermodynamic data for normal conditions (room temperature and atmospheric pressure) are used to determine the equation of state and the unloading isentrope (5.88). The isentrope (5.88) and Eq. (5.19) can be used to calculate a more accurate value of the mass velocity u_r in the unloading wave and the first-approximation value of $u_1 = u_{fs} - u_r$. The new values of u_1 and u_s are again used to plot the Hugoniot curve, to determine the equation of state, and to find the isentrope (5.88); the value of u_r is once more calculated using this isentrope and Eq. (5.19), and the whole process is repeated.

A detailed study, reported by Walsh and Christian [6], showed that this method gives rapidly converging results and even the values of the ratio u_r/u_1 obtained in the first approximation are sufficiently accurate. The free-surface velocity approximation was used successfully later [7] in investigations of a large number of metals at pressures up to 500 kbar. It was found that the ratio u_r/u_1 lies, in the majority of cases, within the range $1 \le u_r/u_1 \le 1.03$, and the corrections to the compression $(1 - x)$ on the Hugoniot curve at a fixed pressure range from 0 to -1%.

The free surface method, or the free-surface velocity approximation, is inapplicable to porous samples, since it gives rise to very high temperatures and the unloading isentrope then deviates considerably from the Hugoniot curve.

Collision Method*

An increase in the shock-wave amplitude produces high pressures and strongly raises the temperature on the Hugoniot curve. This temperature rise makes the unloading isentrope depart considerably from the Hugoniot curve so that the free-surface velocity approximation of Eq. (5.115) is too rough. Al'tshuler et al. [9] suggested a general method for the determination of the mass velocity behind the front of a shock wave, which can be used at any pressures. In this method an explosive charge accelerates a striker plate to a velocity w, which then collides with a target made of the same material. Since the striker and target plates are made of the same material, identical shock waves are generated in the striker and the target. Consequently, the pressures P and the mass velocities u_1 on both sides of the contact boundary are identical until the shock waves emerge at the free surfaces of the striker and the target (the pattern is then distorted by the unloading waves).

We can easily see that the mass velocity u_1 behind the front of a shock wave in the target is equal to half the velocity of the striker: $u_1 = w/2$. Before the collision, the target is at rest and, therefore, the mass velocity in the target after the impact is equal to the velocity of the contact boundary between the colliding bodies. On the other hand, before the collision, the matter in the striker moves at the velocity w but after the collision and the formation of a shock wave, this velocity decreases by u_1, i.e., it becomes $w - u_1$. At the contact boundary, the mass

*The authors call this the "deceleration" or "breaking" method; it is also sometimes called the "momentum transfer" method.

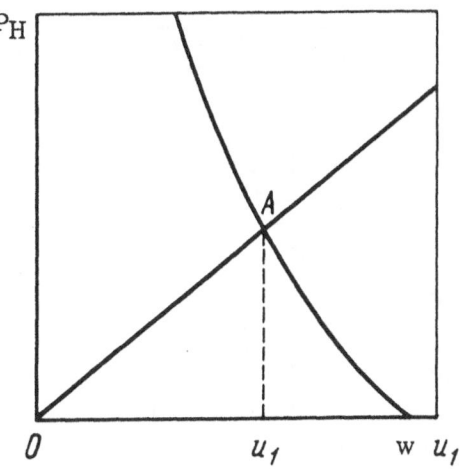

Fig. 5.15. Graphical determination of the mass velocity of a sample u_1 in the collision method.

velocities of the colliding bodies are equal, i.e., $w - u_1 = u_1$ and $u_1 = w/2$.

The collision method makes it possible to determine also the wave velocity u_S in the target. This velocity is determined either by the method of electrical contacts or by an optical method (using high-speed photography). The experimental values of u_1 and u_S make it possible to determine the Hugoniot curve in the (P, x) plane using Eqs. (5.27b) and (5.29b).

The collision method can be generalized to the case of strikers and targets made of different materials. This generalization is convenient because we can calibrate very carefully the Hugoniot curve of a standard striker and then carry out relative measurements which, other conditions being equal, are always more accurate. In the generalized collision method, the condition of equality of the pressures at the contact boundary still applies, but not the condition of equality of the mass velocity in the striker $w - u_1$ and the mass velocity in the target u_1, i.e., we now have $w - u_1 \neq u_1$. Nevertheless, the known Hugoniot curve of a standard striker, expressed in terms of the pressure and the mass velocity, $P = f(w - u_1)$, can be used to determine the mass velocity u_1 in the target.

The pressure in the target is related to the mass velocity u_1 which, as before, is equal to the velocity of motion of the contact boundary [Eq. (5.29b)]:

$$P = \rho_0 u_s u_1.$$

The equality of the pressures gives the required formula

$$f(w - u_1) = \rho_0 u_s u_1, \qquad (5.118)$$

which makes it possible to determine u_1 by measuring the velocity of the striker w and the wave velocity in the target u_S.

The solution of Eq. (5.118) can be obtained most easily by a graphical method (Fig. 5.15), plotting the data in the (P, u_1) plane. The straight line starting from the origin, $P = \rho_0 u_s u_1$, gives the shock pressure in the target for a given wave velocity u_S. The Hugoniot curve of the striker originates on the abscissa at a point $u_1 = w$, i.e., at $P = 0$ the mass velocity of the striker is equal to its free flight velocity. The two curves intersect at a point A, which gives the required mass velocity u_1 [Eq. (5.118)]. If the target and the striker are made of the same material, $u_1 = w/2$ and the abscissa of A is the exact midpoint of the segment 0w.

Calibrated Reflection Method*

Let us now consider the following arrangement. A plate (which shall be called the "screen") with a known equation of state and, consequently, with a known Hugoniot curve and known isentropes is used as a base for the mounting of samples, one of which is made of the same material as the screen. Let us assume that a shock wave enters the screen (the screen is in contact with an explosive charge or a shock wave is generated in it by the impact of a

---

*The authors term it the "reflection" method; it is also called the "impedance match" method.

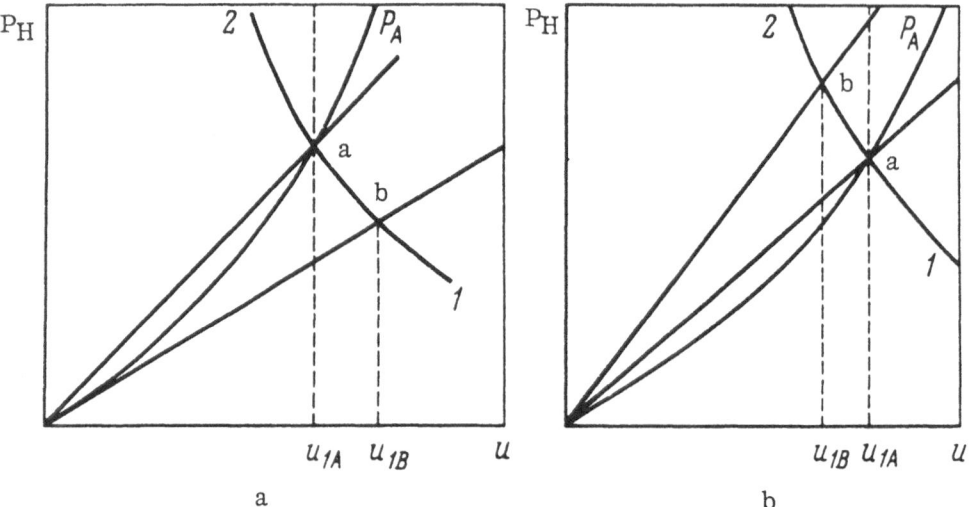

Fig. 5.16. Graphical determination of the mass velocity of a sample u_{1B} in the calibrated reflection method. a) Shock "hardness" of the sample less than the shock "hardness" of the screen $\rho_{0B}u_{sB} < \rho_{0A}u_{sA}$; b) shock "hardness" of the sample greater than the shock "hardness" of the screen $\rho_{0B}u_{sB} > \rho_{0A}u_{sA}$.

striker) and then penetrates into the samples. The wave velocities in the screen and in the samples are measured experimentally. The laws of conservation and the boundary conditions can then be used to determine the mass velocities u_1 in the samples and to plot, on the basis of these velocities, the experimental Hugoniot curves.

The calibrated reflection method can be explained best by a graphical plot in the (P_H, u) plane (Fig. 5.16). The curve P_A is the known Hugoniot curve of the screen. Let us assume that the wave velocity in the screen is equal to u_{sA}. Then, according to Eq. (5.29b), the mass velocity behind the shock front in the screen can be found from the intersection of the straight line $P = \rho_{0A}u_{sA}u = u_{1A}$ with the Hugoniot curve P_A. The abscissa of the point of intersection a is $u = u_{1A}$ and the ordinate is $P = P_{1A}$. When the shock wave emerges at the boundary between the screen and the sample, it is reflected back and if the sample is "softer" than the screen, the reflection follows (as in the case of a free surface) the unloading isentrope. However, if the sample is "harder" than the screen, the reflected wave is also a shock wave whose initial parameters are (u_{1A}, P_{1A}). Consequently, during the unloading, the pressure in the screen falls and the mass velocity rises; conversely, when the reflected wave is a shock wave, the pressure in the screen rises and the mass velocity falls. Measurements of the wave velocity in the sample u_{sB} can be used to draw the straight line $P_B = \rho_{0B}u_{sB}u$ in the (P, u) plane [Eq. (5.29b)], which represents the state of the sample behind the shock front.

According to the boundary conditions, we must find those states in the screen and the sample for which the pressures and mass velocities are equal in both. These states are realized at the point of intersection of the straight line $P_B = \rho_{0B}u_{sB}u$ with the unloading isentrope if $\rho_{0B}u_{sB} < \rho_{0A}u_{sA}$ and with the reflected shock wave if $\rho_{0B}u_{sB} > \rho_{0A}u_{sA}$. In the former case, the sample is regarded as softer than the screen, and in the latter case as harder. Knowing the equation of state of the screen, we can determine the unloading isentrope as well as the Hugoniot curve for the reflected shock wave, and we can thus find the required values of the mass velocities in the samples. In practice, the procedure is usually simpler, because the intersection points b lie close to the points a (Fig. 5.16). An analysis shows that in this case the lower and upper branches of the reflected wave can be found sufficiently accurately by "reflecting" the Hugoniot curve of the screen P_A in a vertical line passing through the point a.

The following three conditions must be satisfied whenever strikers are used in the collision and calibrated reflection methods [8].

1. The pressure in the striker plate at the moment of collision with the target or the screen should be uniform. The required uniformity of the pressure in a striker plate 0.5 cm thick, accelerated by an explosive charge 10 cm thick, is reached after a free flight of only a few centimeters.

2. The accelerated plate (the striker) should be sufficiently thick to ensure that the initial shock wave in the screen and in the samples is not overtaken by the strong rarefaction wave generated at the rear side of the striker facing the explosive charge. The condition for the maximum ratio r of the screen thickness (including the samples) to the thickness of the striker plate is given in [8]:

$$r = \frac{[u_s^{-1} + (xc)^{-1}]_{str}}{[u_s^{-1} - (xc)^{-1}]_{tgt}},$$

where c is the hydrodynamic velocity of sound behind the shock-wave front; $x = V/V_0$ is the ratio of the specific volumes on the two sides of the shock front; the subscripts "str" and "tgt" refer to the striker and target plates, respectively. If the striker and the target are made of brass, the ratio r varies with pressure in the following manner:

P, Mbar	0.5	1.0	1.5	2.0
r	7 7	5.7	5.0	4.5

3. The samples mounted on the screen should have a sufficiently large width — thickness ratio in order to avoid the influence of the rarefaction waves from the edges on the shock front. The ratio of the width l to the thickness h should not be less than [8]

$$l/h = 2\left[\left(\frac{c}{x u_s}\right)^2 - 1\right]^{1/2}.$$

Thus, for compressions $x^{-1} = 1.6$, one could use $l/h = 2.5$, which is fully satisfactory [8].

There are some limitations to the dynamic methods. As the shock pressures are increased, the experimental difficulties and errors increase rapidly. L. V. Al'tshuler, the leading Soviet authority on the dynamic methods in the physics of high pressures, is of the opinion [31] that the dynamic methods are not suitable at pressures greater than 10–15 Mbar.

Experimental Data

The experimental values of u_s and u_1 are described using the formulas

$$u_s = C + S u_1 \tag{5.98}$$

and

$$u_s = C + S u_1 + K u_1^2, \tag{5.119}$$

in which the coefficients are found by the method of least squares. The calculated probable experimental error for the linear equation is less than that for the quadratic formula. Moreover, the coefficient of the quadratic term is always small. Consequently, it is usual to employ the linear formula (5.98), which is the fundamental relationship in the dynamic methods used at high pressures.

We shall consider also the coefficient C_B, which is the hydrodynamic velocity of sound at normal temperature and pressure:

TABLE 5.2

Substance	Init. dens., g/cm³	Range of pressures, kbar	C, cm / μsec	s	Transition pressure, kbar	Refs.	Remarks
			Elements				
Li	0.53	35—698	0.445	1.115		[70]	
Na	0.97	7—32				[40]	
		42—966	0.255	1.262		[70]	
K	0.86	33—860	0.200	1.172		[70]	
Cu	8.90	216—1444	0.396	1 50		[7, 8, 10, 18, 41]	
	8.93	450—9070				[10—12, 14]	
Ag	10.49	216—4010	0.324	1.59		[7, 8, 10]	
Au	19.24	269—5130	0.308	1.56		[7, 8, 10, 41]	
Be	1.85	142—311	0.798	1.09		[7, 41]	
Mg	1.725	50—260	0.449	1.27		[7, 41—45]	
Zn	7.14	186—1403	0.305	1.56		[7, 8, 41]	
	7.14	350—7860				[10, 14]	
Cd	8.64	183—1351	0.244	1.67		[7, 8, 46]	
	8.64	360—8410				[10, 14]	
Hg	13.5	226—464	0.12	2.4		[47]	
Al	2.79	20—1970	0.525	1.39		[7, 11, 12, 18, 42—46, 48—51]	
In	7.27	214—405	0.237	1.61		[7]	
Tl	11.84	213—1517	0.186	1.52		[7, 8, 41]	
Ti	4.51	168—1063	0.474	1.09		[7, 8, 40, 51]	
Zr	6.49	208—407	0.377	0.93		[7]	
C	1.7—2.2	3—850			180—400; 600	[53, 54]	Two transitions
Sn	7.28	330—3100	0.264	1.48		[7, 8, 10, 14]	
Pb	11.34	195—1390	0.203	1.52		[7, 8, 10, 42]	
	11.34	390—9150				[11, 12, 14, 18]	Different porosities
V	6.1	204—1244	0.511	1.21		[8]	
Nb	8.60	244—482	0.445	1.21		[7]	
Ta	16.46	272—547	0.337	1.16		[7, 52]	
N	0.808	29,6—404	0.16	1.4		[55]	
Sb	6.69	248—1175	0.20	1.60	27.2 at 19° C	[8]	
Bi	9.79	17.6—3450	0.20	1.34	31.3 at 28° C, 17.6 at 236° C	[7, 8, 10, 25, 56]	Linear relationships valid above 350 kbar
Cr	7.13	233—1379	0.522	1.47		[7, 8, 41]	
Mo	10.20	254—1633	0.516	1.24		[7, 8, 41, 52]	
W	19.17	394—2074	0.400	1.27		[8, 41]	
S	2.1	60—200	0.22	1.0	67—106	[57, 58]	Below transition
			0.32	0.8			Above transition
I	4.93	50—1100	0.135	1.7	~700	[59]	Below transition
			0.36	0.7			Above transition
Fe	7.84	38—1730	0.380	1.58	130, 19 at 885°C, 150 at 195° C	[7, 8, 41, 42, 50, 56]	Linear relationships valid above 300 kbar
	7.85	415—8700				[9, 11, 12, 14]	

TABLE 5.2 (continued)

substance	Init. dens., g/cm^3	Range of pressures, kbar	C, $\frac{cm}{\mu sec}$	s	Transition pressure, kbar	Refs.	Remarks
Co	8.82	244—1603	0.475	1.33		[7, 8, 41]	
Rh	12.42	279—551	0.468	1.65		[7]	
Ni	8.86	235—9180	0.465	1.45		[7, 8, 14, 41]	
Pd	11.95	263—372	0.379	1.92		[7]	
Pt	21.4	295—586	0.367	1.41		[7]	
Th	11.68	203—1405	0.213	1.28		[7, 8]	
U	18.9	335—6450	0.255	1.504		[45]	
Alkali halides							
CsBr	4.43	146—328	0.31	0.3		[60]	
	4.45	110—5450				[39]	Different porosities
CsCl	3.95	60—318	0.22	0.5		[60]	
CsI	4.51	64—5540				[60, 15, 61]	
KBr	2.75	21—4380			20—70	[60, 15, 39]	
KCl	1.98	40—229	0.18	1.8		[60]	
	1.99	20—3790			20—70	[15, 39, 62]	Different porosities
KF	2.49	117—266	0.24	1.6		[60]	
KI	3.1	110—278	0.18	1.4		[60]	
LiBr	3.30	136—300	0.26	1.4		[60]	
LiCl	2.05	121—263	0.41	1.5		[60]	
LiF	2.62	155—331	0.50	1.6		[60]	
	2.65	79—4850	0.50	1.6		[39, 15]	Different porosities
LiI	4.01	205—320	0.28	0.9		[60]	
NaBr	3.16	58—305	0.26	1.3		[60]	
NaCl	2.15	52—791	0.34	1.37		[13, 60]	
	2.165	276—4030				[15, 39, 62]	Different porosities
NaI	3.64	134—312	0.20	1.6		[60]	
	3.67	62—1009				[15]	
RbBr	3.30	112—286	0.14	1.6		[60]	
RbCl	2.70	109—268	0.15	1.6		[60]	
RbI	3.5	117—279	0.14	1.5		[60]	
Various minerals and alloys							
Albite	2.61	410—904				[63]	
Andalusite	3.06—3.10	594—1158				[63]	
Anorthite	2.70—2.79	331—952				[63]	
Bronzitite	3.27—3.30	551—1090				[63]	
Gabbro	3.0	150—720				[64]	
	3.15	314—2325				[65]	
Hematite	4.90—5.05	896—1420				[63]	
Granite	2.63	406—919				[63]	
Dunite	3.25	180—2300				[64, 66]	
	2.9	298—1021	0.459	1.290		[65]	
	2.96	307—2240	0.454	1.370		[65]	
	3.32	734—1123				[63]	
	3.68—3.85	653—1190				[63]	

TABLE 5.2 (continued)

Substance	Init. dens., g/cm³	Range of pressures, kbar	C, cm/μsec	s	Transition pressure, kbar	Refs.	Remarks
Diabase	3.13	131—2280	0.448	1.326		[65]	
	2.97—3.02	326—1029				[63]	
Dialogite	3.01	316—1052	0.533	1.120		[65]	
Jadeite	3.33—3.35	269—1147				[63]	
Cassiterite	6.45—6.75	992—1410				[63]	
Quartz	2.66	40—200			144; 400	[28, 67]	Fused quartz
	2.204	0—623			250	[28]	
Corundum	3.83—4.02	310—1312	0.86	1.1		[41, 63]	
Magnetite	5.01—5.20	620—1310	0.60	1.5		[41, 63]	
Marble	2.70	51—518	0.35	1.95	150	[68]	
Olivenite	3.31	136—5050	0.508	1.287		[65]	
	3.21	327—2360	0.523	1.270		[65]	
	3 69	356—1239	0.496	1.324		[65]	
Periclase	3 58	202—1258	0.68	1.25		[41, 63]	
Peridotite	3.22	319—2369	0.450	1.400		[65]	
Pyroxenite	3.29	136—2400				[65]	
Pyrolusite	4.19—4.42	183—1202				[63]	
Rutile	4 25	1033—1235	0.70	2.2		[41, 63]	
Sillimanite	3.07—3.15	584—1099				[63]	
Serpentine	2.76—2.84	575—940				[63]	
Fayalite	4.18—4.30	577—1137				[63]	
Ferrosilicon	6.86—6.94	480—3310				[69]	
Forsterite	3.03—3.07	664—1035				[63]	
Spinel	3.40—3.43	678—1158				[63]	
Eclogite	3.39—3.59	439—1125				[63]	
Enstatite	2.71	300—637				[63]	
Brass	8.41	167—1764	0.38	1.42		[7, 8]	

$$C_B^2 = -V_0^2 \left(\frac{\partial P}{\partial V} \right)_s. \qquad (5.120)$$

The values of C, deduced from the dynamic data using Eq. (5.98), can be compared with the static values of C_B calculated from the Bridgman data using Eq. (5.120). For the investigated metals [8] it is found that the dynamic and static values of the "velocities of sound" at atmospheric pressure (C and C_B) are in agreement to within 1.5%; in nine of the investigated cases, $C < C_B$; and in seven cases, $C > C_B$, i.e., we can assume that the agreement is satisfactory. Equation (5.98) can be used to express the experimental values of the Hugoniot pressure P_H and energy E_H as explicit functions of the volume x and of the empirical constants C and S:

$$P_H = \frac{\rho_0 C^2 (1-x)}{[1-S(1-x)]^2},$$ (5.99)

$$E_H = E_0 + \frac{1}{2}\left[\frac{C(1-x)}{1-S(1-x)}\right]^2.$$ (5.121)

The limiting compression $\bar{x}$ is then found from Eq. (5.99):

$$1 - S(1-\bar{x}) = 0, \qquad \bar{x} = (S-1)/S.$$ (5.122)

It should always be remembered that Eq. (5.122) is obtained by the extrapolation of the linear formula (5.98) to infinite pressures, which may not be justified. We should also mention that the average probable error of the values of the coefficients C and S is 0.5% and 0.9%, respectively.

Extensive experimental data are now available on the dynamic compression of solids. The upper limit of pressures investigated by this method is currently about 10^4 kbar (10 Mbar). These pressures have been reached only by Soviet investigators. The pressures achieved by non-Soviet workers usually do not exceed $2 \cdot 10^3$ kbar and most investigations are limited to $5 \cdot 10^2$ kbar.

The available experimental data on the shock-wave studies of solids are presented in Table 5.2. This table is based on that given by Duvall and Fowles [38], but it has been modified appreciably and new data are included.

5.6. Determination of the Equation of State

from Dynamic Data

We shall now consider the practical aspects of the détermination of the equations of state for solids. We shall consider the potential term and the high-temperature corrections. As long as the temperature on the Hugoniot curve is not too high, we can ignore terms proportional to T^2. Then, the potential term is the only unknown part of the equation of state. As mentioned earlier, this applies at relatively small compressions (pressures of the order of 10^2 kbar). Consequently, we shall determine the potential of a solid using only the lower part of the Hugoniot curve.

The theoretical expression for the Hugoniot curve

$$P_H = \frac{P_p(x) + \frac{\rho_0 \gamma}{x}[E_0 - E_p(x)]}{1 - \frac{\gamma(1-x)}{2x}}$$ (5.75)

contains only two unknown functions P_p and γ. The experimental data usually yield only the function $P_H(x)$, which prevents us from determining the functions P_p and γ independently. Therefore, as in the static case (Chapter 4), we shall assume a functional relationship between P_p and γ. We shall postulate that this relationship is of the form

$$\gamma = -\frac{x}{2}\frac{\partial^2(P_p x^{2m/3})/\partial x^2}{\partial(P_p x^{2m/3})/\partial x} + \frac{1}{3}(m-2) + \delta,$$ (2.45)

where the normalization constant δ is found (as in Chapter 4) from the thermodynamic value γ_0.

Now we find that Eq. (5.75) has three unknown parameters A, b, and K since $P_p(x) = -\delta E_p/\delta x$ and $E_p(x)$ is given by one of the expressions in Sec. 1.3, suitable for a given solid. Let us take $E_p(x)$ in the form of Eq. (1.7). We shall calculate the parameter K using the condition that $P_H = 0$ when $x = 1$. Then, Eqs. (5.75), (2.45), and (1.7) yield

$$K = A + \rho_0 \gamma_0 \frac{R}{\mu} \left[\frac{9}{8} \theta_0 + 3T_0 D\left(\frac{\theta_0}{T_0}\right) \right]. \qquad (5.123)$$

We have derived Eq. (5.123) by assuming additionally that E_0 (the energy of a unit mass of matter in front of a shock wave) can be calculated using the Debye theory (Appendix B.2). The zero subscript in Eq. (5.123) is used for the values of the various properties before the arrival of a shock wave.

The other two parameters A and b can be found by the method of least squares, i.e., from the condition for a minimum of the function

$$\mathcal{M} = \sum_{j=1}^{l} [P_{Hej} - P_H(x_j)]^2 \frac{1}{\sigma_j^2}, \qquad (5.124)$$

where l is the number of experimental points; $P_H(x_j)$ and P_{Hej} are the theoretical and experimental values of the pressure on the Hugoniot curve of Eq. (5.75) at a point x_j; σ_j is the absolute error at the same point x_j. If all the points are determined with the same relative error η, it follows that $\sigma_j = P_{Hej}\eta$ and Eq. (5.124) becomes

$$\mathcal{M} = \frac{1}{\eta^2} \sum_{j=1}^{l} \left[1 - \frac{P_H(x_j)}{P_{Hej}} \right]^2.$$

The equations for the determination of A and b are found by minimizing $\mathcal{M}$ with respect to A and b:

$$\frac{\partial \mathcal{M}}{\partial A} = 0, \qquad \frac{\partial \mathcal{M}}{\partial b} = 0,$$

or, more explicitly,

$$\left. \begin{array}{l} \displaystyle\sum_{j=1}^{l} \frac{1}{\sigma_j^2} [P_{Hej} - P_H(x_j)] \frac{\partial P_H(x_j)}{\partial A} = 0, \\[4mm] \displaystyle\sum_{j=1}^{l} \frac{1}{\sigma_j^2} [P_{Hej} - P_H(x_j)] \frac{\partial P_H(x_j)}{\partial b} = 0. \end{array} \right\} \qquad (5.125)$$

The derivatives of $P_H(x_j)$ with respect to A and b are given in Appendix D. The system of equations (5.125) is fairly complex and, therefore, we shall solve it by the method of successive approximations. The method for solving the system (5.125) is described in detail in Appendix D, which also gives all the necessary formulas.

The parameters for the equations of state of various solids, calculated in this way, are listed in Table 5.3. This table names the substance and lists the literature from which the experimental values of P_{He} were taken. Three sets of parameters A, b, K, and δ, are given for each substance: these three sets correspond to the values m = 0, 1, and 2 in Eq. (2.45) for the Grüneisen parameter. The first row of figures corresponds to m = 0, the second to m = 1, and the third to m = 2.

Figures 5.17–5.33 show the equations of state for some solids. The quantities given in these figures were calculated using the parameters A, b, K, and δ taken from the middle row of figures (m = 1) in Table 5.3.

TABLE 5.3

Substance	$A \cdot 10^{-5}$, bar	b	$K \cdot 10^{-5}$, bar	δ
Noble metals				
Cu [7,8]	5.2196	10.0094	5.4404	—0.292
	5.3191	9.8052	5.5398	0.100
	5.4328	9.6111	5.6536	0.482
Ag [7,8]	3.8306	10.6513	4.0110	0.024
	3.9401	10.3651	4.1205	0.432
	4.0592	10.0981	4.2396	0.828
Au [7,8]	6.7416	10.0098	6.9630	0.727
	6.8611	9.8357	7.0825	1.108
	6.9809	9.6810	7.2023	1.480
Divalent metals				
Be [7]	14.2362	4.5709	14.5015	—0.153
	14.1815	4.5451	14.4468	0.203
	14.2066	4.5261	14.4720	0.548
Mg [7,45]	1.9956	7.3883	2.0762	—0.417
	2.1730	6.9383	2.2535	0.024
	2.3435	6.5877	2.4241	0.435
Zn [7,8]	2.1189	11.0863	2.3148	—0.182
	2.2360	10.5465	2.4319	0.295
	2.3800	10.0195	2.5759	0.752
Cd [7]	1.3286	12.9825	1.4572	—0.614
	1.4485	12.1505	1.5770	—0.088
	1.5974	11.3252	1.7260	0.418
Trivalent metals				
Al [7,45]	3.8771	8.0755	4.0518	0.128
	4.0380	7.7860	4.2127	0.540
	4.2041	7.5377	4.3788	0.934
In [7]	4.1181	6.0214	4.2227	0.622
	4.9311	5.3958	5.0357	1.096
	5.5445	5.0475	5.6491	1.506
Tetravalent metals				
Sn [7]	1.3618	11.7748	1.4563	—0.626
	1.4718	11.0932	1.5664	—0.139
	1.6008	10.4390	1.6953	0.330
Pb [7]	5.6139	4.9428	5.7256	1.378
	9.6042	3.8204	9.7159	1.989
	10.8725	3.6318	10.9842	2.379
Transition elements				
Ti [7]	4.9890	7.7348	5.0769	—0.735
	5.0990	7.5847	5.1869	—0.364
	5.2027	7.4602	5.2906	—0.002
Zr [7]	17.7278	3.6118	17.7694	—0.294
	18.4181	3.5502	18.4598	0.060
	18.4667	3.5454	18.5083	0.396
V [8]	9.1314	7.1766	9.2992	—0.521
	9.2284	7.1227	9.3461	—0.170
	9.2927	7.0718	9.4104	0.177
Nb [7]	20.6043	4.6471	20.7230	0.356
	21.2230	4.5655	21.3417	0.713
	21.4926	4.5318	21.6113	1.055

Substance	$A \cdot 10^{-5}$, bar	b	$K \cdot 10^{-5}$, bar	δ
Ta [7]	9.3749	7.6925	9.4905	—0.213
	9.5038	7.5979	9.6194	0.144
	9.6159	7.5229	9.7315	0.496
Mo [7,8]	15.6789	7.1619	15.9118	—0.224
	15.8570	7.0914	15.9899	0.127
	16.0093	7.0369	16.1422	0.473
W [8]	17.1170	7.3983	17.2408	—0.305
	17.3243	7.3234	17.4481	0.046
	17.4946	7.2660	17.6185	0.392
Fe [27, 56]	5.5105	10.8970	5.7001	—0.796
	5.3870	10.9992	5.5766	—0.462
	5.3235	11.0372	5.5132	—0.123
Co [7,8]	9.4444	8.2997	9.6804	—0.031
	9.5134	8.2144	9.7495	0.332
	9.6041	8.1322	9.8402	0.689
Rh [7]	26.8330	5.4510	27.0498	0.777
	27.3303	5.3840	27.5471	1.130
	27.6149	5.3457	27.8317	1.474
Ni [7]	14.1636	6.3736	14.3912	0.240
	14.4904	6.2576	14.7181	0.606
	14.7468	6.1755	14.9744	0.960
Pd [7]	2.9550	18.3446	3.1415	—1.571
	2.9719	18.1434	3.1583	—1.176
	2.9990	17.9185	3.1855	—0.782
Pt [7]	13.4996	8.4276	13.7164	0.592
	13.6590	8.3348	13.8758	0.951
	13.8092	8.2559	14.0259	1.304
Actinides				
Th [7,8]	2.4462	8.5078	2.4878	—0.926
	2.6096	8.1274	2.6512	—0.515
	2.7653	7.8136	2.8069	—0.121
U [45]	4.3967	10.3871	4.5045	—0.550
	4.4319	10.2740	4.5397	—1.184
	4.4784	10.1587	4.5861	0.178
Minerals and rocks				
Fe_3O_4 [63]	3.3038	10.1210	3.4937	—0.965
	3.7818	9.2528	3.9717	—0.448
	4.3506	8.4656	4.5405	0.406
Al_2O_3 [63]	38.1679	4.4514	38.4422	0.318
	41.4459	4.2629	41.7202	0.701
	43.1222	4.1787	43.3964	1.056
MgO [63]	7.3066	3.6902	7.5489	—0.697
	7.5987	8.4008	7.8410	—0.293
	7.9164	8.1358	8.1587	0.098
Olivine [63]	11.7193	5.0472	11.9081	—0.218
	11.8206	4.9920	12.0094	0.140
	11.8873	4.9616	12.6761	0.486

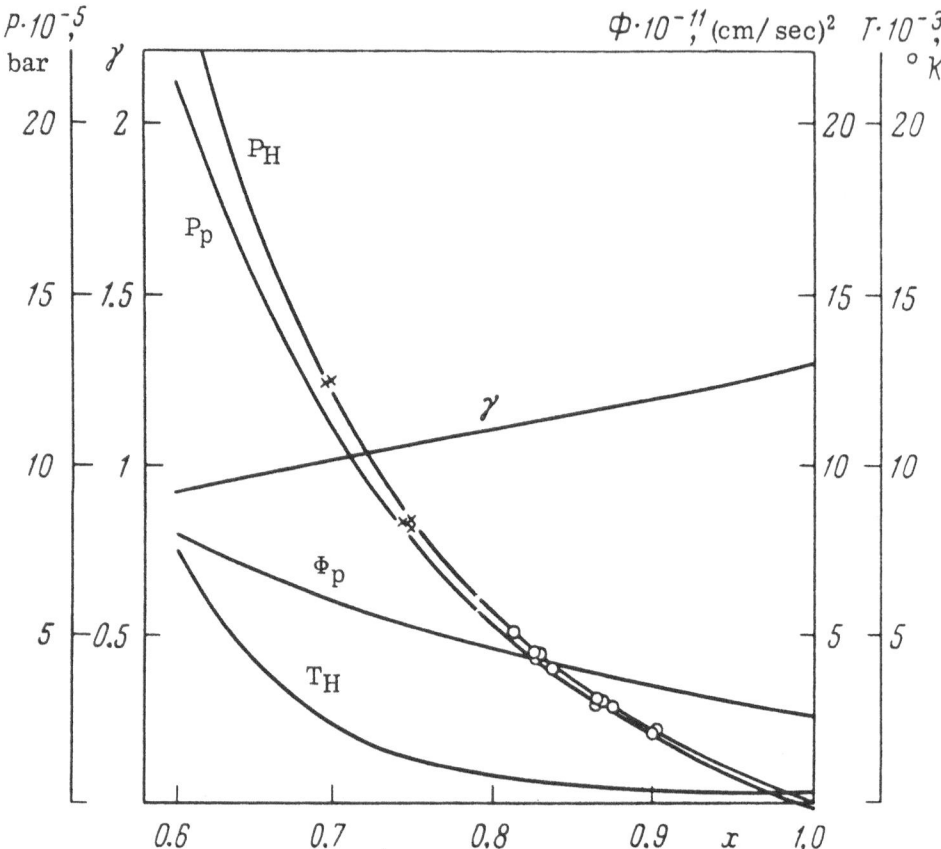

Fig. 5.17. Equation of state of vanadium: $\circ$ — P_{He} points from [8], used in calculations; $\times$ — other P_{He} points from [8].

The potential pressure P_p was calculated using Eq. (1.9), the Grüneisen parameter γ was found by substituting m = 1 in Eq. (4.3), and the quantity $\Phi_p = K_p/\rho$ (K_p is the potential term in the bulk modulus) was found using Eq. (1.11). The Hugoniot curve P_H and the temperature along this curve T_H were calculated employing formulas (5.75) and (5.77). Whenever experimental points were avilable in the pressure range above ~500 kbar, they were also included in these figures.

It is evident from Figs. 5.17–5.33 that the theoretical Hugoniot curves usually agree well with the experimental curves. The only exception is lead, whose temperature increases rapidly along the Hugoniot curve (Fig. 5.21) and whose theoretical Hugoniot curve deviates considerably from the experimental points. This can be explained by the contribution of the thermally excited conduction electrons. Qualitatively, this follows from Eq. (5.111) and from Table 5.1.

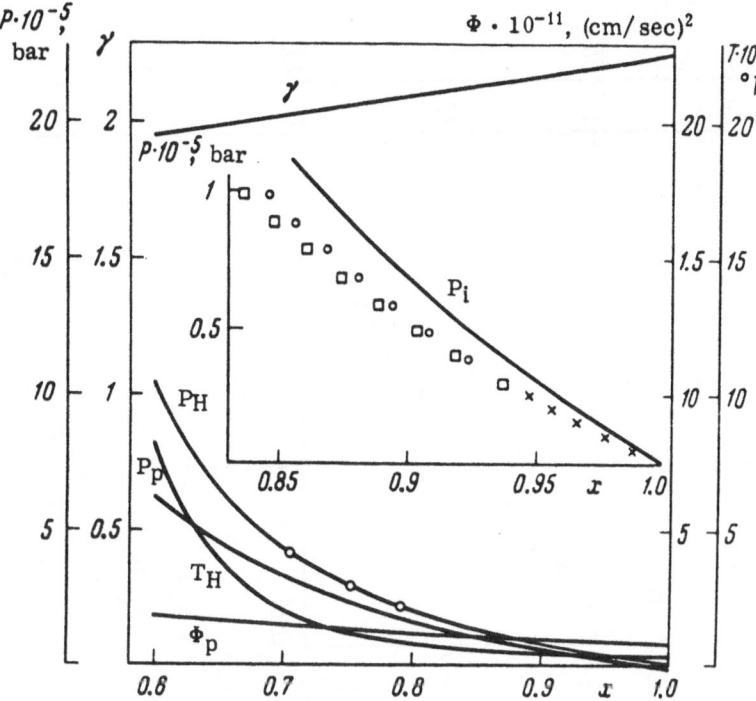

Fig. 5.18. Equation of state of indium: O — P_{He} points from [7], used in calculations; $\times$ — P_{ie} points from [9] of Chapter 4; O — from [16]; $\square$ — from [7] of Chapter 4.

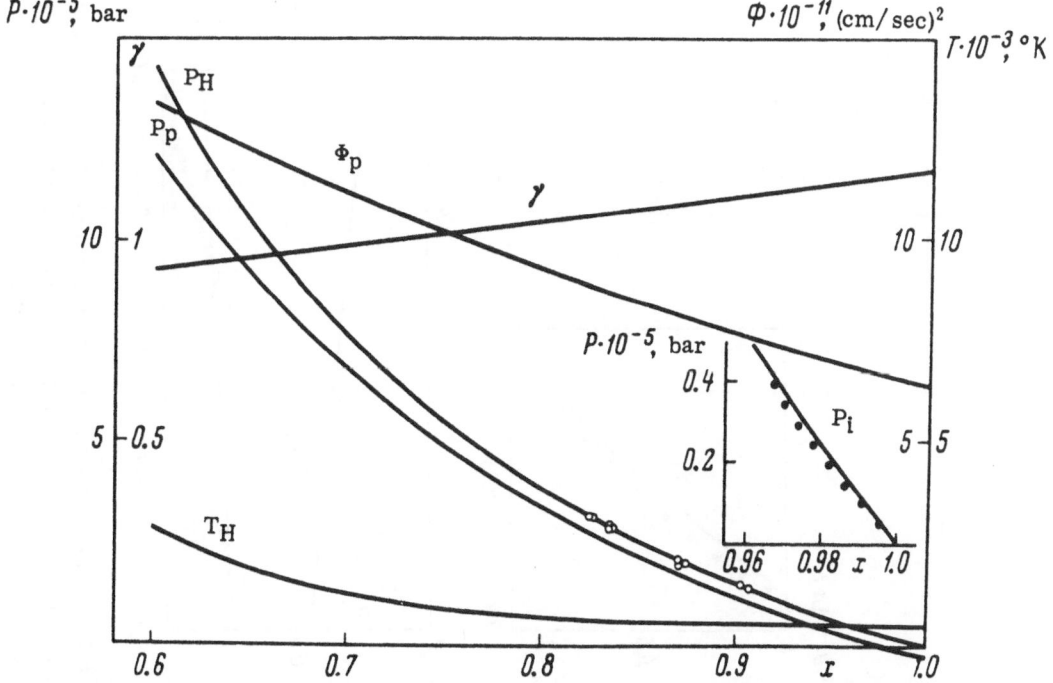

Fig. 5.19. Equation of state of beryllium: O — P_{He} points from [7], used in calculations; $\bullet$ — P_{ie} points from [19] of Chapter 4.

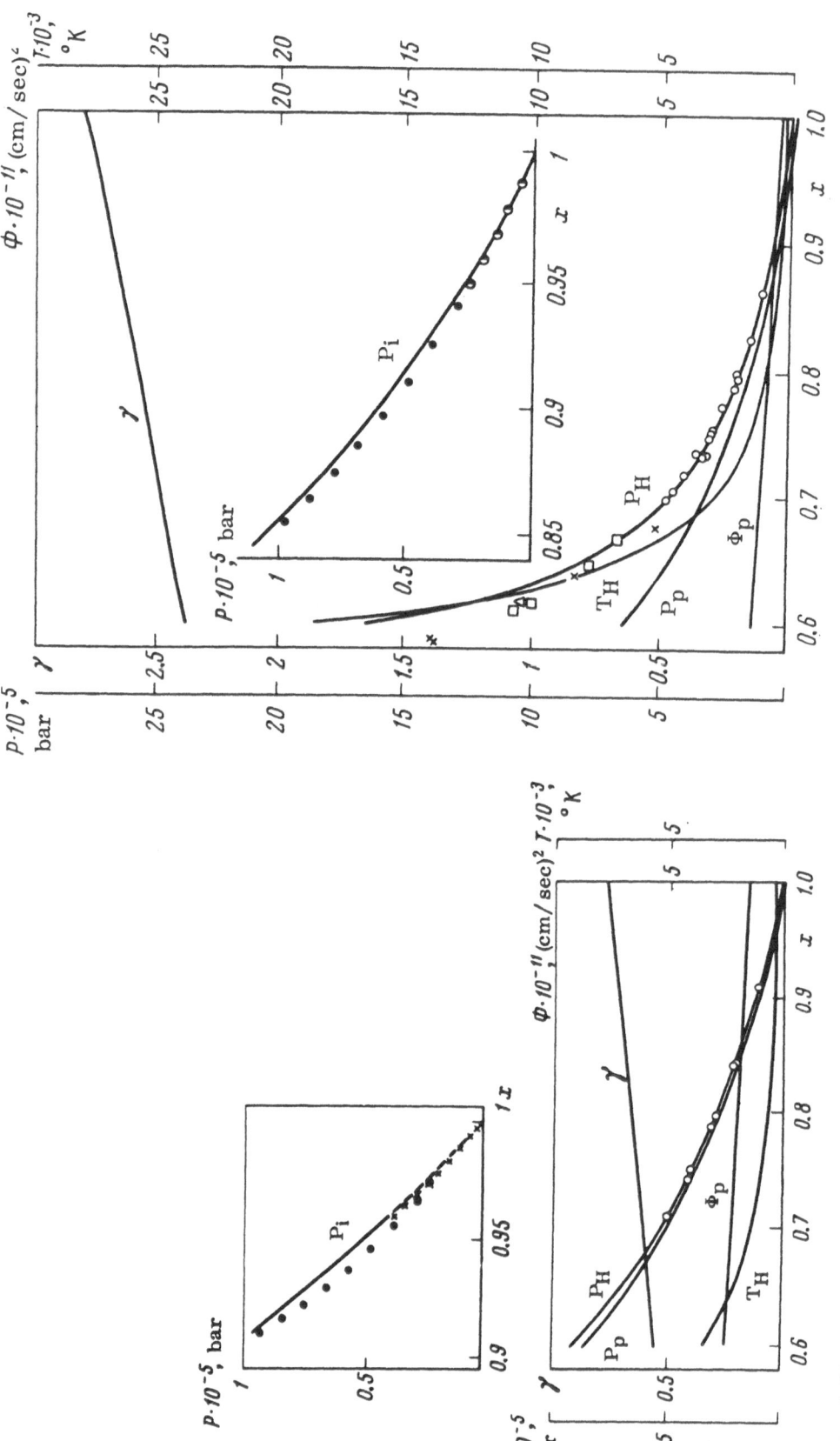

Fig. 5.21. Equation of state of lead: $\bigcirc$ — P_{He} points from [7], used in calculations; $\times$ — P_{He} points from [8], $\triangle$ — from [11], $\square$ — from [12]; $\bullet$ — P_{ie} points from [8] of Chapter 4, $\bigcirc$ — from [9] of Chapter 4.

Fig. 5.20. Equation of state of zirconium: $\bigcirc$ — P_{He} points from [7], used in calculations; $\times$ — P_{ie} points from [19] of Chapter 4, $\bullet$ — from [10] of Chapter 4.

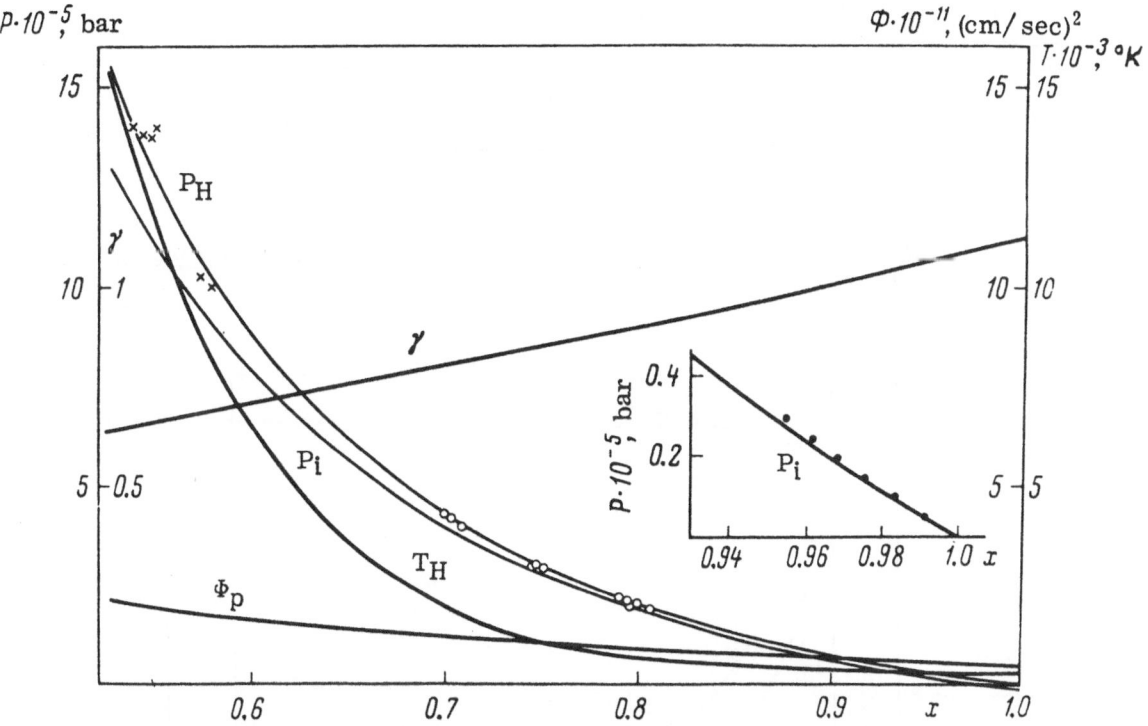

Fig. 5.22. Equation of state of thorium: $\bigcirc$ — P_{He} points from [7, 8] used in calculations; $\times$ — P_{He} points from [8]; $\bullet$ — P_{ie} points from [21] of Chapter 4.

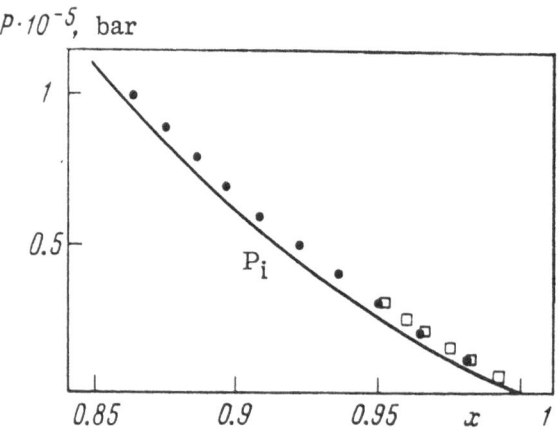

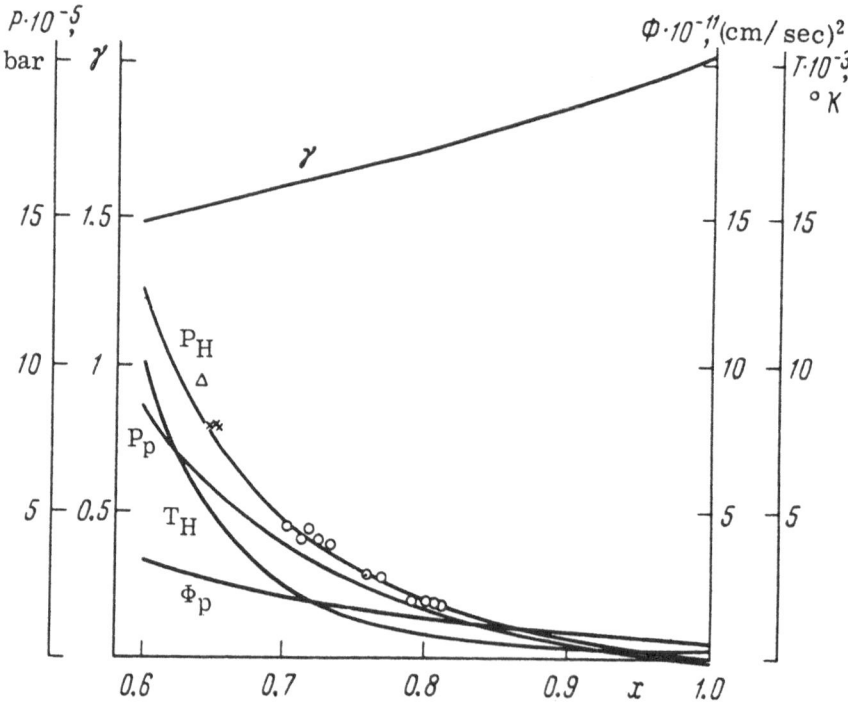

Fig. 5.23. Equation of state of tin: $\bigcirc$ — P_{He} points from [7], used in calculations; $\times$ — P_{He} points from [8], $\triangle$ — from [10]; $\bullet$ — P_{ie} points from [7] of Chapter 4, $\square$ — from [21] of Chapter 4.

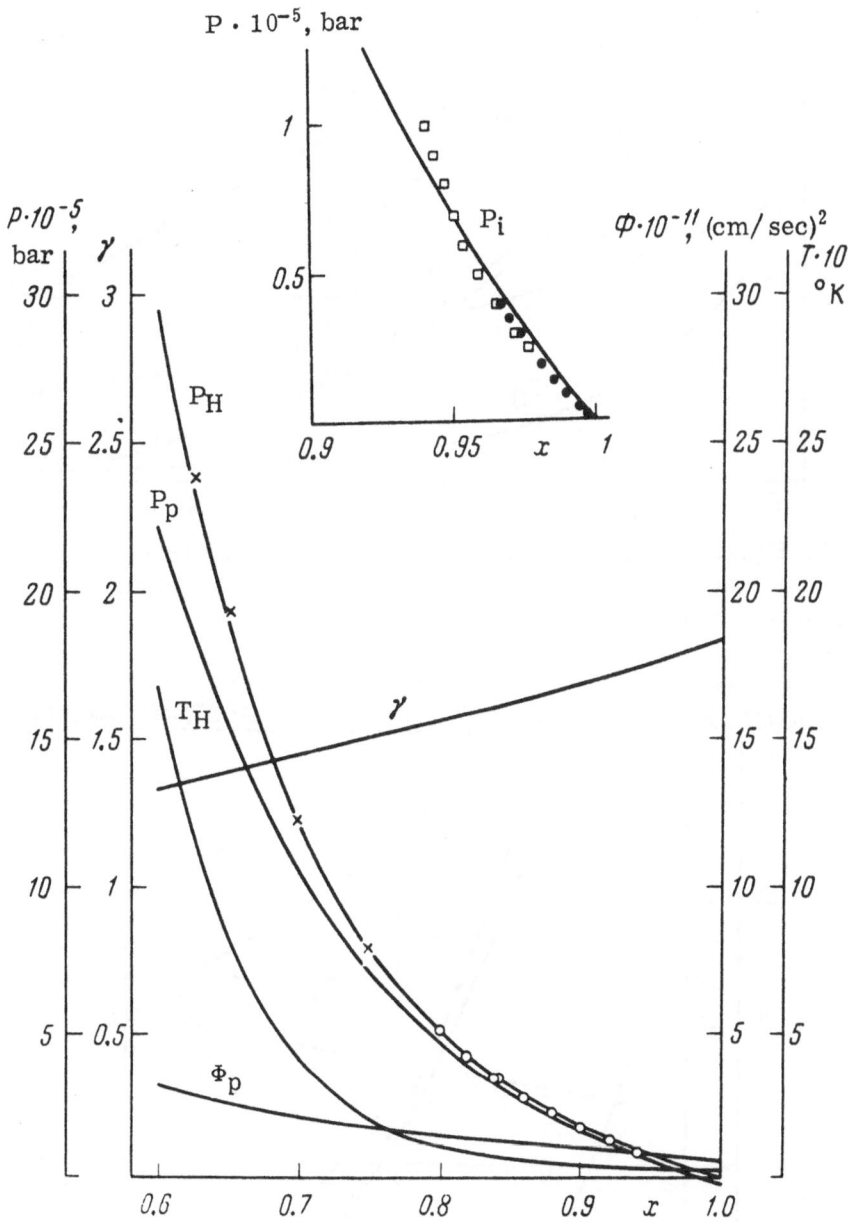

Fig. 5.24. Equation of state of uranium: $\bigcirc$ — P_{He} points from [45], used in calculations; $\times$ — other P_{He} points from [45]; $\bullet$ — P_{ie} points from [19] of Chapter 4, $\square$ — from [10] of Chapter 4.

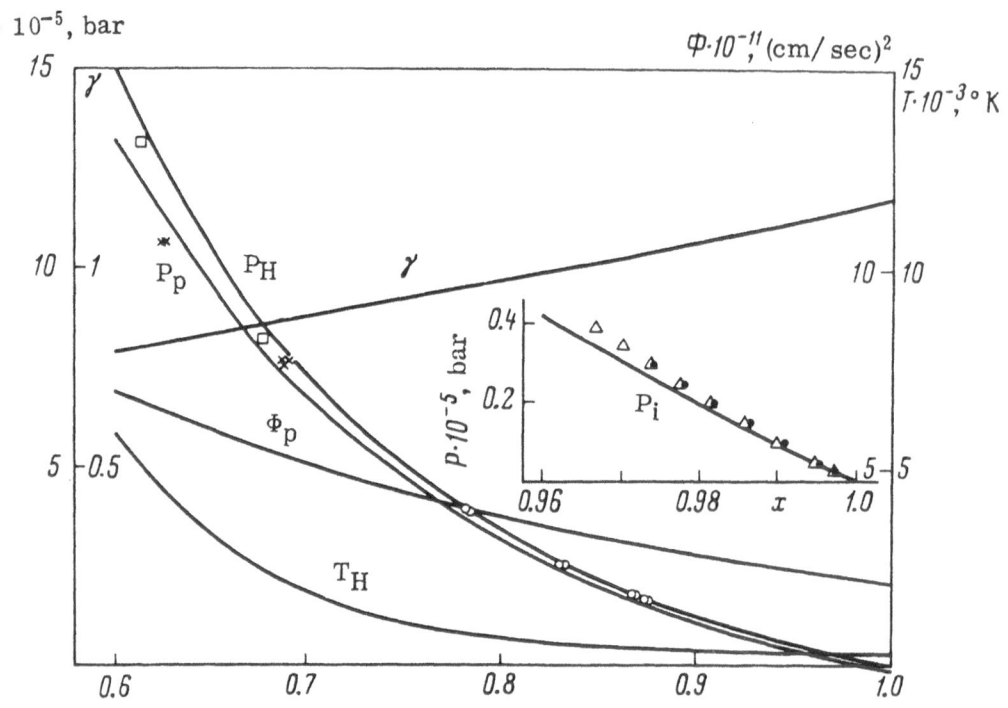

Fig. 5.25. Equation of state of titanium: O — P_{He} points from [7], used in calculations; $\times$ — P_{He} points from [8], $\square$ — from [52]; $\bullet$ — P_{ie} points from [21] of Chapter 4, $\triangle$ — from [19] of Chapter 4.

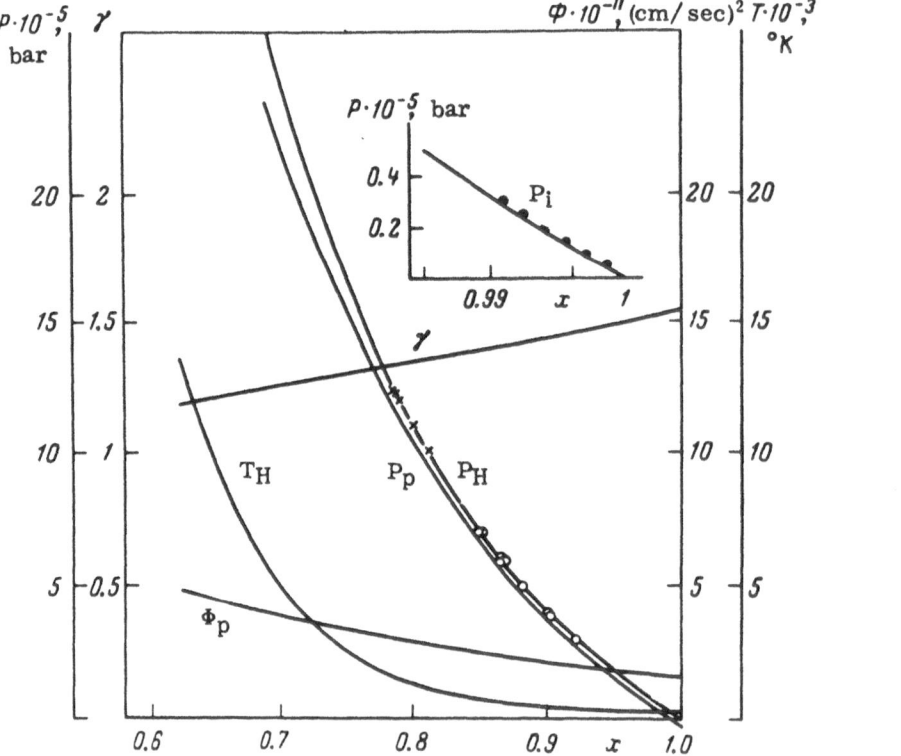

Fig. 5.26. Equation of state of tungsten: O — P_{He} points from [8], used in calculations; $\times$ — other P_{He} points from [8]; $\bullet$ — P_{ie} points from [21] of Chapter 4.

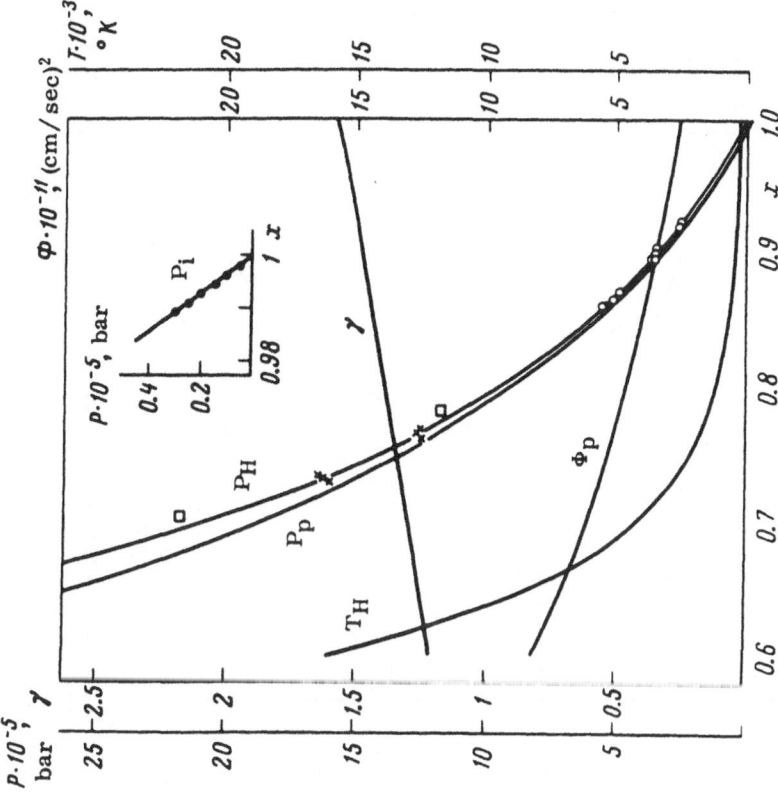

Fig. 5.28. Equation of state of molybdenum: $\bigcirc$ — P_{He} points from [7], used in calculations; $\times$ — P_{He} points from [8]; $\square$ — from [52]; $\bullet$ — P_{ie} points from [21] of Chapter 4.

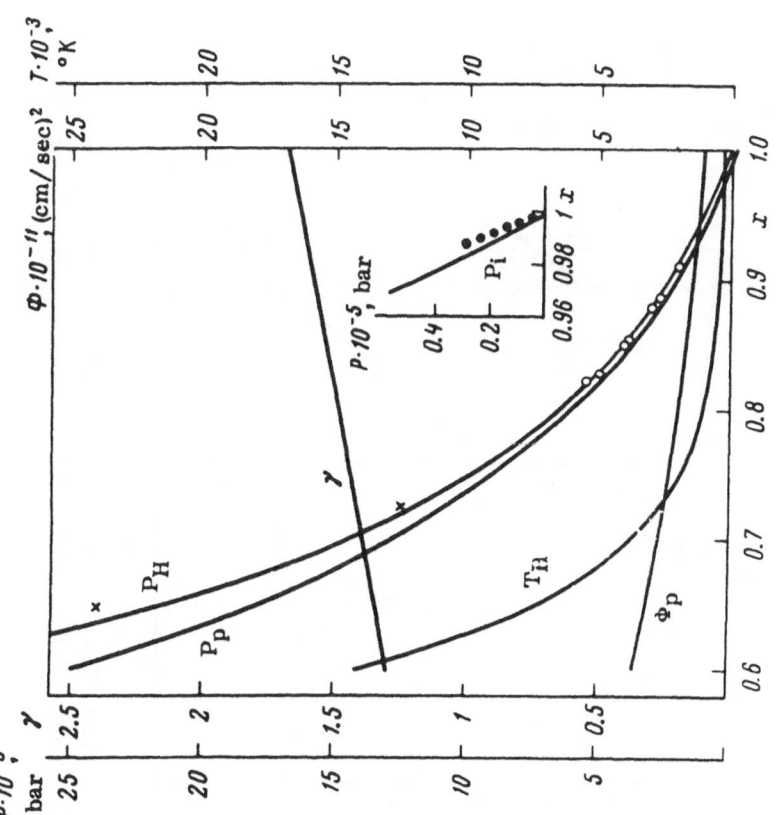

Fig. 5.27. Equation of state of tantalum: $\bigcirc$ — P_{He} points from [7], used in calculations; $\times$ — P_{He} points from [52]; $\bullet$ — P_{ie} points from [21] of Chapter 4.

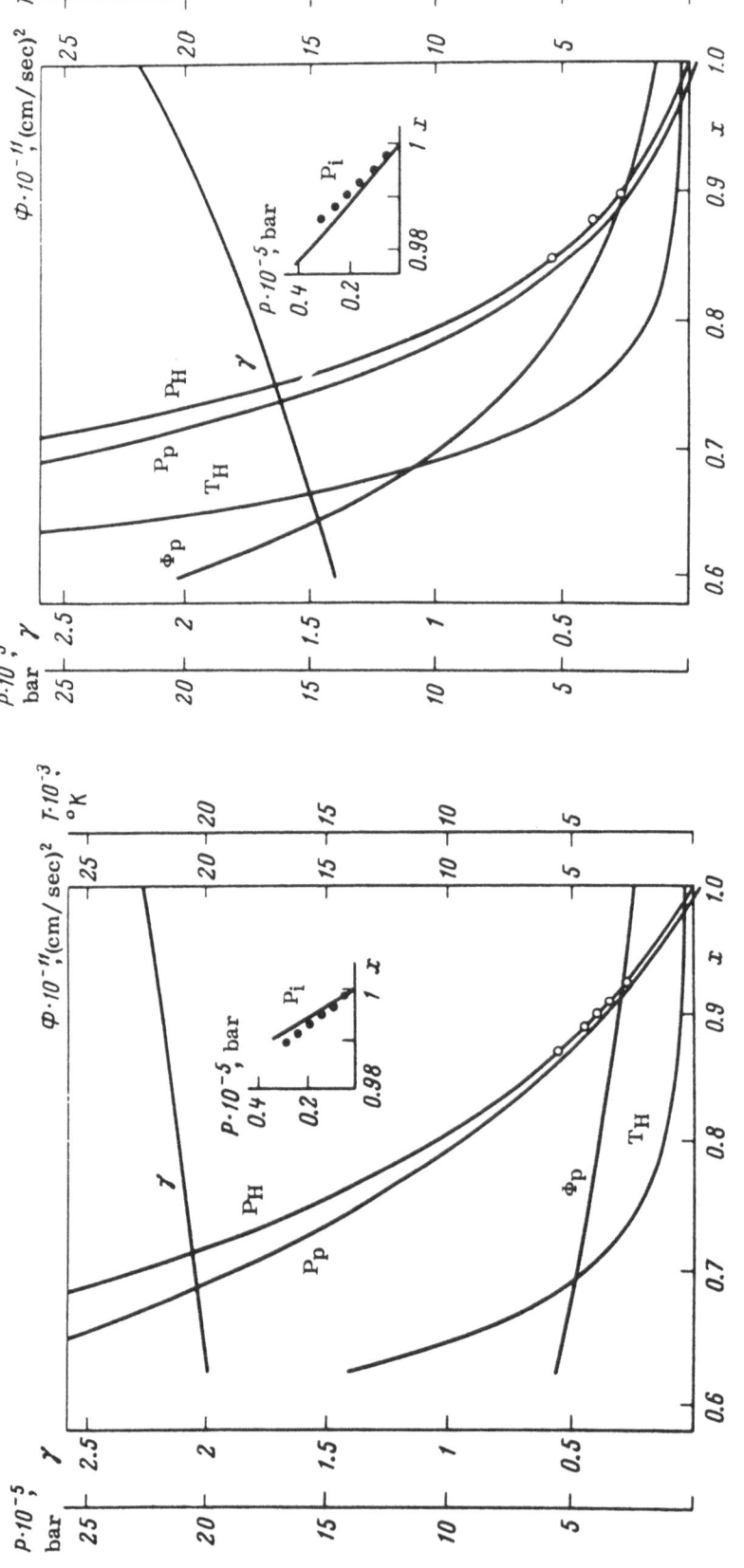

Fig. 5.30. Equation of state of palladium: ○ — P_{He} points from [7], used in calculations; ● — P_{ie} points from [21] of Chapter 4.

Fig. 5.29. Equation of state of rhodium: ○ — P_{He} points from [7], used in calculations; ● — P_{ie} points from [21] of Chapter 4.

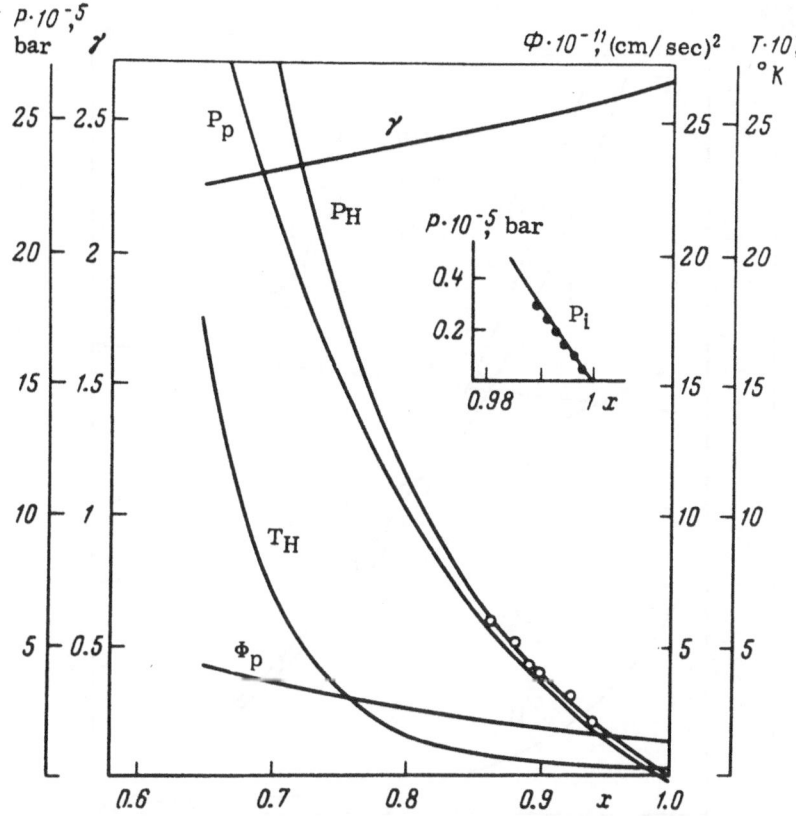

Fig. 5.31. Equation of state of platinum: ○ — P_{He} points from [7], used in calculations; ● — P_{ie} points from [21] of Chapter 4.

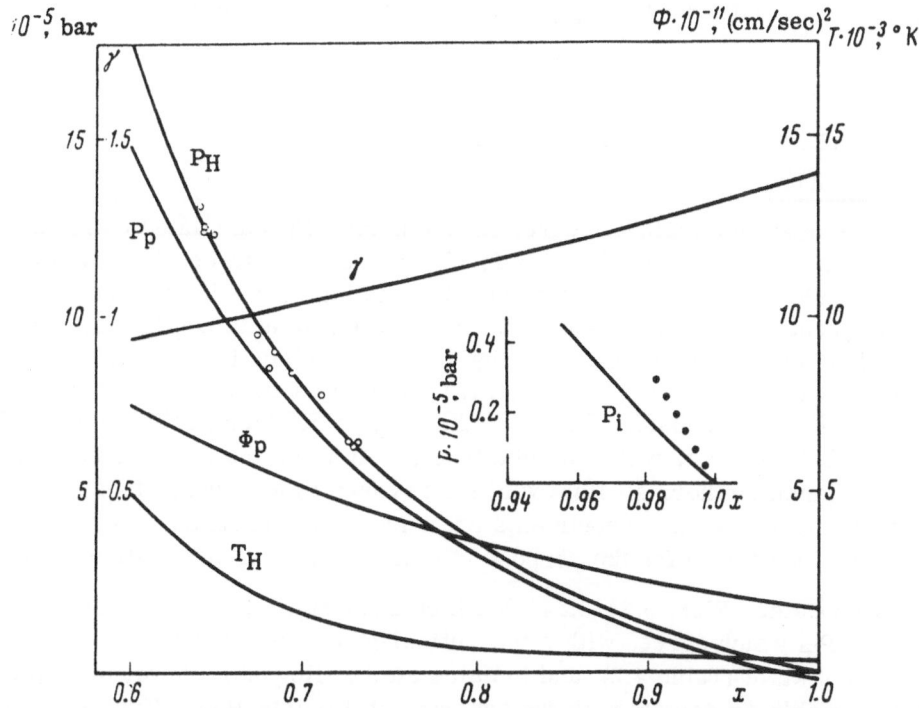

Fig. 5.32. Equation of state of magnetite; ○ — P_{He} points from [63], used in calculations; ● — P_{ie} points from [21] of Chapter 4.

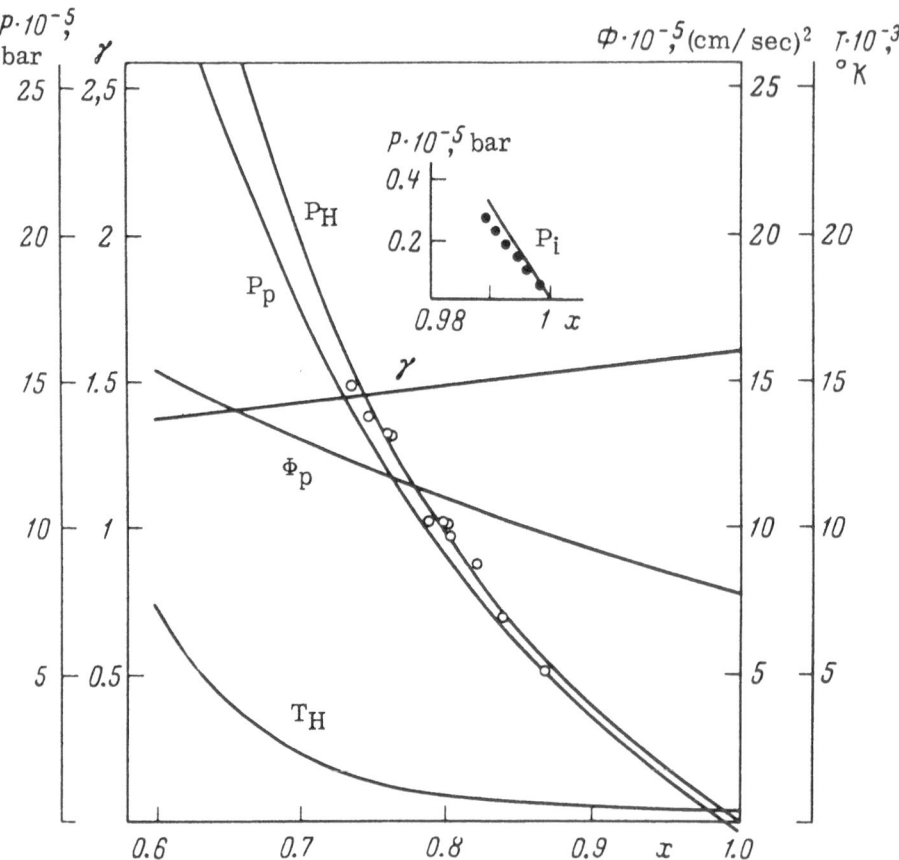

Fig. 5.33. Equation of state of corundum: $\bigcirc-$ P_{He} points from [63], used in calculations; $\bullet-$ P_{ie} points from [21] of Chapter 4.

5.7. Equations of State from Static

and Dynamic Data

Figures 5.18–5.44 also include theoretical isotherms P_i, calculated using the parameters A, b, K, and δ taken from Table 5.3 for m = 1. Each of these figures also gives the experimental points alongside the theoretical curves. At first sight, the discrepancy between the theory and experiment is — for the majority of substances — quite negligible. However, when the experimental points P_{ie}, represented by the various symbols in Figs. 5.18–5.33, are used to determine the equation of state (Chapter 4), it is found that the corresponding theoretical Hugoniot curves differ very considerably from the P_{He} points included in the same figures. Thus, we can see that an equation of state obtained solely from the dynamic data describes quite well the Bridgman isotherms. However, the reverse is not always true. This is because changes of the volume x in static experiments are small and measurements must be highly accurate before they can be used for the purpose of extrapolation of an equation of state.

Let us now consider Figs. 5.34–5.44. Each of these figures has at least one set of experimental points P_{ie} which agrees with a theoretical P_i curve. Using these points to determine the equation of state (Table 4.2), and to calculate the Hugoniot curve, we obtain results which are in reasonable agreement with the experimental points P_{He}. Thus, for each of the substances in these figures, we have two equations of state, each of which describes well the experimental data in the range of its validity.

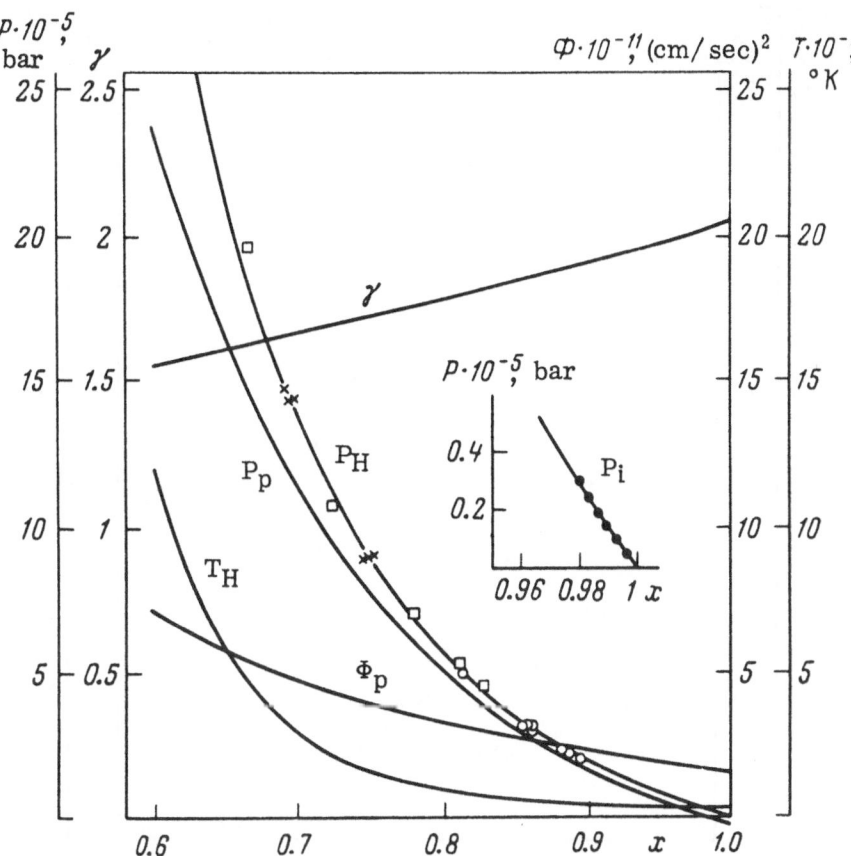

Fig. 5.34. Equation of state of copper: ● — P_{ie} points from [20] of Chapter 4; ○ — P_{He} points from [7, 8], used in calculations; × — P_{He} points from [8], □ — from [11, 12].

We recall that the dynamic data in the lower part of the Hugoniot curve, as well as the static data, describe basically the potential part of the equation of state. Therefore, it is reasonable to use both the dynamic and the static data in order to find the optimum values of the parameters in an expression for the potential. The method used in such calculations is a combination of the methods described in Secs. 4.3 and 5.6. In other words, the parameters A and b are found from the minimum of the function

$$M + \mathscr{M} = \sum_{i=1}^{n}[P_{e_i} - P(x_i)]^2 \frac{1}{\sigma_i^2} + \sum_{j=1}[P_{He_j} - P_H(x_j)]^2 \frac{1}{\sigma_j^2} \qquad (5.126)$$

The other stages of the calculation have been described earlier (see Appendix D). In particular, the parameter K is again found from

$$K = A + \rho_0\gamma_0 \frac{R}{\mu}\left[\frac{9}{8}\theta_0 + 3T_0 D\left(\frac{\theta_0}{T_0}\right)\right].$$

The values of the parameters calculated by this combined method are listed in Table 5.4. We start with the same experimental data as those employed to obtain the results listed in Tables 4.2 and 5.3. Table 5.4 lists three sets of values of the parameters for each substance: the top row corresponds to m = 0 in Eq. (2.45), the middle row to m = 1, and the bottom row to m = 2.

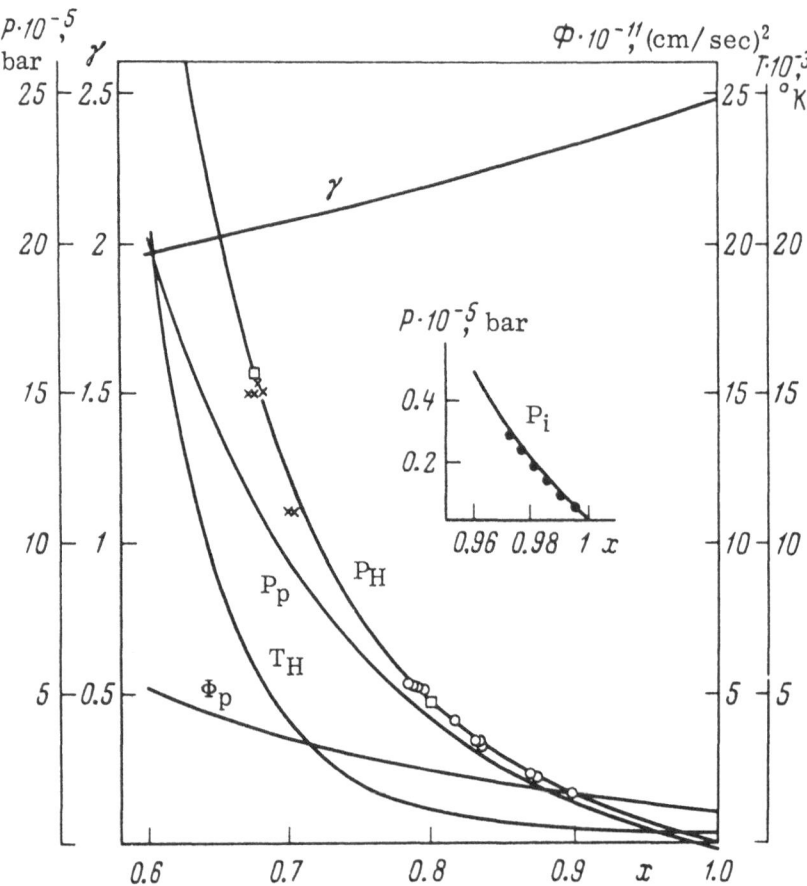

Fig. 5.35. Equation of state of silver: • — P_{ie} points from [21] of Chapter 4; ○ — P_{He} points from [7, 8], used in calculations; × — P_{He} points from [8], □ — from [10].

Figures 5.34–5.46 give the theoretical values of the same quantities as Figs. 5.17–5.33. These quantities are obtained using the parameters A, b, K, and δ taken from Table 5.4 (middle row of values corresponding to m = 1). The correctness of the obtained equations of state can be judged by the agreement between the theoretical Hugoniot curve and the experimental points P_{He}. These points are usually found in the pressure range above 500 kbar and they are the points which had not been used in the calculation of the theoretical Hugoniot curve. In the majority of cases, the agreement is so good that the high-temperature corrections to the equation of state cannot be determined from the experimental data. These corrections would require experimental data obtained at still higher pressures for which the temperatures on the Hugoniot curves are very high. Very few such measurements have yet been carried out and the determination of the high-temperature corrections to the equation of state is a matter for the future.

The parameters A, b, K, and δ listed in Tables 5.3 and 5.4 for m = 0 and m = 2 shift the P_i and P_p curves only slightly, within the limits of the error of the experimental data P_{He}.

TABLE 5.4

Substance	$A \cdot 10^{-3}$, bar	b	$K \cdot 10^{-3}$, bar	δ	Substance	$A \cdot 10^{-3}$, bar	b	$K \cdot 10^{-3}$, bar	δ
Noble metals					Pb	1.9633	8.7562	2.0750	0.649
						2.1555	8.1366	2.2672	1.125
Cu	5.6053	9.5524	5.8261	—0.211		2.3856	7.5548	2.4973	1.581
	5.5819	9.5021	5.8026	0.153					
	5.6002	9.4248	5.8210	0.515	Transition elements				
Ag	3.1328	12.0830	3.3132	—0.228	Ti	9.7080	5.2351	9.7959	—0.267
	3.1495	11.9180	3.3299	0.163		9.7676	5.1966	9.8555	0.081
	3.1864	11.7287	3.3668	0.549		9.8310	5.1672	9.9189	0.424
Au	5.8698	10.9810	6.0911	0.557					
	5.9159	10.8552	6.1372	0.931	Zr	9.5491	4.7875	9.5908	—0.579
	5.9753	10.7302	6.1966	1.300		9.7368	4.7222	9.7784	—0.229
						9.8542	4.6852	9.8959	0.114
Divalent metals									
					Nb	12.6719	6.1079	12.7907	0.065
Be	4.8040	8.8590	5.0693	—0.978		12.7746	6.0594	12.8933	0.414
	4.5354	9.0593	4.8008	—0.648		12.8657	6.0223	12.9845	0.758
	4.4109	9.1189	4.6762	—0.303					
					Fe I	26.3094	3.9593	26.4991	0.514
Mg	2.0861	7.2015	2.1666	—0.383		26.3796	3.9454	26.5693	0.858
	2.1448	6.9851	2.2254	0.015		26.4000	3.9411	26.5897	1.195
	2.2166	6.7872	2.2972	0.398					
					Co	8.0309	9.2461	8.2669	—0.199
Zn	2.1023	11.1411	2.2982	—0.192		8.0000	9.2138	8.2361	0.156
	2.1130	10.9190	2.3078	0.230		8.0160	9.1603	8.2520	0.510
	2.1516	10.6433	2.3475	0.645					
					Ni	8.1912	9.0337	8.4188	—0.241
Cd	1.2509	13.3360	1.3795	—0.679		8.1740	8.9905	8.4016	0.116
	1.2554	13.0634	1.3840	—0.248		8.2020	8.9293	8.4296	0.470
	1.2787	12.7098	1.4073	0.183					
					Pd	5.6651	11.9849	5.8516	—0.476
Trivalent metals						5.6386	11.9559	5.8250	—0.122
						5.6389	11.9012	5.8253	0.233
Al	3.0701	9.3207	3.2448	—0.098					
	3.1010	9.1416	3.2757	0.299	Minerals and rocks				
	3.1601	8.9456	3.3348	0.687	Fe$_3$O$_4$	68.0680	2.7555	68.2580	0.713
In	1.2516	11.5090	1.3563	—0.386		67.9612	2.7540	68.1511	1.051
	1.2886	11.1422	1.3932	0.054		64.5303	2.7932	64.7202	1.355
	1.3510	10.6853	1.4556	0.495					
					MgO	8.9809	7.7037	9.2232	—0.519
						9.0628	7.5862	9.3050	—0.148
Tetravalent metals						9.2018	7.4646	9.4441	0.217
Sn	3.0638	7.4340	3.1583	0.154	Olivine	18.7741	3.9718	18.9629	0.027
	3.1921	7.1736	3.2867	0.554		18.7756	3.9548	18.9644	0.375
	3.3059	6.9742	3.4004	0.935		18.7648	3.9504	18.9536	0.713

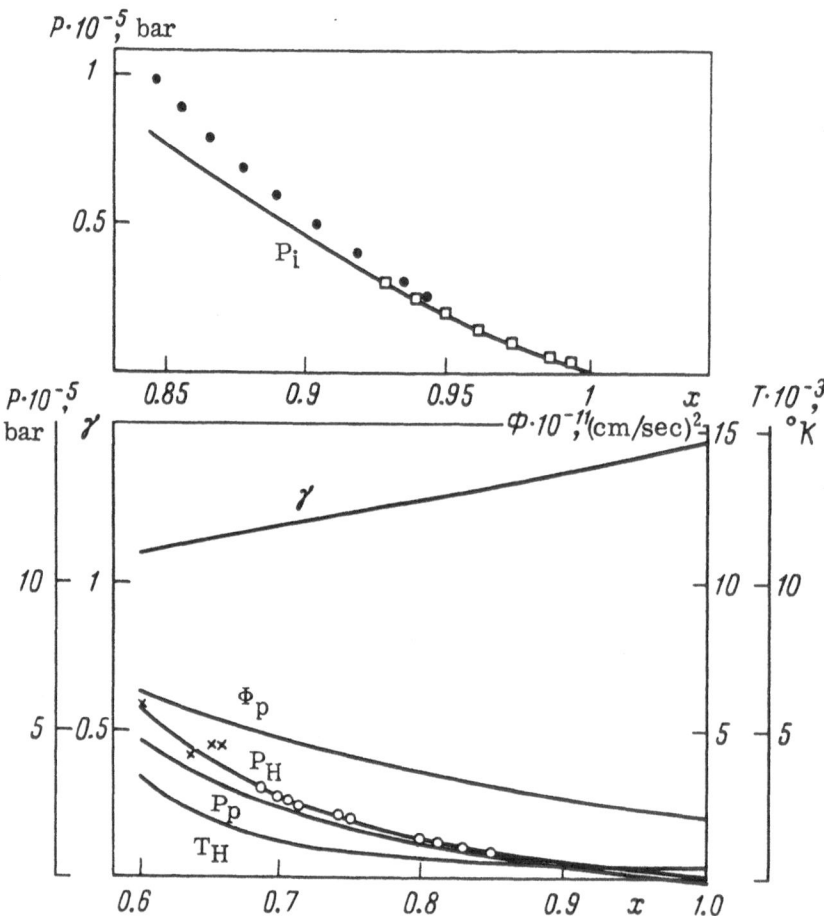

Fig. 5.36. Equation of state of magnesium: □ — P_{ie} points from Chapter 4,
● — from [10] of Chapter 4; ○ — P_{He} points from [7, 45], used in calculations;
× — P_{He} points from [45].

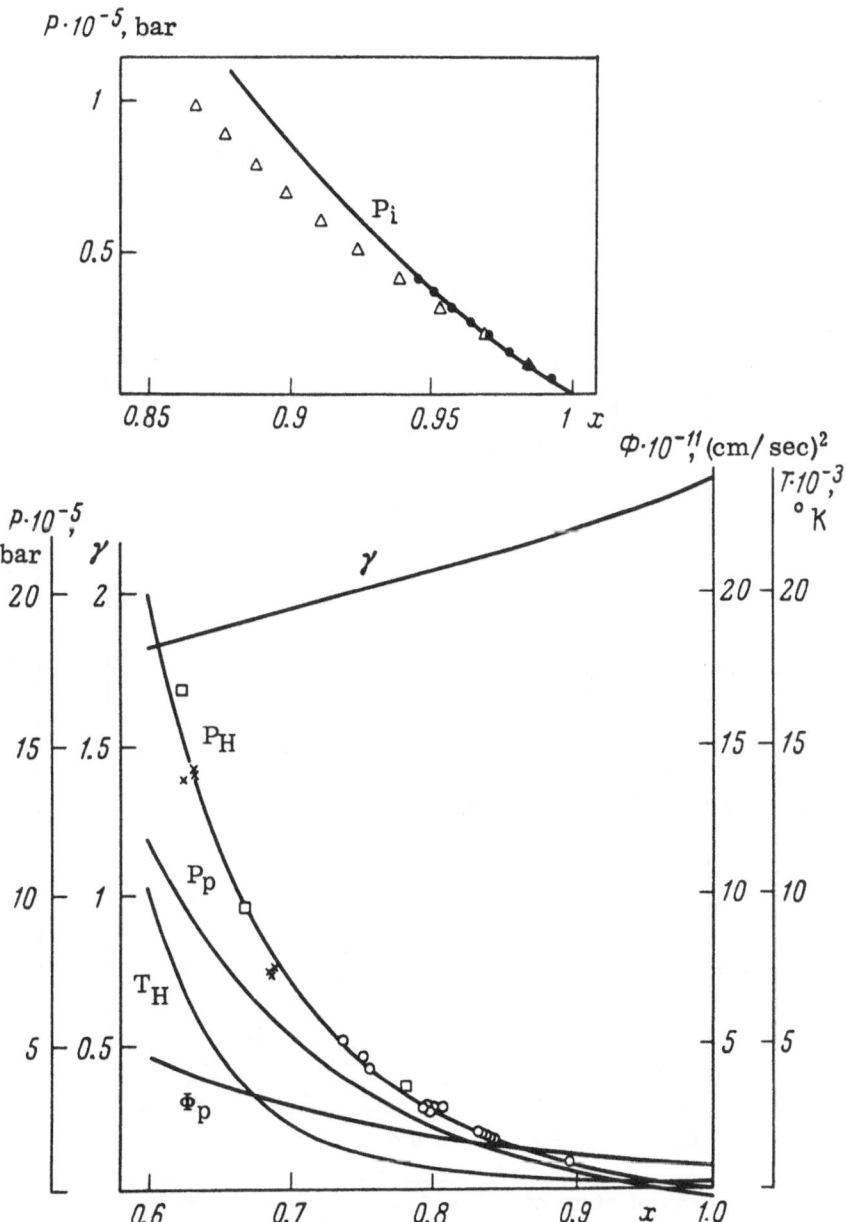

Fig. 5.37. Equation of state of zinc: $\bullet$ — P_{ie} points from [19, 21] of Chapter 4, $\triangle$ — from [7] of Chapter 4; $\bigcirc$ — P_{He} points from [7, 8], used in calculations; $\times$ — P_{He} points from [8], $\square$ — from [14].

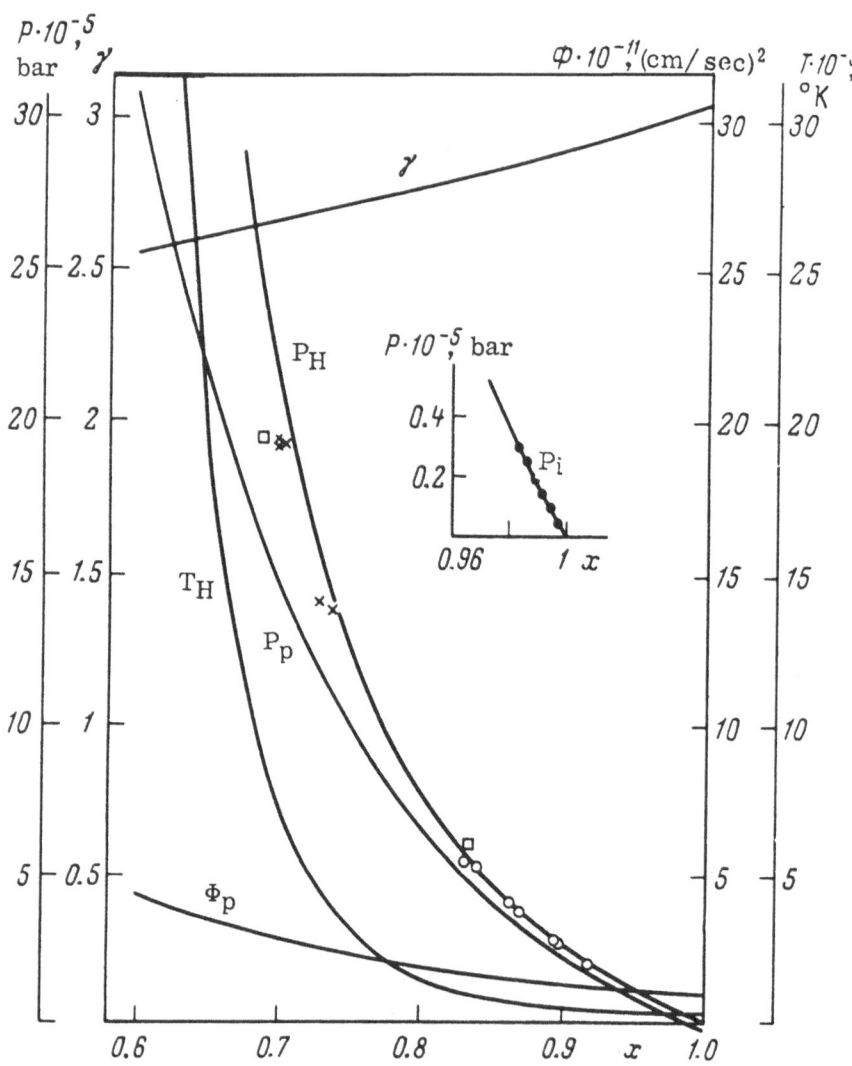

Fig. 5.38. Equation of state of gold: ● — P_{ie} points from [21] of Chapter 4; ○ — P_{He} points from [7, 8], used in calculations; × — P_{He} points from [8], ☐ — from [10].

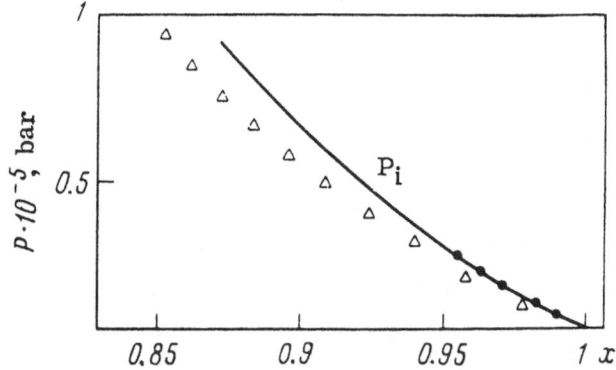

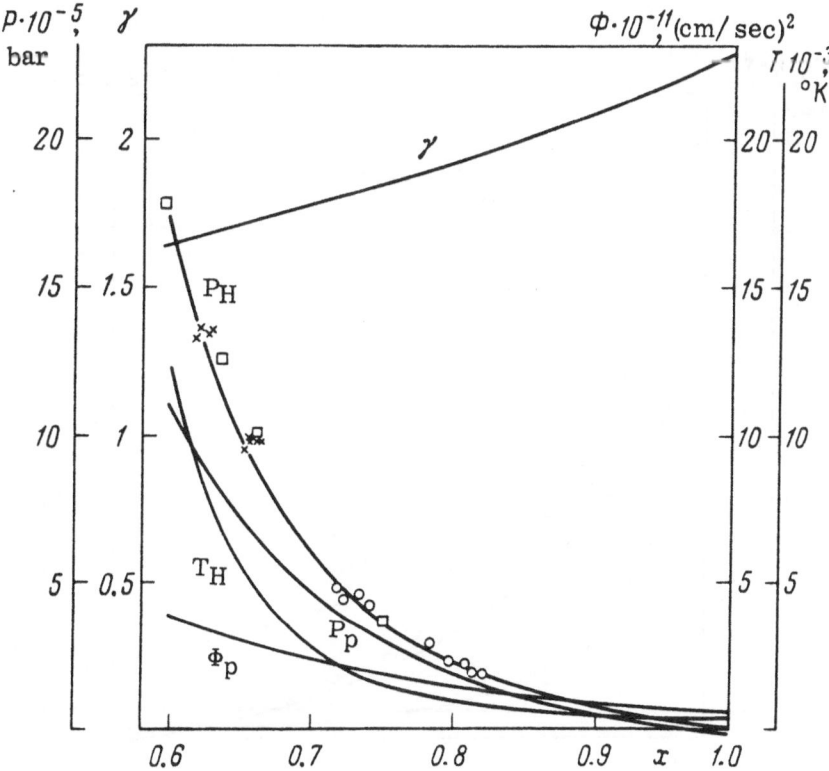

Fig. 5.39. Equation of state of cadmium: ● — P_{ie} points from [8] of Chapter 4, △ — from [7] of Chapter 4; ○ — P_{He} points from [7] used in calculations; × — P_{He} points from [8], □ — from [14].

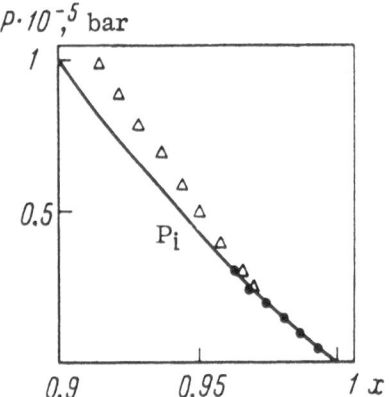

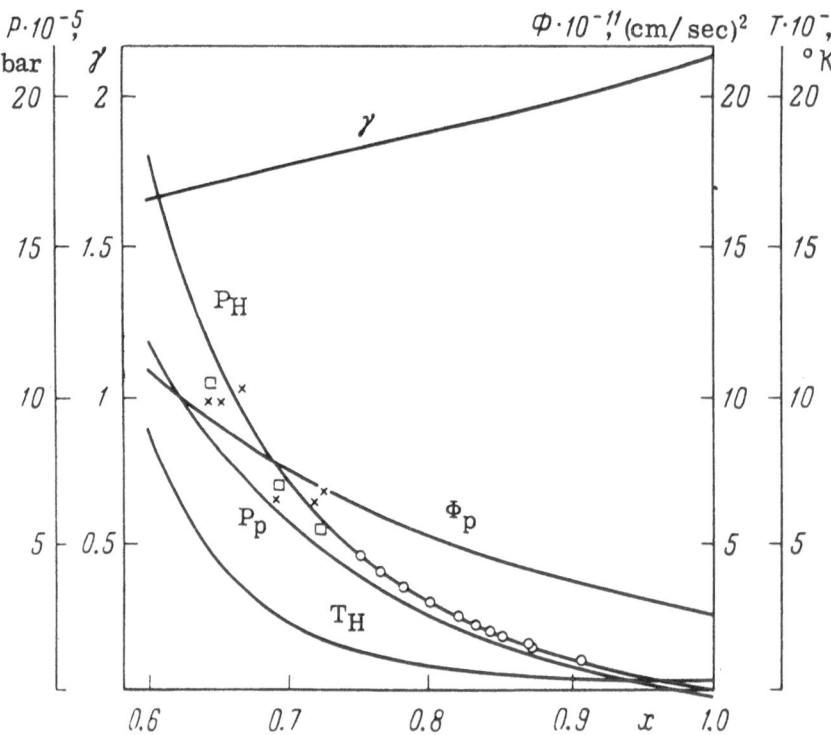

Fig. 5.40. Equation of state of aluminum: ● — P_{ie} points from [21] of Chapter 4, △ — from [10] of Chapter 4; ○ — P_{He} points from [7, 45], used in calculations; × — P_{He} points from [45], □ — from [11, 12].

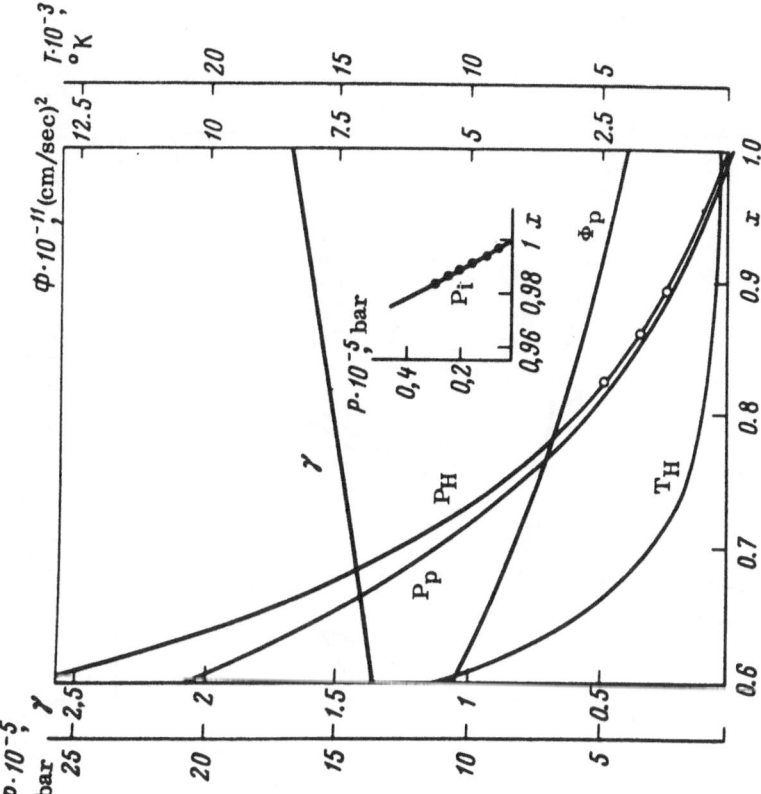

Fig. 5.42. Equation of state of niobium: ● − P_{ie} points from [21] of Chapter 4; ○ − P_{He} points from [7] used in calculations.

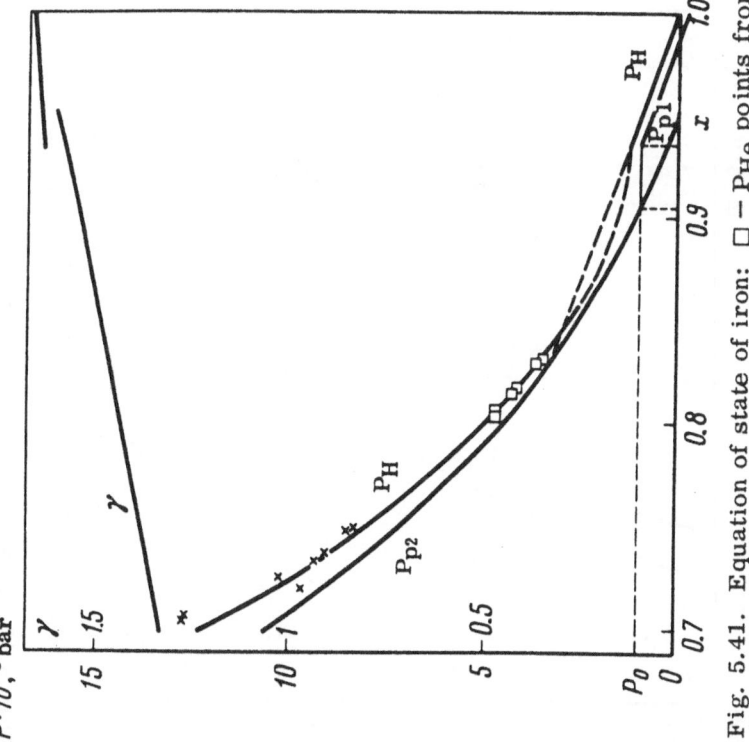

Fig. 5.41. Equation of state of iron: □ − P_{He} points from [7, 8] used in calculations; × − P_{He} points from [8].

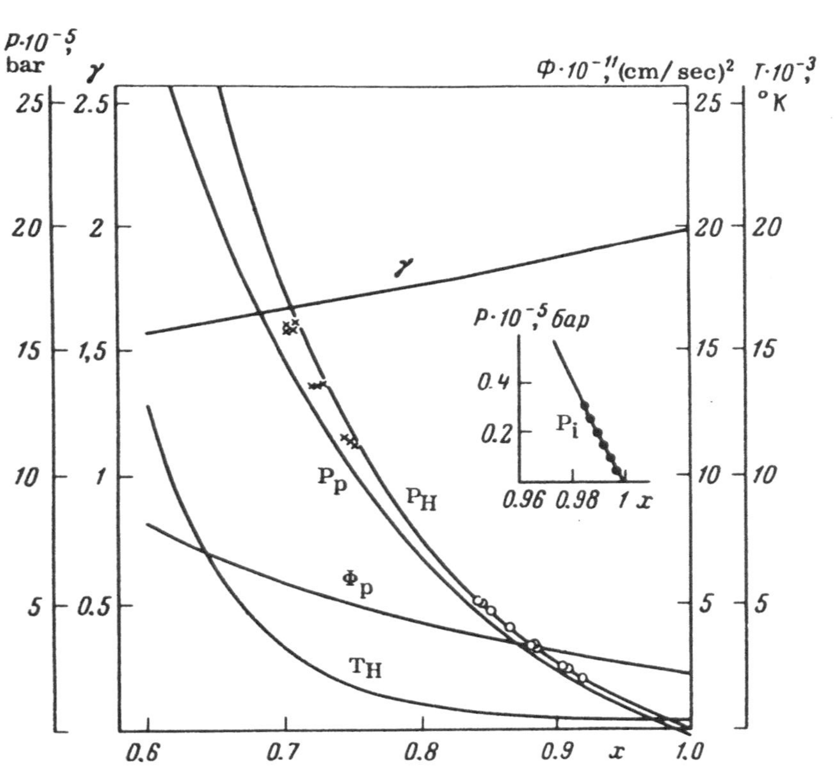

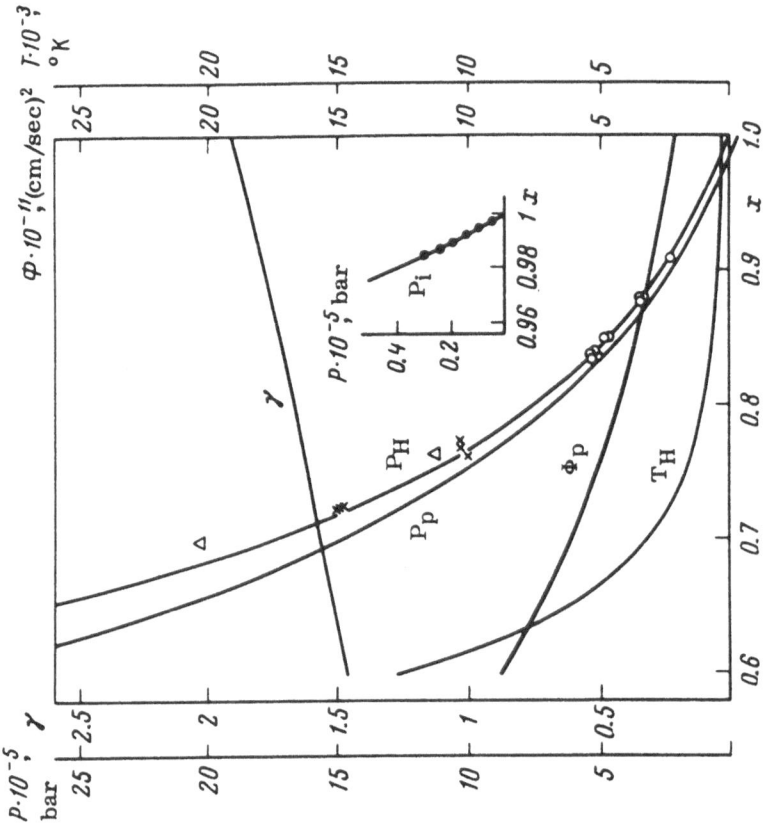

Fig. 5.44. Equation of state of nickel; ● — P_{ie} points from [21] of Chapter 4; ○ — P_{He} points from [7], used in calculations; × — P_{He} points from [8], △ — from [14].

Fig. 5.43. Equation of state of cobalt: ● — P_{ie} points from [21] of Chapter 4; ○ — P_{He} points from [7, 8] used in calculations; × — P_{He} points from [8].

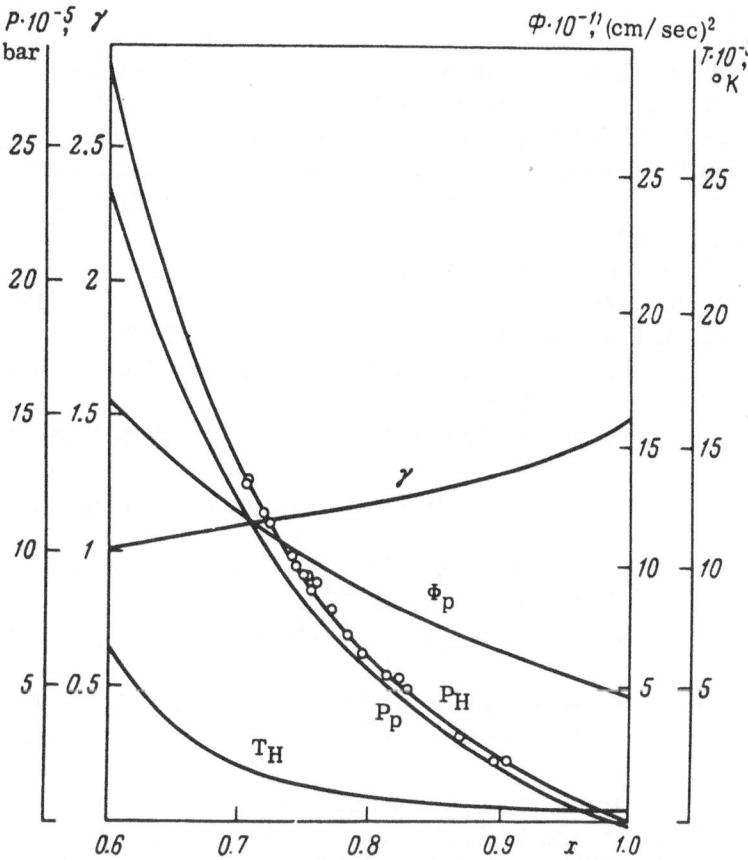

Fig. 5.45. Equation of state of periclase: $\bigcirc$ — P_{He} points from [63], used in calculations.

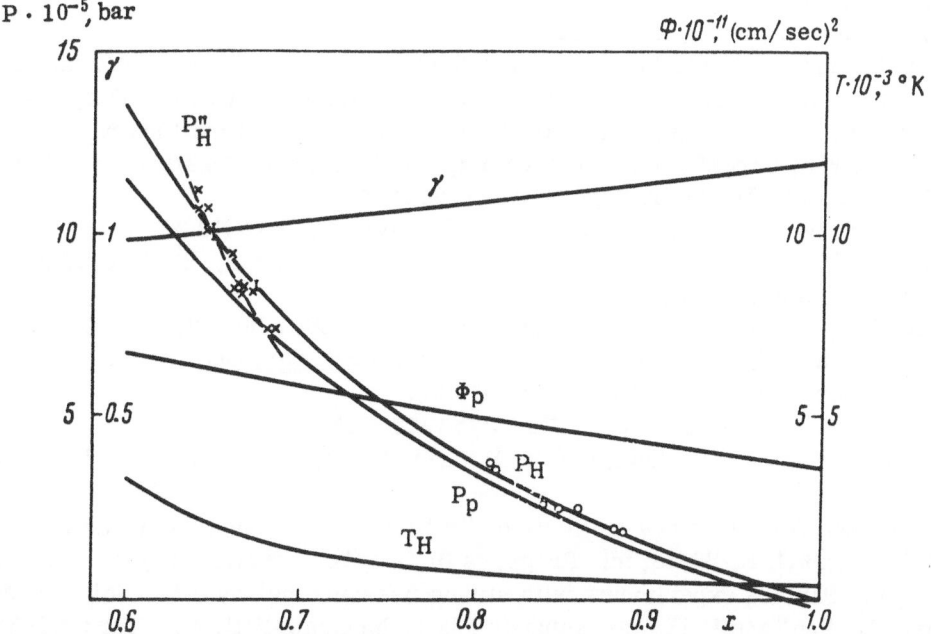

Fig. 5.46. Equation of state of olivine: $\bigcirc$ — P_{He} points from [63], used in calculations; $\times$ — P_{He} points from [63]; P_H^{π} — Hugoniot curve of the high-pressure phase.

Literature Cited

1. V. I. Davydov, Izv. Akad. Nauk SSSR, Ser. Geofiz., No. 12, p. 1411 (1956).
2. V. A. Kalinin, Trudy Inst. Fiziki Zemli Akad. Nauk SSSR, No. 11(178), p. 67 (1960).
3. V. N. Zharkov and V. A. Kalinin, Dokl. Akad. Nauk SSSR, 135:811 (1960).
4. V. N. Zharkov and V. A. Kalinin, Dokl. Akad. Nauk SSSR, 145:551 (1962).
5. V. N. Zharkov, in: Solids at Pressures and Temperatures in Earth's Interior [in Russian], Nauka, Moscow (1964), p. 41.
6. J. M. Walsh and R. H. Christian, Phys. Rev., 97:1544 (1955).
7. M. H. Rice, R. G. McQueen, and J. M. Walsh, Solid State Phys., 6:1 (1958).
8. R. G. McQueen and S. P. Marsh, J. Appl. Phys., 31:1253 (1960).
9. L. V. Al'tshuler, K. K. Krupnikov, B. I. Ledenev, V. I. Zhuchikhin, and M. I. Brazhnik, Zh. Éksp. Teor. Fiz., 34:874 (1958).
10. L. V. Al'tshuler, K. K. Krupnikov, and M. I. Brazhnik, Zh. Éksp. Teor. Fiz., 34:886 (1958).
11. L. V. Al'tshuler, S. B. Kormer, A. A. Bakanova, and R. F. Trunin, Zh. Éksp. Teor. Fiz., 38:790 (1960).
12. L. V. Al'tshuler, S. B. Kormer, M. I. Brazhnik, L. A. Vladimirov, M. P. Speranskaya, and A. I. Funtikov, Zh. Éksp. Teor. Fiz., 38:1061 (1960).
13. L. V. Al'tshuler, L. V. Kuleshova, and M. N. Pavlovskii, Zh. Éksp. Teor. Fiz., 39:16 (1960).
14. L. V. Al'tshuler, A. A. Bakanova, and A. F. Trunin, Zh. Éksp. Teor. Fiz., 42:91 (1962).
15. L. V. Al'tshuler, M. N. Pavlovskii, L. V. Kuleshova, and G. V. Simakov, Fiz. Tverd. Tela, 5:279 (1963).
16. S. B. Kormer and V. D. Urlin, Dokl. Akad. Nauk SSSR, 131:542 (1960).
17. S. B. Kormer, V. D. Urlin, and L. T. Popova, Fiz. Tverd. Tela, 3:2131 (1961).
18. S. B. Kormer, A. I. Funtikov, V. D. Urlin, and A. N. Kolesnikova, Zh. Éksp. Teor. Fiz., 42:686 (1962).
19. Ya. B. Zel'dovich and Yu. P. Raizer, Physics of Shock Waves and High-Temperature Hydrodynamic Phenomena, Vol. 2, Academic Press, New York (1966), Ch. XI.
20. V. P. Koryavov, Zh. Prikl. Mekh. Tekhn. Fiz., No. 5, p. 123 (1964).
21. L. D. Landau and E. M. Lifshitz, Fluid Mechanics, Pergamon Press, London (1959).
22. R. Courant and K. O. Friedrichs, Supersonic Flow and Shock Waves, Wiley, New York (1957).
23. A. H. Cottrell, Dislocations and Plastic Flow in Crystals, Oxford University Press (1953).
24. H. G. van Bueren, Imperfections in Crystals, North-Holland, Amsterdam (1960).
25. R. E. Duff and F. S. Minshall, Phys. Rev., 108:1207 (1957).
26. P. Johnson, B. Stein, and R. S. Davis, J. Appl. Phys., 33:557 (1962).
27. D. Bancroft, E. L. Peterson, and F. S. Minshall, J. Appl. Phys., 27:291 (1956).
28. J. Wackerle, J. Appl. Phys., 33:922 (1962).
29. V. N. Zharkov, Trudy Inst. Fiziki Zemli Akad. Nauk SSSR, No. 20(187), p. 3 (1962).
30. V. N. Zharkov and V. A. Kalinin, Izv. Akad. Nauk SSSR, Ser. Geofiz., No. 3, p. 298 (1962).
31. L. V. Al'tshuler, Usp. Fiz. Nauk, 85:197 (1965).
32. Ya. B. Zel'dovich, Zh. Éksp. Teor. Fiz., 32:1577 (1957).
33. K. K. Krupnikov, M. I. Brazhnik, and V. P. Krupnikova, Zh. Éksp. Teor. Fiz., 42:675 (1962).
34. Physics of Low Temperatures (collection) [in Russian], IL, Moscow (1959).
35. J. G. Collins and G. K. White, in: Progress in Low Temperature Phys., 4:450 (1964).
36. B. J. Alder, "Physics experiments with strong pressure pulses," in: Solids under Pressure (W. Paul and D. M. Wareschauer, eds.), McGraw-Hill, New York (1963), p. 385.
37. F. A. Baum, K. P. Stanyukovich, and B. I. Shekhter, Physics of Explosions [in Russian], Fizmatgiz, Moscow (1959).
38. G. E. Duvall and G. R. Fowles, in: High Pressure Physics and Chemistry (R. S. Bradley, ed.), Vol. 2, Academic Press, New York (1963), p. 209.

39.   Yu. S. Stepanov, Modern Methods for Determination of Equations of State for Solids [in Russian], Moskovsk. Gos. Universitet, Moscow (1961).

39a. S. B. Kormer, M. V. Sinitsyn, A. I. Funtikov, V. D. Urlin, and A. V. Blinov, Zh. Éksp. Teor. Fiz., 47:1202 (1964).

40.   J. Dapoigny, J. Kieffer, and B. Vodar, J. Rech. Cent. Nat. Rech. Sci., 6:260 (1955).

41.   R. G. McQueen, J. N. Fritz, and S. P. March, J. Geophys. Res., 69:2947 (1964).

42.   H. Lawton and I. C. Skidmore, Discussions Faraday Soc., 22:188 (1956).

43.   R. Schall, in: Proc. Third Intern. Congress on High-Speed Photography held in London, 1956 (R. B. Collins, ed.), Butterworths, London (1956), p. 228.

44.   R. Schall, Explosivstoffe, 6:120 (1958).

45.   I. C. Skidmore and E. Morris, Thermodynamics of Nuclear Materials, International Atomic Energy Agency, Vienna (1962).

46.   R. W. Goranson, D. Bancroft, L. B. Blendin, T. Blechar, E. E. Houston, E. F. Gittings, and S. A. Landeen, J. Appl. Phys., 26:1472 (1955).

47.   J. M. Walsh and M. H. Rice, J. Chem. Phys., 26:815 (1957).

48.   D. G. Doran, G. R. Fowles, and G. A. Peterson, Phys. Rev. Letters, 1:402 (1958).

49.   G. R. Fowles, J. Appl. Phys., 32:1475 (1961).

50.   S. Katz, D. G. Doran, and D. R. Curran, J. Appl. Phys., 30:568 (1959).

51.   H. D. Mallory, J. Appl. Phys., 26:555 (1955).

52.   K. K. Krupnikov, A. A. Bakanova, M. I. Brazhnik, and R. F. Trunin, Dokl. Akad. Nauk SSSR, 148:1302 (1963).

53.   B. J. Alder and R. H. Christian, Phys. Rev. Letters, 7:367 (1961).

54.   D. G. Doran, J. Appl. Phys., 34:844 (1963).

55.   V. N. Zubarev and G. S. Telegin, Dokl. Akad. Nauk SSSR, 142:309 (1962).

56.   D. S. Hughes, L. E. Gourley, and M. F. Gourley, J. Appl. Phys., 32:624 (1961).

57.   J. Berger, S. Joigneau, and G. Battet, Compt. Rend., 250:4331 (1960).

58.   J. Berger et al., Les Ondes de Détonation, Paris (1961).

59.   B. J. Alder and R. H. Christian, Phys. Rev. Letters, 4:450 (1960).

60.   R. H. Christian, The Equation of State of the Alkali Halides, Univ. California Res. Lab., Livermore, Calif. (1957).

61.   M. N. Pavlovskii, B. Ya. Vashchenko, and G. V. Simakov, Fiz. Tverd. Tela, 7:1212 (1965).

62.   S. B. Kormer, M. V. Sinitsyn, G. A. Kirillov, and V. D. Urlin, Zh. Éksp. Teor. Fiz., 48:1033 (1965).

63.   R. McQueen and S. P. Marsh, in: Handbook of Physical Constants (S. P. Clark, Jr., ed.), Geological Society of America, New York (1961).

64.   D. S. Hughes and R. G. McQueen, Trans. Am. Geophys. Union, 39:959 (1958).

65.   R. F. Trunin, V. I. Gon'shakova, G. V. Simakov, and N. E. Galgin, Izv. Akad. Nauk SSSR, Fizika Zemli, No. 9, p. 1 (1965).

66.   L. V. Al'tshuler and S. B. Kormer, Izv. Akad. Nauk SSSR, Ser. Geofiz., No. 1, p. 33 (1961).

67.   G. R. Fowles, Doctoral Thesis, Dept. of Geophysics, Stanford University, Calif. (1962).

68.   G. A. Adadurov, D. B. Balashov, and A. N. Dremin, Izv. Akad. Nauk SSSR, Ser. Geofiz., No. 5, p. 712 (1961).

69.   S. B. Kormer and A. I. Funtikov, Izv. Akad. Nauk SSSR, Fizika Zemli, No. 5, p. 1 (1965).

70.   A. A. Bakanova, I. P. Dudoladov, and R. F. Trunin, Fiz. Tverd. Tela, 7:1615 (1965).

CHAPTER 6

INFLUENCE OF PRESSURE ON THE ANHARMONICITY, AND THE ELASTIC MODULI, AND THEIR TEMPERATURE DEPENDENCES FOR CRYSTALS WITH PAIR INTERACTIONS

6.1. Introduction

The classical form of the method of potentials was introduced to solid-state physics by Born. Born assumed that in some solids — for example, in ionic crystals — the interaction between particles (ions) can be described quite accurately as being of the pair type. Then, having specified the actual nature of the pair potential between ions, we cannot only calculate theoretically the equation of state but also predict many other properties of ionic crystals which are of great interest in the physics of high pressures and in geophysics.

It must be stressed that the assumption of the pair interaction is a hypothesis, which is sometimes difficult to justify. Therefore, one would hardly expect the Born theory of ionic crystals to give very good agreement with the experimental data. On the other hand, it is well known that the Born theory of ionic crystals describes correctly, in the qualitative sense, many properties of these crystals. Therefore, it seems reasonable to extend the Born theory to high pressures. This should give information on the high-pressure behavior of such important parameters as the elastic moduli, their temperature dependences, and the anharmonicity.

Our calculations will show that the temperature coefficients of the elastic moduli and the anharmonicity decrease exponentially with decreasing lattice constants, i.e., with increasing pressure. This result can be understood quite easily. The contribution of the anharmonicity to the free energy F is proportional to T^2 and is of the form given by Eqs. (6.36)-(6.38):

$$F_a = F_3 + F_4, \quad F_3 = -\frac{9}{4} s^3 N (kT)^2 \hat{P}, \quad F_4 = \frac{9}{8} s^2 N (kT)^2 \hat{Q},$$

where only quantities $\hat{P}$ and $\hat{Q}$ depend on the lattice constants: the value of $\hat{P}$ is proportional to the third-order terms in the expansion of the potential energy, and the value of $\hat{Q}$ is proportional to the fourth-order terms. The main conclusion which follows from the calculations given in Secs. 6.2 and 6.3 can be summarized by the formula

$$\hat{P}, \hat{Q} \propto \exp(\varkappa y),$$

where $\varkappa$ is an average value of the exponent in the repulsive potential; $y = l / l_0$ is the relative value of the lattice constant. In fact,

$$\hat{P} \propto \frac{|\Phi^{III}|^2}{(\overline{\omega^2})^3}, \qquad \hat{Q} \propto \frac{\Phi^{IV}}{(\overline{\omega^2})^2},$$

where Φ^{III}, Φ^{IV} are the third and fourth derivatives of the potential energy; $\overline{\omega^2}$ is the average value of the frequency squared, given by Eq. (6.4a). The values of ω^2, Φ^{III}, and Φ^{IV} are proportional to the second, third, and fourth derivatives of the potential, respectively. The exponential term dominates the second derivative and becomes even stronger in the higher-order derivatives. This proves our conclusion. Thus, the anharmonicity of crystals with exponential repulsive forces decreases rapidly with decreasing volume. This increases the value of the "anharmonic analog" of the Grüneisen parameter γ_a (Chapter 3):

$$\gamma_a = \left(\frac{\partial \ln F_a}{\partial \ln x} \right)_T \propto \frac{\varkappa}{3}.$$

Similarly, we can show that the temperature coefficients of the elastic moduli decrease exponentially. The treatment which will be given here is based on that in [1].

6.2.   General Relationships

We shall now consider the effect of pressure on the anharmonicity of molecular and ionic crystals. The main forces acting in these crystals can be regarded quite satisfactorily as central forces with known potentials. The theory of anharmonicity is very cumbersome and, therefore, we shall extend various relationships to the case of high pressures without deriving the relevant formulas [2].

We shall use $\mathbf{m}(m_1, m_2, m_3)$ and $\mathbf{n}(n_1, n_2, n_3)$ as integral identifying numbers of the atoms in a crystal lattice. In the case of complex lattices, we shall also use μ and ν to number the atoms or ions in a unit cell. In this case, each particle is then denoted by a double index $\mathbf{m}\mu$ or $\mathbf{n}\nu$. Sometimes, for the sake of brevity, the indices μ and ν will be omitted (but only in those cases where this will cause no misunderstanding). Using the interaction potential $\varphi(|\mathbf{R}^m - \mathbf{R}^n|)$, where $\mathbf{R}$ is the radius vector of a particle, we can express the total potential energy in the form

$$\Phi = \frac{1}{2} \sum_{m,n} \varphi(|\mathbf{R}^m - \mathbf{R}^n|). \tag{6.1}$$

It follows from the form of Φ in Eq. (6.1) that the derivatives of Φ with respect to the coordinates of three or more particles are equal to zero. Introducing the relative radius vector $\mathbf{R}^h = \mathbf{R}$ and the operator $O = d/RdR$, we obtain the following relationships for $\mathbf{h} \neq 0$ [2]:

$$\left.\begin{aligned}
&\Phi_{ij}^{0h} = -(O^2\varphi) X_i X_j - (O\varphi) \delta_{ij}, \qquad \mathbf{R} = \{X_1, X_2, X_3\}, \\
&\Phi_{ijk}^{00h} = (O^3\varphi) X_i X_j X_k + (O^2\varphi)\{X_i\delta_{jk} + X_j\delta_{ik} + X_k\delta_{ij}\}, \\
&\Phi_{ijkl}^{00hh} = -\Phi_{ijkl}^{00h} = (O^4\varphi) X_i X_j X_k X_l + (O^3\varphi)\{X_i X_j\delta_{kl} + X_i X_k\delta_{jl} + \\
&\quad + X_i X_l\delta_{jk} + X_j X_k\delta_{il} + X_j X_l\delta_{ik} + X_k X_l\delta_{ij}\} + (O^2\varphi)\{\delta_{ij}\delta_{kl} + \delta_{ik}\delta_{jl} + \delta_{il}\delta_{jk}\},
\end{aligned}\right\} \tag{6.2}$$

where

$$\left.\begin{aligned}
&O\varphi = \frac{\varphi^I}{R}, \qquad O^2\varphi = \frac{\varphi^{II}}{R^2} - \frac{\varphi^I}{R^3}, \\
&O^3\varphi = \frac{\varphi^{III}}{R^3} - \frac{3\varphi^{II}}{R^4} + \frac{3\varphi^I}{R^5}, \\
&O^4\varphi = \frac{\varphi^{IV}}{R^4} - \frac{6\varphi^{III}}{R^5} + \frac{15\varphi^{II}}{R^6} - \frac{15\varphi^I}{R^7}.
\end{aligned}\right\} \tag{6.2a}$$

If Δ is used to denote the radial part of the Laplace operator, we find that

$$
\left.
\begin{aligned}
\Delta\varphi &= \varphi^{II} + \frac{2\varphi^{I}}{R}, \qquad \Delta\Delta\varphi = \varphi^{IV} + \frac{4\varphi^{III}}{R}, \\
O\Delta\varphi &= \frac{\varphi^{III}}{R} + \frac{2\varphi^{II}}{R^2} - \frac{2\varphi^{I}}{R^3}, \\
O^2\Delta\varphi &= \frac{\varphi^{IV}}{R^2} + \frac{\varphi^{III}}{R^3} - \frac{6\varphi^{II}}{R^4} + \frac{6\varphi^{I}}{R^5},
\end{aligned}
\right\}
\tag{6.2b}
$$

where the upper index of the potential φ indicates the order of the derivative.

The derivatives should not be affected by a vanishingly small translation. This gives the following relationships:

$$
\sum_{n,\,\nu} \Phi^{mn}_{\substack{\mu\nu \\ ij}} = 0,
\tag{6.3a}
$$

$$
\sum_{p,\,\varkappa} \Phi^{mnp}_{\substack{\mu\nu\varkappa \\ ijk}} = 0.
\tag{6.3b}
$$

In the calculations of the various complex anharmonic quantities, the square of the frequency ω^2 is averaged independently using the following relationship [2]:

$$
\overline{\omega^2} = \frac{1}{3sN} \sum_{k,\,\lambda} \omega^2(k,\lambda) = \frac{1}{3s} \sum_{\mu,\,i} \frac{\Phi^{00}_{\substack{\mu\mu \\ ii}}}{M_\mu}.
\tag{6.4}
$$

According to Eq. (6.3a),

$$
\overline{\omega^2} = -\frac{1}{3s} \sum_{\mu,\,i} \frac{1}{M_\mu} \left(\sum_{\substack{h \\ \nu \,(\neq\,{}^0_\mu)}} \Phi^{0h}_{\substack{\mu\nu \\ ii}} \right),
\tag{6.4a}
$$

where s is the number of particles in one unit cell; M_μ is the mass of a μ-th particle; N is the number of unit cells in a crystal; k is the wave vector; λ is an index used for numbering the phonon modes and the phonon polarization.

We shall now generalize the relationships, which describe the influence of the anharmonicity on the elastic moduli, to the case of high pressures.

The potential energy, per unit cell, is of the form

$$
\Phi_z = \frac{1}{2} \sum_{h,\,\mu\nu}{}' \varphi_{\mu\nu}(R^h_{\mu\nu}).^{*}
\tag{6.5}
$$

where $R^h_{\mu\nu} = |Ah + R^\mu - R^\nu|$ is the distance between the μ-th and ν-th particles in unit cells separated by a vector Ah. Using Eqs. (6.4) and (6.5), the expression for the mean-square frequency can be expressed in the form

---

*A primed summation sign means that the point $h = 0$, $\mu = \nu$ is excluded, i.e., it means that the "zeroth" energy and its derivatives are excluded. We are not speaking here of the energy of the zero-point vibrations of a crystal but of the infinite intrinsic energy of a crystal which has no physical meaning.

$$3s\overline{\omega^2} = \sum_{\nu,\,i} \frac{1}{M_\nu} \Phi^{00}_{ii} = -{\sum_{\substack{h,\,i \\ \mu,\,\nu}}}' \frac{1}{M_\nu} \Phi^{h0}_{ii} = {\sum_{\substack{h \\ \mu,\,\nu}}}' \frac{1}{M_\nu} \Delta\varphi_{\mu\nu}(R^h_{\mu\nu}). \tag{6.6}$$

It is known [2, 3] that the radius vector after deformation R is related to the radius vector before deformation $\overline{R}$ by the expression

$$R^2 - \overline{R}^2 = 2(\overline{R}, u\overline{R}) = 2\sum_{ik} \overline{X}_i \overline{X}_k u_{ik}, \tag{6.7}$$

where u_{ik} is the strain tensor. Consequently, any function $G(R)$ can be expanded using the relationship

$$G(R) = G(\overline{R}) + \sum_{ik} (OG)\overline{X}_i\overline{X}_k u_{ik} + \frac{1}{2}\sum_{ikjl} (O^2G)\overline{X}_i\overline{X}_k\overline{X}_j\overline{X}_l u_{ik}u_{jl} + \ldots, \tag{6.8}$$

where, as before,

$$O = 2\frac{\partial}{\partial(R^2)} = \frac{1}{R}\frac{\partial}{\partial R}. \tag{6.9}$$

We shall introduce the Grüneisen parameter

$$\gamma_{ik} = -\frac{1}{2}\frac{\partial \ln \overline{\omega^2}}{\partial u_{ik}}. \tag{6.10}$$

Using Eq. (6.6), we shall transform γ_{ik} to a form convenient for calculations

$$\gamma_{ik} = -\frac{1}{2}\frac{\sum' (O\Delta\varphi)(X_i X_k/M_\nu)}{\sum' (\Delta\varphi/M_\nu)}. \tag{6.10a}$$

Finally, in the case of cubic crystals,

$$\gamma_{ik} = \gamma\delta_{ik}, \qquad X_i X_k = \frac{1}{3}R^2\delta_{ik}, \tag{6.11}$$

$$\gamma = -\frac{1}{6}\frac{\sum' \dfrac{R^2}{M_\nu}(O\Delta\varphi)}{\sum' \dfrac{1}{M_\nu}\Delta\varphi}. \tag{6.10b}$$

It follows from Eq. (6.10b) that the Grüneisen parameter depends only on the volume and, therefore, it can be calculated using the cold-compression isotherm. The quantities referring to the cold-compression isotherm will be denoted by a superscript zero. We shall determine the change in the lattice constant δl due to the transition from an isotherm for a temperature T to the cold-compression isotherm for T = 0°K. To do this, we shall write the free energy in the "quasiharmonic" approximation:

$$F_{qh} = \Phi_z + \frac{1}{N\beta}\sum_{k,\lambda} \ln\left[2\,\sinh\,\frac{1}{2}\beta h|\omega|\right], \qquad \omega = \omega(k,\lambda). \tag{6.12}$$

The approximation is called quasiharmonic rather than harmonic because the frequencies ω depend on the volume. The quantity Φ_z in Eq. (6.12) is the zero-point energy (the zero-point vibrations can be ignored in the classical limit of high temperatures), and the second term in this equation describes the thermal energy of a crystal in the harmonic approximation.

The condition for equilibrium at a point P, T is

$$\frac{\partial F_{qh}}{\partial u_{ik}} = -V_z P \delta_{ik}.$$ (6.13)

The condition for equilibrium at a point P, T = 0 is

$$\frac{\partial F_{qh}}{\partial u_{ik}} = -V_z{}^0 P \delta_{ik},$$ (6.13a)

where, as already mentioned, the superscript zero indicates the quantities referring to the cold-compression isotherm, and

$$V_z - V_z{}^0 = V_z{}^0 u_{ii}, \qquad u_{ii} = u_{11} + u_{22} + u_{33}.$$ (6.14)

Repeated italic indices indicate summation in Eq. (6.14) and in all the subsequent equations.

Using Eqs. (6.8) and (6.9), we transform Eq. (6.13) to

$$\frac{\partial \Phi_z}{\partial u_{ik}} = \frac{1}{2} {\sum_{h,\mu\nu}}' (O\varphi) X_i X_k = 3s\gamma_{ik}\bar{\varepsilon} - V_z P \delta_{ik},$$ (6.15)

where

$$\bar{\varepsilon} = \frac{1}{N} \sum_k \varepsilon(\omega, T), \qquad \varepsilon = \hbar\omega\left(\bar{n} + \frac{1}{2}\right), \qquad \bar{n} = \frac{1}{e^{\beta\hbar\omega} - 1}.$$ (6.16)

At high temperatures,

$$\bar{\varepsilon} = kT.$$ (6.17)

Those quantities in Eq. (6.15) which refer to a general isotherm are related to the corresponding quantities of the cold-compression isotherm:

$$X_i - X_i{}^0 = u_{ij} X_j{}^0,$$ (6.18)

$$\left. \begin{array}{l} (O\varphi) = (O\varphi)^0 + (O^2\varphi)^0 X_i{}^0 X_k{}^0 u_{ik}, \\[2mm] (O\varphi) X_i X_k = (O\varphi)^0 X_i{}^0 X_k{}^0 + 2u_{kl}(O\varphi)^0 X_i{}^0 X_l{}^0 + u_{jl}(O^2\varphi)^0 X_i{}^0 X_k{}^0 X_j{}^0 X_l{}^0. \end{array} \right\}$$ (6.19)

The elastic moduli on the cold-compression isotherm are given by

$$C^0_{ik\,jl} = \frac{1}{2V_z{}^0} {\sum_{h,\mu\nu}}' (O^2\varphi)^0 X_i{}^0 X_k{}^0 X_j{}^0 X_l{}^0 - P(\delta_{ij}\delta_{kl} + \delta_{il}\delta_{jk} - \delta_{ik}\delta_{jl}).$$ (6.20)

Substituting Eqs. (6.18)-(6.20) into Eq. (6.15), we obtain

$$C^0_{ik,jl} u_{jl} = \frac{3s\gamma_{ik}}{V_z{}^0}\bar{\varepsilon}.$$ (6.21)

We shall now apply Eq. (6.21) to cubic crystals. For these crystals, the strain is spherically symmetrical:

$$u_{ik} = \frac{\delta l}{l^0}\delta_{ik}, \qquad V_z = V_z^0\left(1 + 3\frac{\delta l}{l^0}\right). \tag{6.22}$$

Using Eq. (6.11), we obtain

$$\frac{\delta l}{l^0} = \frac{s\gamma\chi^0}{V_z^0}\overline{\varepsilon}, \qquad \frac{1}{\chi^0} = \frac{C_{11,jj}^0}{9}, \tag{6.23}$$

where χ^0 is the compressibility on the cold-compression isotherm.

We have derived Eq. (6.23) using the following well-known relationships for cubic crystals:

$$\left.\begin{array}{l} C_{11,11} = C_{22,22} = C_{33,33} = c_{11}, \\ C_{11,22} = C_{22,33} = C_{33,11} = c_{12}, \\ C_{12,12} = C_{23,23} = C_{31,31} = c_{44}, \end{array}\right\} \tag{6.24}$$

$$\chi = \frac{3}{c_{11} + 2c_{12}}. \tag{6.25}$$

The quantity γ in Eq. (6.23) refers to a general isotherm but the difference between γ and γ^0 is of the second order of smallness, and therefore we can use γ in the expressions relating to the cold-compression isotherm.

According to Leibfried and Ludwig [2], the adiabatic elastic moduli are given by the formula

$$C_{ik,jl}^{ad} = \frac{1}{V_z}\frac{\partial^2\Phi_z}{\partial u_{ik}\partial u_{jl}} + \frac{3s}{2\overline{\omega}^2 V_z}\left(\frac{\partial^2\overline{\omega^2}}{\partial u_{ik}\partial u_{jl}} - \frac{1}{2\overline{\omega}^2}\frac{\partial\overline{\omega}^2}{\partial u_{ik}}\frac{\partial\overline{\omega}^2}{\partial u_{jl}}\right)\overline{\varepsilon} - P(\delta_{ij}\delta_{kl} + \delta_{il}\delta_{jk} - \delta_{ik}\delta_{jl}), \tag{6.26}$$

where all the quantities in the second term can be calculated using the cold-compression isotherm because this term is small; however, the first term must be calculated with an accuracy to within δl.

Using Eqs. (6.8), (6.9), (6.18), and (6.22), we easily obtain the following relationship for cubic crystals

$$\frac{1}{V_z}\frac{\partial^2\Phi_z}{\partial u_{ik}\partial u_{jl}} = \frac{1}{2V_z}\sum_{h,\mu\nu}'\langle O^2\varphi\rangle\overline{X}_i\overline{X}_k\overline{X}_j\overline{X}_l = C_{ik,jl}^* + \frac{\delta l}{l}\left[C_{ik,jl}^* + \frac{1}{2V_z^0}\sum_{h,\mu\nu}' R^{0^2}(O^3\varphi)\,{}^0X_i{}^0X_k{}^0X_j{}^0X_l\right], \tag{6.27}$$

where the following symbol (which will be convenient in later calculations) is introduced:

$$C_{ik,jl}^* = \frac{1}{2V_z^0}\sum_{h,\mu\nu}'(O^2\varphi)\,{}^0X_i{}^0X_k{}^0X_j{}^0X_l. \tag{6.27a}$$

According to Eqs. (6.26) and (6.27), $C_{ik,jl}^*$ are related to $C_{ik,jl}^0$ by the expression

$$C_{ik,jl}^0 = C_{ik,jl}^* - P(\delta_{ij}\delta_{kl} + \delta_{il}\delta_{jk} - \delta_{jk}\delta_{jl}). \tag{6.26a}$$

Using Eqs. (6.8) and (6.9), we shall now expand $\overline{\omega^2}$ of Eq. (6.6) up to terms quadratic in strain:

$$\overline{\omega^2} = \frac{1}{3s}\left\{\sum_{h,\mu\nu}'\frac{1}{M_\nu}\left[\Delta\varphi + (O\Delta\varphi)X_iX_ku_{ik} + \frac{1}{2}(O^2\Delta\varphi)X_iX_kX_jX_lu_{ik}u_{jl}\right]\right\}. \tag{6.28}$$

Since ε is a small quantity, it is sufficient to retain only the first nonvanishing expressions in the derivatives of the mean-square frequency (we shall denote these expressions by a zero superscript):

$$\left.\begin{aligned}
\frac{\partial \overline{\omega^2}}{\partial u_{ik}} &= \frac{1}{3s} \sum_{h,\mu\nu}{}' \frac{1}{M_\nu} (O\Delta\varphi)^0 X_i^0 X_k^0, \\
\frac{\partial^2 \overline{\omega^2}}{\partial u_{ik} \partial u_{jl}} &= \frac{1}{3s} \sum_{h,\mu\nu}{}' \frac{1}{M_\nu} (O^2\Delta\varphi)^0 X_i^0 X_k^0 X_j^0 X_l^0.
\end{aligned}\right\} \tag{6.29}$$

Collecting our results, we now obtain

$$C_{ik,jl}^{\mathrm{ad}} = C_{ik,jl}^{\bullet} + \left[C_{ik,jl}^{\bullet} + \frac{1}{2V_z^0} \sum_{h,\mu\nu}{}' R^{0^2} (O^3\varphi)^0 X_i^0 X_k^0 X_j^0 X_l^0 \right] \frac{s\gamma\chi^0\overline{\varepsilon}}{V_z^0} +$$

$$+ \frac{3s}{4V_z^0 \sum{}' \frac{1}{M}(\Delta\varphi)^0} \left\{ \sum{}' \frac{2}{M_\nu} (O^2\Delta\varphi)^0 X_i^0 X_k^0 X_j^0 X_l^0 - \right.$$

$$\left. - \frac{\left(\frac{1}{M_\nu}(O\Delta\varphi)^0 X_i^0 X_k^0 \right)\left(\sum{}' \frac{1}{M_\nu}(O\Delta\varphi)^0 X_j^0 X_l^0 \right)}{\sum{}' \frac{1}{M_\nu}(\Delta\varphi)^0} \right\} \overline{\varepsilon} - P(\delta_{ij}\delta_{kl} + \delta_{il}\delta_{jk} - \delta_{ik}\delta_{jl}). \tag{6.30}$$

Using more compact notation, Eq. (6.30) can be represented in the form [see Eq. (6.24)]

$$c_{\alpha\beta}^{\mathrm{ad}} + P^0(\delta_{ij}\delta_{kl} + \delta_{il}\delta_{jk} - \delta_{ik}\delta_{jl}) = c_{\alpha\beta}^{\bullet} \{1 - D_{\alpha\beta}\overline{\varepsilon}\}, \tag{6.30a}$$

$$\alpha = (ik), \qquad \beta = (jl),$$

where

$$c_{\alpha\beta}^{\bullet} D_{\alpha\beta} = \frac{3s}{V_z^0} (\gamma\Gamma_{\alpha\beta} - \Gamma_{\alpha\beta}'), \tag{6.31}$$

$$-\frac{P^0}{3} + \frac{1}{\chi^{\mathrm{ad}}} = \frac{1}{\chi^{\bullet}} - \frac{3s}{V_{z_i}^0}(\gamma\Gamma - \Gamma')\overline{\varepsilon} = \frac{1}{\chi^{\bullet}}(1 - D_\chi \overline{\varepsilon}), \tag{6.32}$$

$$\left.\begin{aligned}
\Gamma_{11} &= -\frac{\chi^0 \sum{}' R^{0^2} X^{0^4} (O^3\varphi)^0}{6V_z^0}, \\
\Gamma_{12} &= \Gamma_{44} = -\frac{\chi^0 \sum{}' R^{0^2} X^{0^2} Y^{0^2} (O^3\varphi)^0}{6V_z^0}, \\
\Gamma_{11}' &= -\gamma^2 + \gamma\chi^0 \frac{c_{11}^{\bullet}}{3} + \frac{\sum \frac{1}{M_\nu} X^{0^4} (O^2\Delta\varphi)^0}{2\sum{}' \frac{1}{M_\nu}(O\varphi)^0}, \\
\Gamma_{12}' &= \Gamma_{44}' - \gamma^2 = -\gamma^2 + \frac{\gamma\chi^0 c_{12}^{\bullet}}{3} + \frac{\sum{}' \frac{1}{M_\nu} X^{0^2} Y^{0^2} (O^2\Delta\varphi)^0}{2\sum{}' \frac{1}{M_\nu}(\Delta\varphi)^0},
\end{aligned}\right\} \tag{6.33a}$$

$$\Gamma = \frac{1}{3}(\Gamma_{11} + 2\Gamma_{12}) = -\frac{\chi^0}{54V_z{}^0}\sum{}' R^{0^4}(O^3\varphi)^0,$$

$$\Gamma' = \frac{1}{3}(\Gamma_{11}' + 2\Gamma_{12}') = -\gamma^2 + \frac{1}{3}\gamma\frac{\chi^0}{\chi^*} + \frac{\sum{}' R^{0^4}\frac{1}{M_\nu}(O^2\Delta\varphi)^0}{18\sum{}'\frac{1}{M_\nu}(O\varphi)^0}. \qquad (6.33b)$$

All the formulas in Eqs. (6.33a) and (6.33b) transform into the expressions given in [2] when P = 0. In this case, the quantities with the asterisk are identical with the quantities denoted by the superscript zero. The adiabatic and isothermal elastic moduli are known to be related by the following thermodynamic equation [2]:

$$C_{ik,jl}^{\ \ \mathrm{ad}} - \mathcal{C}_{ik,ji} = \frac{3s}{V_z{}^0}\gamma_{ik}\gamma_{jl}Tc_V{}^s, \qquad (6.34)$$

where c_V^s is the specific heat of one s-th particle,

$$c_V = 3sNc_V{}^s.$$

For cubic crystals,

$$c_{11}{}^{\mathrm{ad}} - c_{11}{}^{\mathrm{is}} = c_{21}{}^{\mathrm{ad}} - c_{12}{}^{\mathrm{is}} = \frac{3s}{V_z{}^0}\gamma^2 Tc_V{}^s,$$

$$c_{44}{}^{\mathrm{ad}} = c_{44}{}^{\mathrm{is}}, \qquad c_{11}{}^{\mathrm{ad}} - c_{12}{}^{\mathrm{ad}} = c_{11}{}^{\mathrm{is}} - c_{12}{}^{\mathrm{is}}. \qquad (6.35)$$

In order to determine the specific heat and other thermodynamic parameters, we must calculate the free energy up to terms including T^2. The relevant expressions for the central forces are given in [2]:

$$F_a = F_3 + F_4,$$

$$F_3 = -\frac{9}{4}s^2N(kT)^2\hat{P}, \qquad F_4 = \frac{9}{8}s^2N(kT)^2\hat{Q}, \qquad (6.36)$$

where

$$\hat{P} = \frac{\sum\limits_{\substack{hg \\ ij \ k\mu\nu x}}\frac{1}{M_\mu M_\nu M_x}\left|\Phi_{\mu\nu x}^{ohg}{}_{ijk}\right|^2}{\sum\limits_h\left[\sum\limits_{\mu i}\frac{1}{M_\mu}\Phi_{\mu\mu}^{oh}{}_{ii}\right]^3} = \frac{\sum\limits_{h,\ \mu\nu}'\frac{3}{2M_\mu M_\nu}\left(\frac{1}{M_\mu}+\frac{1}{M_\nu}\right)\sum\limits_{ijk}\left|\Phi_{\nu\nu\mu}^{oh}{}_{ijk}\right|^2}{(3s\overline{\omega^2})^3 + \sum\limits_{h(\neq 0)}\left[\sum\limits_{iv}\frac{1}{M_\nu}\Phi_{\nu\nu}^{oh}{}_{ii}\right]^3}, \qquad (6.37)$$

and

$$\hat{Q} = \frac{\sum\limits_{h,\ \mu\nu ij}\Phi_{\mu\mu\nu\nu}^{oohh}{}_{iijj}\frac{1}{M_\mu M_\nu}}{\left[\sum\limits_{\mu,\ i}\frac{1}{M_\mu}\Phi_{\mu\mu}^{oo}{}_{ii}\right]^2} = \frac{\sum\limits_{h,\ \mu\nu}'\frac{1}{2}\left(\frac{1}{M_\mu}+\frac{1}{M_\nu}\right)^2\sum\limits_{ij}\Phi_{\nu\nu\mu\mu}^{oohh}{}_{iijj}}{(3s\overline{\omega^2})^2}. \qquad (6.38)$$

Equations (6.36)–(6.38) show that the fourth-order terms contribute to the first approximation of $\hat{Q}$ and the third-order terms contribute to the second approximation of $\hat{P}$. Using the formulas in Eq. (6.2), we can easily calculate the sums over (i, j, k) which occur in Eqs. (6.37) and (6.38):

$$
\left.
\begin{aligned}
\sum_{ijk} \left| \Phi_{vv\mu \atop ijk}^{00h} \right|^2 &= (\varphi_{\mu\nu}^{III})^2 + 6R^2 (O^2 \varphi_{\mu\nu})^2, \qquad \overline{\omega^2} = \frac{1}{3s} \sum_{h, \mu\nu} \frac{\Delta \varphi_{\mu\nu}}{\overline{M_\mu}}, \\
\sideset{}{'}\sum_{ij} \Phi_{vv\mu\mu \atop iijj}^{00hh} &= \Delta\Delta\varphi_{\mu\nu}, \qquad \sum_i \Phi_{vv \atop ii}^{0h} = -\Delta\varphi_{vv}.
\end{aligned}
\right\}
\tag{6.39}
$$

The anharmonic correction to c_V is calculated using Eq. (6.36):

$$
c_{V^a} = -T(F_a)_{TT} = 3sNk^2T \left(\frac{3}{2} s^2 \hat{P} - \frac{3}{4} s\hat{Q} \right).
\tag{6.40}
$$

It is usual to measure experimentally the specific heat at constant pressure c_P. This quantity is related to c_V by the well-known thermodynamic equation

$$
c_P - c_V = 3sNk^2Ts\hat{P}', \qquad \hat{P}' = \frac{3\chi^{is}\gamma^2}{V_z},
\tag{6.41}
$$

or by

$$
\frac{c_P}{k} = 1 + kT \left[3s^2 \frac{\hat{P}}{2} - \frac{3}{4} s\hat{Q} + s\hat{P}' \right].
\tag{6.42}
$$

Equation (6.42) is valid for the high-temperature specific heat, $T > Q$, even without the inclusion of the quantum corrections. The quantum correction for the specific heat in the temperature range $T > \theta_\infty$ (θ_∞ is the Debye temperature at high temperatures) is given by the well-known expression [2]:

$$
\frac{c_P}{k} = \dots - \frac{1}{20} \left(\frac{\theta_\infty}{T} \right)^2.
\tag{6.43}
$$

The term in Eq. (6.42) proportional to temperature is small and, therefore, the volume-dependent quantities $\hat{P}$, $\hat{Q}$, and $\hat{P}'$ can be calculated approximately using the cold-compression isotherm.

In order to investigate the influence of pressure on the anharmonicity, it is necessary to specify the nature of the potential of the interaction between particles. The next two sections will deal with the influence of pressure on the anharmonicity of van der Waals and ionic crystals, i.e., of crystals whose interaction potential is known quite well.

6.3. Van der Waals Crystals

Let us consider the influence of pressure on the anharmonicity of inert-gas crystals. These crystals have the fcc structure. The attractive forces in these crystals can be represented by a series of reciprocals of R^2, beginning from R^{-6}. The repulsive forces are the usual exponential overlap forces.

The potential of the attractive forces will be represented by a single term proportional to R^{-6}. Higher-order terms of this potential may be important at high pressures but they can be ignored for the following reasons.

We effectively include the higher-order terms in the attractive-force potential by selecting the equation of state for any particular substance on the basis of the experimental data, i.e., by calculating the coefficients in the van der Waals potential from the experimental data.

We shall show later that the anharmonicity decreases with increasing pressure. The inclusion of the higher-order attractive terms would have the result of increasing the rate of decrease of the anharmonicity with increasing pressure.

We shall assume that the potential representing the interaction of two atoms is

$$\psi = -A_m y^{-m} \left(\frac{l}{R}\right)^m + A^{\varkappa} \exp\left(-\varkappa y \frac{R}{l}\right), \tag{6.44}$$

where $y = l/l_0$; l_0 and l are the shortest distances between these two atoms at P = 0, T = 0 and at P, T, respectively; R is the distance between any two atoms in the lattice. Assuming that the energy per atom is

$$\Phi_{z0} = -\eta \quad \text{when} \quad y = 1, \quad T = 0 \tag{6.45}$$

and that the pressure is P = 0 when y = 1, T = 0, we obtain

$$A_m = \frac{2\eta\varkappa}{z_m(\varkappa-m)}, \qquad A_{\varkappa} = \frac{2\eta\, m\, \exp\varkappa}{z(\varkappa-m)}, \tag{6.46}$$

where z always denotes the number of nearest neighbors (in our case, the number is 12),

$$z_m = \sum_h{}' \left(\frac{l}{R^h}\right)^m,$$

$$z_m{}^{\alpha\beta} = 3 \sum_h{}' \frac{1}{(R^h)^4} (X_\alpha{}^h X_\beta{}^h)^2 \left(\frac{l}{R^h}\right)^m, \quad \alpha, \beta = 1, 2, 3, \tag{6.47}$$

We shall find it convenient to use $z_m^{44} = z_m^{12}$. The quantities z_m and $z_m^{\alpha\beta}$ are calculated in [4].

The lattice sums for the fcc structure are:

z	12	z_{13}	12.088	z_{25}	12.001	z_8^{11}	6.6019
z_6	14.454	z_{15}	12.040	z_{32}	12	z_{12}^8	3.1200
z_8	12.802	z_{17}	12.020	z_6^{11}	7.6896	l_{10}/l_0	$2/\sqrt{2}$
z_{10}	12.311	z_{18}	12.013	z_6^{12}	3.3820	$\tau = V_{z0}/l_0^3$	$\sqrt{2}/2$

The constant $\varkappa$ in Eqs. (6.44) and (6.46) is assumed to be known. In practice, this constant can be determined from independent experimental data, for example, from the compressibility χ. The values calculated for T = 0°K, i.e., for the cold-compression isotherm, will be denoted by the superscript zero:

$$P^0 = \frac{1}{6V_{z0}^0}(\varkappa z A_{\varkappa} y^{-2}\exp(-\varkappa y) - m z_m A_m y^{-(m+3)}), \tag{6.48}$$

$$\frac{1}{\chi^0} = \frac{1}{18V_{z0}^0}\left[z\varkappa^2 A_{\varkappa} y^{-1}\left(1+\frac{2}{\varkappa y}\right)\exp(-\varkappa y) - m(m+3)z_m A_m y^{-(m+3)}\right], \tag{6.49}$$

$$\Phi_z = -\frac{1}{2}[z_m A_m y^{-m} - z A_{\varkappa}\exp(-\varkappa y)], \tag{6.50}$$

$$\varphi^{0s} = \frac{d^s\varphi}{dR^s} = \frac{(-1)^s}{l_0^s}\left[\varkappa^s A_{\varkappa}\exp\left(-\varkappa y\frac{R}{l}\right) - \frac{(m+s-1)!}{(m-1)!}A_m y^{-(m+s)}\left(\frac{l}{R}\right)^{m+s}\right]. \tag{6.51}$$

Using Eqs. (6.44), (6.51), (6.2), (6.2a), and (6.2b), we can calculate all the required quantities. The Grüneisen parameter of Eq. (2.10b) is

$$\gamma = \frac{1}{6}\left\{ z\varkappa^3 A_\varkappa y\left(1 - \frac{2}{\varkappa y} - \frac{2}{\varkappa^2 y^2}\right)\exp(-\varkappa y) - (m+2)m(m-1)z_{m+2}A_m y^{-(m+2)}\right\} \times$$

$$\times \left\{ z\varkappa^2 A_\varkappa \left(1 - \frac{2}{\varkappa y}\right)\exp(-\varkappa y) - m(m-1)z_{m+2}A_m y^{-(m+2)}\right\}^{-1}. \tag{6.52}$$

The quantities c^* in Eqs. (6.27a) and (6.24) are given by

$$c_{11}^* = \frac{1}{2V_{z0}^0}\left\{ 2\varkappa^2 A_\varkappa y^{-1}\left(1 + \frac{1}{\varkappa y}\right)\exp(-\varkappa y) - \frac{z_m^{11}}{3}m(m+2)A_m y^{-(m+3)}\right\}, \tag{6.53}$$

$$c_{12}^* = c_{44}^* = \frac{1}{2V_{z0}^0}\left\{ \varkappa^2 A_\varkappa y^{-1}\left(1 + \frac{1}{\varkappa y}\right)\exp(-\varkappa y) - \frac{z_m^{12}}{3}m(m+2)A_m y^{-(m+3)}\right\}, \tag{6.54}$$

$$(\chi^*)^{-1} = \frac{1}{3}(c_{11}^* + 2c_{12}^*) = \frac{1}{18V_{z0}^0}\left\{ z\varkappa^2 A_\varkappa y^{-1}\left(1 + \frac{1}{\varkappa y}\right)\exp(-\varkappa y) - m(m+2)z_m A_m y^{-(m+3)}\right\}. \tag{6.55}$$

The quantities with the superscript zero can be expressed in terms of the quantities with an asterisk using Eq. (6.26a):

$$c_{11}^0 = c_{11}^* - P, \qquad c_{12}^0 = c_{12}^* + P, \qquad c_{44}^0 = c_{44}^* - P, \tag{6.56}$$

$$\frac{1}{\chi^0} = \frac{1}{\chi^*} + \frac{P}{3}. \tag{6.57}$$

The quantities Γ, defined by the formulas in Eq. (6.33) now become

$$\left.\begin{aligned}
\Gamma_{11} &= \frac{\chi^0}{6V_{z0}^0}\left\{ 2\varkappa^3 A_\varkappa\left(1 + \frac{3}{\varkappa y} + \frac{3}{\varkappa^2 y^2}\right)\exp(-\varkappa y) - \frac{m}{3}(m^2 + 6m + 8)A_m z_m^{11} y^{-(m+3)}\right\}, \\[2mm]
\Gamma_{12} &= \Gamma_{44} = \frac{\chi^0}{6V_{z0}^0}\left\{ A_\varkappa \varkappa^3\left(1 + \frac{3}{\varkappa y} + \frac{3}{\varkappa^2 y^2}\right)\exp(-\varkappa y) - \frac{m}{3}(m^2 + 6m + 8)A_m z_m^{12} y^{-(m+3)}\right\}, \\[2mm]
\Gamma_{11}' &= -\gamma^2 + \gamma\chi^0\frac{c_{11}^*}{3} + \left\{ 2\varkappa^4 A_\varkappa y^2\left(1 - \frac{1}{\varkappa y} - \frac{6}{\varkappa^2 y^2} - \frac{6}{\varkappa^3 y^3}\right)\exp(-\varkappa y) - \frac{m}{3}(m+2)[(m+\right. \\
&\quad + 2)(m+1) - 6]A_m z_{m+2}^{11} y^{-(m+2)}\bigg\}\frac{1}{2}\left[z\varkappa^2 A_\varkappa\left(1 - \frac{2}{\varkappa y}\right)\exp(-\varkappa y) - m(m-1)z_{m+2}A_m y^{-(m+2)}\right]^{-1}, \\[2mm]
\Gamma_{12}' &= \Gamma_{44}' - \gamma^2 = -\gamma^2 + \frac{\gamma\chi^0 c_{12}^*}{3} + \left\{ \varkappa^4 A_\varkappa y^2\left(1 - \frac{1}{\varkappa y} - \frac{6}{\varkappa^2 y^2} - \frac{6}{\varkappa^3 y^3}\right)\exp(-\varkappa y) - \frac{m}{3}(m+2)[(m+2)\times\right. \\
&\quad \times (m+1) - 6]A_m z_{m+2}^{12} y^{-(m+2)}\bigg\}\frac{1}{2}\left[z\varkappa^2 A_\varkappa\left(1 - \frac{2}{\varkappa y}\right)\exp(-\varkappa y) - m(m-1)z_{m+2}A_m y^{-(m+2)}\right]^{-1}.
\end{aligned}\right\} \tag{6.58}$$

The quantities Γ and Γ' can be calculated using the known values of Γ_{ik} and Γ_{ik}' [see Eq. (6.33b)], and they are given by the following expressions:

$$\left.\begin{aligned}
\Gamma &= \frac{\chi}{54V_{z0}^0}\left\{ z\varkappa^3 A_\varkappa\left(1 + \frac{3}{\varkappa y} + \frac{3}{\varkappa^2 y^2}\right)\exp(-\varkappa y) - z_m m(m^2 + 6m + 8)A_m y^{-(m+3)}\right\}, \\[2mm]
\Gamma' &= -\gamma^2 + \frac{\gamma}{3}\frac{\chi^0}{\chi^*} + \left\{ z\varkappa^2 A_\varkappa y^2\left(1 - \frac{1}{\varkappa y} - \frac{6}{\varkappa^2 y^2} - \frac{6}{\varkappa^3 y^3}\right)\exp(-\varkappa y) - m(m+2)[(m+2)\times\right. \\
&\quad \times (m+1) - 6]A_m z_{m+2}y^{-(m+2)}\bigg\}\frac{1}{18}\left[z\varkappa^2 A_\varkappa\left(1 - \frac{2}{\varkappa y}\right)\exp(-\varkappa y) - m(m-1)z_{m+2}A_m y^{-(m+2)}\right]^{-1}.
\end{aligned}\right\} \tag{6.58a}$$

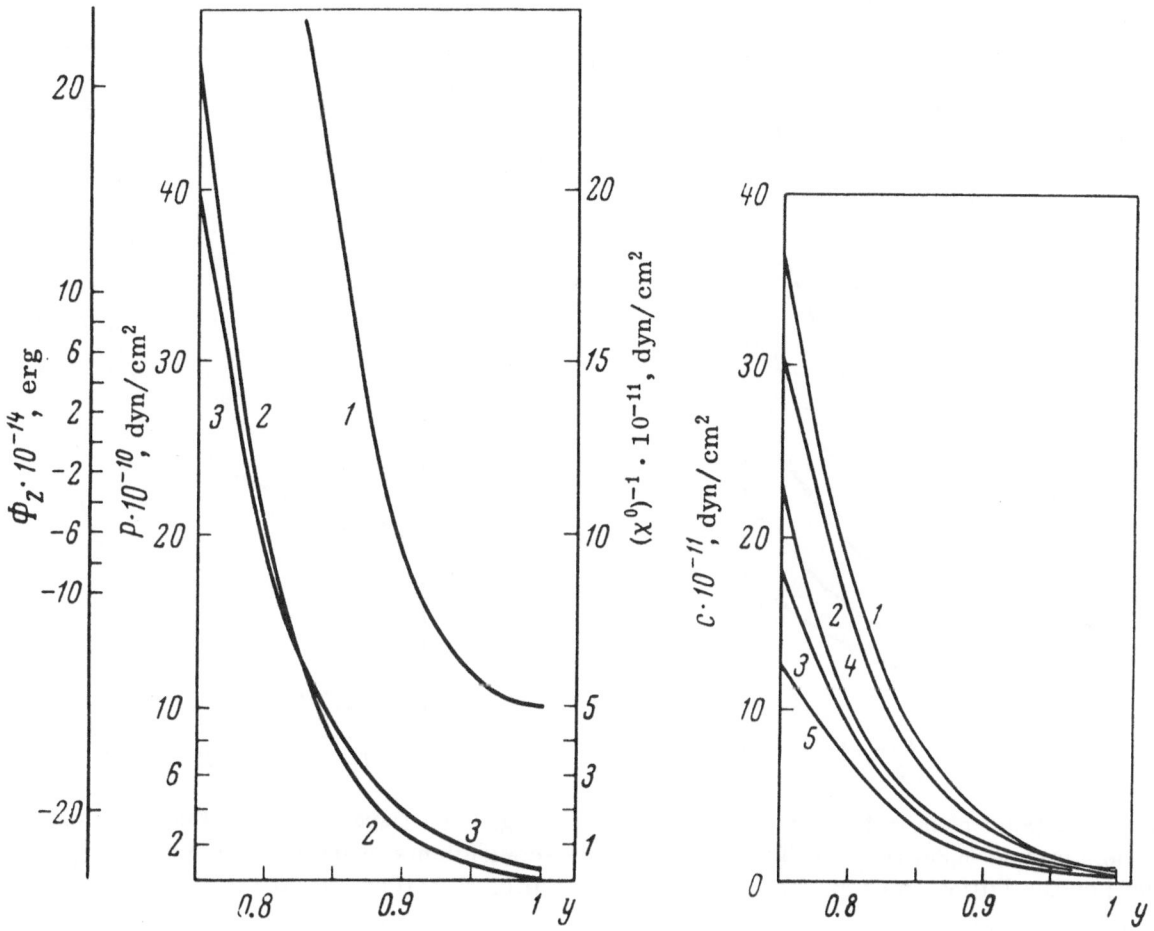

Fig. 6.1. Dependences of the energy, pressure, and the reciprocal of compressibility on the lattice constant for Ar-type crystals. 1) Φ_Z; 2) P; 3) $(\chi^0)^{-1}$.

Fig. 6.2. Dependences of the elastic moduli on the lattice constant for Ar-type crystals. 1) c_{11}^{*}; 2) c_{11}^{0}; 3) c_{12}^{*}; 4) c_{12}^{0}; 5) c_{44}^{0}.

Having determined the volume dependences of the quantities Γ_{ik}, Γ_{ik}', we can then use Eq. (6.30), (6.31), and (6.56) to find the temperature dependences of the elastic moduli at high pressures. The quantities $\hat{P}$ and $\hat{Q}$ of Eqs. (6.37) and (6.38) can be calculated using Eqs. (6.44), (6.51), (6.2), (6.2a), and (6.2b). We thus obtain

$$\hat{P} = \frac{3}{z^2\Omega^3}\left\{ x^6 A_x^2 \left(1 + \frac{6}{x^2 y^2} + \frac{12}{x^3 y^3} + \frac{6}{x^4 y^4} \right) \exp(-2xy) - \right.$$

$$-2m(m+1)(m+2)x^3 A_x A_m \exp(-xy)y^{-(m+3)}\left[1 + \frac{6(1+xy)}{(m+1)x^2 y^2} \right] +$$

$$\left. + \frac{z_{2(m+3)}}{z}m^2(m+2)^2[(m+1)^2+6]A_m^2 y^{-2(m+3)} \right\}, \tag{6.59}$$

$$\hat{Q} = \frac{2}{z\Omega^2}\left\{ x^4 A_x \exp(-xy)\left(1 - \frac{4}{xy} \right) - \frac{z_{m+4}}{z}m(m-1)(m+1)(m+2)A_m y^{-(m+4)} \right\}, \tag{6.60}$$

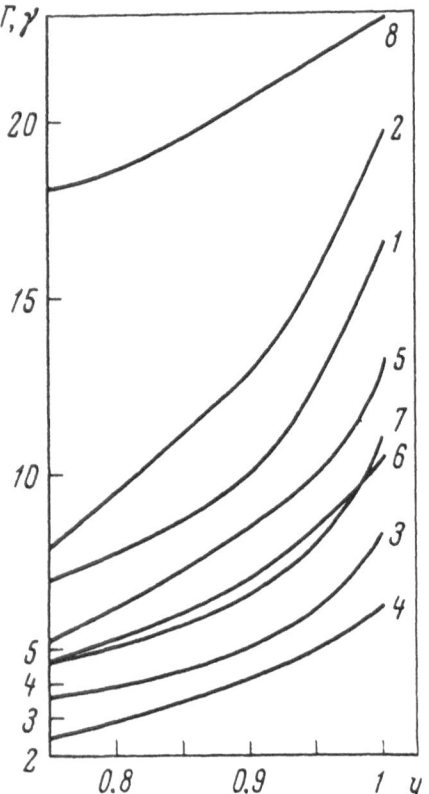

Fig. 6.3. Dependences of the quantities Γ [Eqs. (6.33a) and (6.33b)] and γ [Eq. (6.10b)] on the lattice constant for Ar-type crystals. 1) Γ_{11}; 2) Γ'_{11}; 3) Γ_{12}; 4) Γ'_{12}; 5) Γ'_{44}; 6) Γ; 7) Γ'; 8) 10γ.

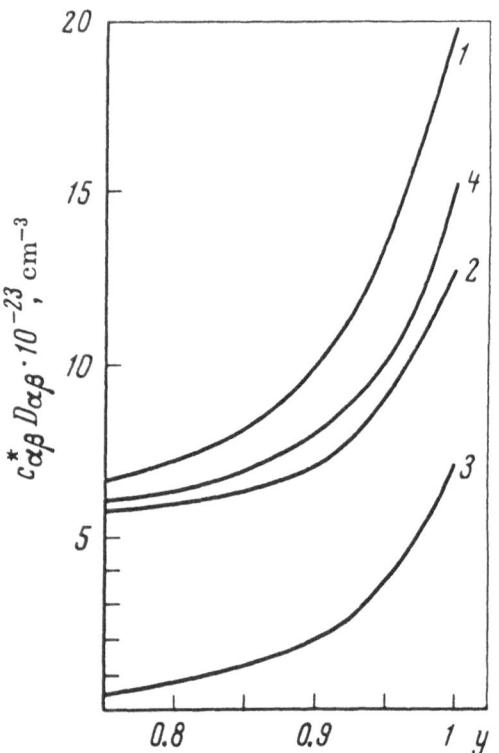

Fig. 6.4. Dependences of the temperature coefficients of the elastic moduli $c^*_{\alpha\beta}D_{\alpha\beta}$ [Eq. (6.31)] on the lattice constant for Ar-type crystals. 1) $c^*_{11}D_{11}$; 2) $c^*_{12}D_{12}$; 3) $c^*_{44}D_{44}$; 4) D_χ/χ^*.

$$\Omega = \left\{ \varkappa^2 A_\varkappa \exp(-\varkappa y)\left(1 - \frac{2}{\varkappa y}\right) - m(m-1)A_m \frac{z_{m+2}}{z} y^{-(m+2)} \right\}. \tag{6.61}$$

Calculation of the denominator of the quantity $\hat{P}$ in Eq. (6.37) shows that

$$\frac{1}{(3\overline{\omega^2})^3}\left(\frac{1}{M^3} \sum_h{}' \left\{ \sum_i \Phi^{0h}_{ii} \right\}^3\right) \approx \frac{1}{z^2}. \tag{6.62}$$

In our case, $z = 12$ and, therefore, the small quantities of the order of z^{-2} are omitted in the denominators of Eqs. (6.59) and (6.60). The results of the calculations are presented in Figs. 6.1-6.7. We must draw attention to the most important points. It is evident from Fig. 6.5 that the anharmonicity parameters $\hat{Q}$ of Eq. (6.38) and $\hat{P}$ of Eq. (6.37) decrease exponentially with decreasing y. This applies also to ionic crystals (Sec. 6.4). Obviously, this result is of general validity because the quantities $\hat{Q}$ and $\hat{P}$ are defined in terms of higher derivatives of the potential, and these derivatives depend mainly on the nature of the repulsive forces. Secondly, we must mention the exponential weakening of the temperature dependence of the elastic moduli with decreasing y (Fig. 6.6) and the corresponding weakening of the temperature dependence of c_P (Fig. 6.7). Finally, according to Fig. 6.7, the inclusion of the anharmonicity gives values of c_V which are smaller than those which follow from the law of Dulong and Petit. In the case of NaCl (Fig. 6.15 in the next section), the situation is the reverse. In general, this conclusion depends on the nature of the potential.

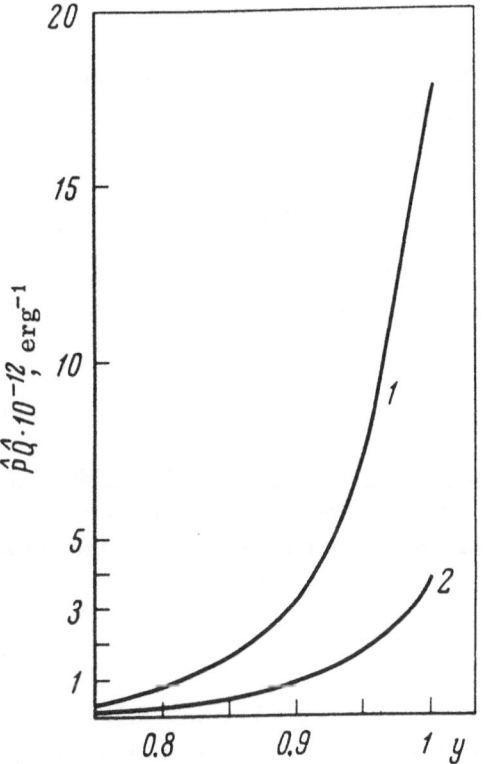

Fig. 6.5. Dependences of the anharmonicity parameters $\hat{P}$ [Eq. (6.37)] and $\hat{Q}$ [Eq. (6.38)] on the lattice constant for Ar-type crystals. 1) $\hat{Q}$; 2) $\hat{P}$.

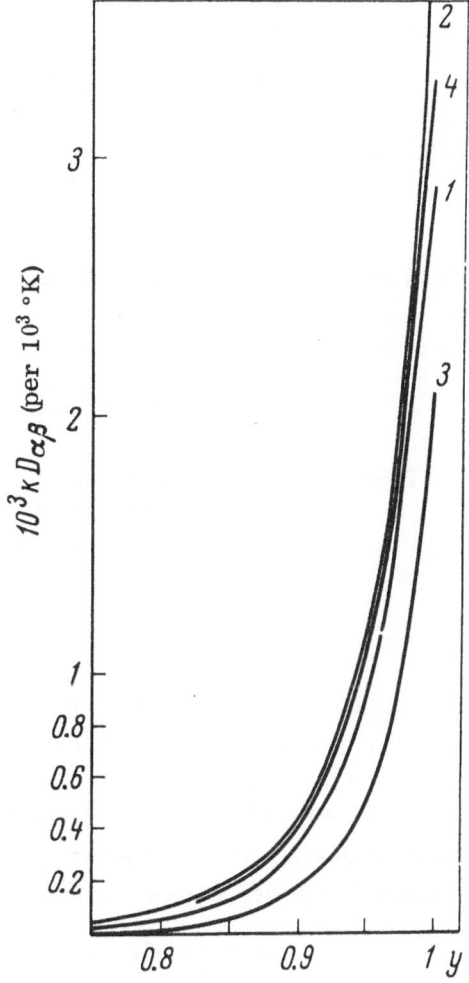

Fig. 6.6. Dependences of the relative temperature coefficients of the elastic moduli $10^3 kD_{\alpha\beta}$ [Eq. (6.31)] on the elastic constant of Ar-type crystals. 1) $10^3 kD_{11}$; 2) $10^3 kD_{12}$; 3) $10^3 kD_{44}$; 4) $10^3 kD_{\chi}$.

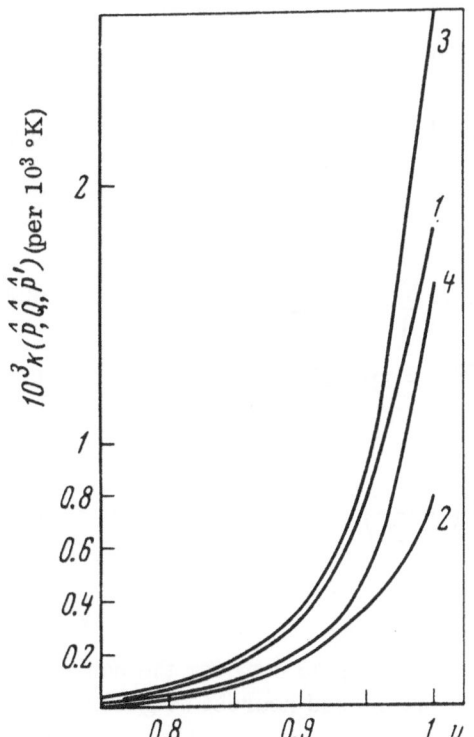

Fig. 6.7. Dependences of the anharmonicity parameters in c_P [Eq. (6.42)] on the lattice constant for Ar-type crystals. 1) $0.75 \cdot 10^3 k\hat{Q}$; 2) $1.5 \cdot 10^3 k\hat{P}$; 3) $10^3 k\hat{P}'$; 4) $10^3 k(\hat{P}' + 1.5\hat{P} - 0.75\hat{Q})$.

6.4.   Ionic Crystals

We shall now consider ionic crystals of the $(I^{-n}I^{+n})$ type. The interaction potential between ions μ, ν (μ, ν = 1, 2 or +, −), with charges $\pm e$ and located at a distance r from one another, will be used in the form

$$\varphi_{\mu\nu}(R) = (-1)^{\mu+\nu}\frac{e^2}{R} + \psi_{\mu\nu}(R) = (-1)^{\mu+\nu}\frac{e^2}{l_0}y^{-1}\left(\frac{l}{R}\right) + \psi_{\mu\nu}(R), \tag{6.63}$$

where l_0 is the shortest distance between unlike (oppositely charged) ions at P = T = 0. We shall use l_{10} to denote the shortest distance between like ions under the same conditions. The potential ψ is of the form

$$\psi_{\mu\nu} = \left\{-A_m^{\mu\nu}y^{-m}\left(\frac{l}{R}\right)^m - A_{m_1}^{\mu\nu}y^{-m_1}\left(\frac{l}{R}\right)^{m_1} + A_{\mu\nu}\exp(-\varkappa_{\mu\nu}y)\right\}, \tag{6.64}$$

where the first two terms describe the van der Waals attraction (m = 6, m_1 = 8) and the last term describes the exponential repulsion. We shall omit the van der Waals terms for the same reasons as those stated at the beginning of Sec. 6.2. The repulsion potential can be found using [5, 6]

$$A_{\mu\nu} = \xi c_{\mu\nu}B\exp\left(\frac{r_\mu + r_\nu}{\rho} - \frac{l_{\mu\nu}}{\rho}\right), \tag{6.65}$$

where r_μ and r_ν are the ionic radii; the Pauling coefficients are c_{+-} = 1, c_{--} = 0.75, c_{++} = 1.25. The factor $\xi \approx 1$ is selected so that the pressure vanishes at T = 0, y = 1. In fact, the pressure should then be equal to the pressure of the zero-point vibrations with its sign reversed. However, since this pressure is small, it can be safely ignored.

We shall carry out our calculations specifically for sodium chloride, which can form NaCl- and CsCl-type structures. In this case,

$$\left.\begin{aligned}
&r_- = 1.81\cdot10^{-8}\text{ cm} \quad r_+ = 0.98\cdot10^{-8}\text{ cm}, \quad \rho = 0.345\cdot10^{-8}\text{ cm,}\\
&B = 0.23\cdot10^{-12}\text{ erg,} \quad e = 4.8025\cdot10^{-10}\text{ cgs esu,} \quad l_0 = 2.81\cdot10^{-8}\text{ cm,}
\end{aligned}\right\} \tag{6.66}$$

$$\varkappa = \frac{l_{0+-}}{\rho} = \frac{l_0}{\rho}, \qquad \varkappa_1 = \frac{l_{0++}}{\rho} = \frac{l_{0--}}{\rho} = \frac{l_{10}}{\rho}. \tag{6.67}$$

We shall also use the following lattice sums:

$$z^{\alpha\beta} = 3\sum_\mu\frac{X^2Y^2}{R^4}, \quad \bar{z}^{\alpha\beta} = 3\sum_{sn}\frac{X^2Y^2}{R^4}, \quad z = \sum_n 1, \quad \bar{z} = \sum_{sn}1; \tag{6.68}$$

$$z_1 = -\sum_{\substack{h\\ \mu\left(\substack{0\\ \neq\\ \nu}\right)}}(-1)^{\mu+\nu}\frac{l}{R}, \quad z_1^{\alpha\beta} = -3\sum_{\substack{h\\ \mu\left(\substack{0\\ \neq\\ \nu}\right)}}(-1)^{\mu+\nu}\frac{l}{R}\frac{X_\alpha^2 X_\beta^2}{R^4}; \tag{6.69}$$

$$z_n^{++} = \sum_{h\neq 0}\left(\frac{l}{R_{11}^h}\right)^n, \quad z_n^{+-} = \sum_h\left(\frac{l}{R_{12}^h}\right)^n; \tag{6.70}$$

$$z_n^{++} + z_n^{+-} = z_n. \tag{6.71}$$

The subscript "n" in Eq. (6.68) denotes that the summation is carried out over the nearest neighbors and the subscript "sn" denotes summation over the second nearest neighbors.

The lattice sums for NaCl- and CsCl-type structures are:

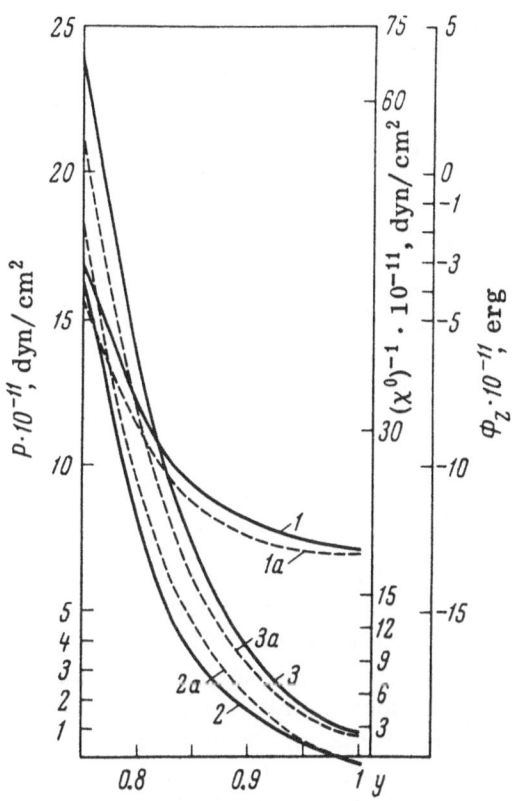

Fig. 6.8. Dependences of the energy, pressure, and the reciprocal of compressibility on the lattice constant of sodium chloride with the NaCl structure. First variant: 1) Φ_z; 2) P; 3) $(\chi^0)^{-1}$. Second variant: 1a) Φ_z; 2a) P; 3a) $(\chi^0)^{-1}$.

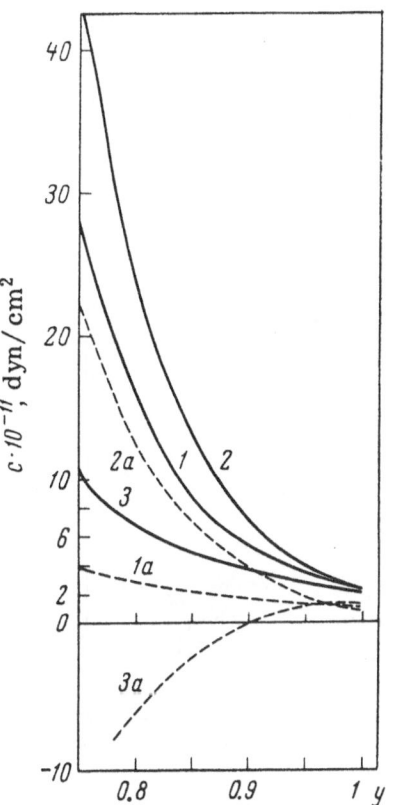

Fig. 6.9. Dependences of the elastic moduli on the lattice constant of sodium chloride with the NaCl structure. First variant: 1) c_{12}^*; 2) c_{12}^0; 3) c_{44}^0. Second variant: 1a) c_{12}^*; 2a) c_{12}^0; 3a) c_{44}^0. The experimental values at normal temperature and pressure [9] are: $c_{12} = 1.23 \cdot 10^{11}$ dyn/cm^2 and $c_{44} = 1.28 \cdot 10^{11}$ dyn/cm^2.

	NaCl	CsCl			NaCl	CsCl
z	6	8		$z_1^{12} = z_1^{44}$	-0.696	1.802
$\bar{z}$	12	6		z_s^{++}	0.800	2.198
z^{11}	6	$8/3$		z_s^{+-}	6.146	8.157
$\bar{z}^{11}$	6	6		z_s	6.946	10.355
$z^{12} = z^{44}$	0	$8/3$		l_{10}/l_0	$\sqrt{2}$	$2/\sqrt{3}$
$\bar{z}^{12} = \bar{z}^{44}$	3	0		$\tau = V_{z0}/l_0^3$	2	$8/3\sqrt{3}$
z_1	1.748	1.763				
z_1^{11}	3.139	-1.841				

As in Sec. 6.3, the quantities calculated for the cold-compression isotherm will be denoted by the superscript zero:

$$P^0 = \frac{1}{3V_{z0}y^2}\left\{ z\varkappa A_{+-}\exp(-\varkappa y) + \frac{\bar{z}}{2}\varkappa_1(A_{++}+A_{--})\exp(-\varkappa_1 y) - \frac{z_1 e^2}{l_0}y^{-2}\right\}, \qquad (6.72)$$

$$(\chi^0)^{-1} = \frac{1}{9V_{z0}y}\left\{ z\varkappa^2 A_{+-}\exp(-\varkappa y)\left(1+\frac{2}{\varkappa y}\right) + \frac{\bar{z}}{2}\varkappa_1^2\left(1+\frac{2}{\varkappa_1 y}\right)(A_{++}+A_{--})\exp(-\varkappa_1 y) - \frac{4z_1 e^2}{l_0}y^{-3}\right\}, \quad (6.73)$$

$$\Phi_z = -\frac{z_1 e^2}{l_0}y^{-1} + zA_{+-}\exp(-\varkappa y) + \frac{\bar{z}}{2}(A_{++}+A_{--})\exp(-\varkappa_1 y), \qquad (6.74)$$

$$\varphi^{0s} = \frac{d^s\varphi}{dR^s} = \frac{(-1)^s}{l_0^s}\left\{ A_{+-}\varkappa^s\exp(-\varkappa y) + (-1)^{\mu+\nu}\frac{e^2}{l_0}y^{-(s+1)}\left(\frac{l}{R}\right)^{s+1}s!\right\} + (-1)^s\left(\frac{\varkappa_1}{l_{10}}\right)^s(A_{++}+A_{--})\exp(-\varkappa_1 y). \quad (6.75)$$

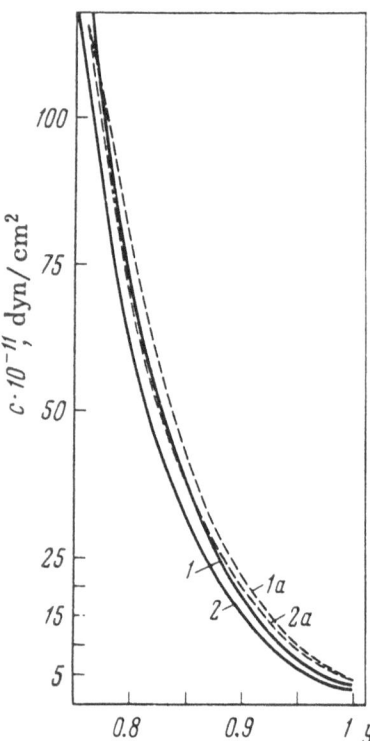

Fig. 6.10. Dependences of the elastic moduli on the lattice constant of sodium chloride with the NaCl structure. First variant: 1) c_{11}^{*}; 2) c_{11}^{0}. Second variant: 1a) c_{11}^{*}; 2a) c_{11}^{0}. The experimental value of c_{11} at normal temperature and pressure is $4.9 \cdot 10^{11}$ dyn/cm^2 [9].

Using Eqs. (6.63), (6.75), (6.2), (6.2a), and (6.2b), we can calculate all the required quantities. The Grüneisen parameter of Eq. (6.10b) is

$$\gamma = \frac{1}{6} \left\{ xy \left(1 - \frac{2}{xy} - \frac{2}{x^2y^2} \right) A_{+-} \exp(-xy) + \right.$$

$$+ \frac{\bar{z}}{z} x_1 y \left(1 - \frac{2}{x_1 y} - \frac{2}{x_1^2 y^2} \right) A_M \exp(-x_1 y) \right\} \times$$

$$\times \left\{ \left(1 - \frac{2}{xy} \right) A_{+-} \exp(-xy) + \right.$$

$$+ \frac{\bar{z}}{z} \left(1 - \frac{2}{x_1 y} \right) A_M \exp(-x_1 y) \right\}^{-1}, \qquad (6.76)$$

where

$$M = \frac{M_+ M_-}{M_+ + M_-}, \quad A_M = M \left(\frac{A_{--}}{M_-} + \frac{A_{++}}{M_+} \right). \quad (6.77)$$

We note that the Coulomb term in the potential does not contribute to the Grüneisen parameter γ of Eq. (6.76). This is because the Grüneisen parameter is defined in Eq. (6.10b) as the logarithmic derivative of the mean-square frequency. This frequency is expressed in terms of the Laplacian of the potential. The application of the Laplacian to the Coulomb term of the potential yields zero as the result. Consequently, we shall calculate the Grüneisen parameter for a more general potential of Eq. (6.64), which includes the van der Waals terms. However, we shall take into account only the van der Waals forces between unlike ions. Then,

$$\Delta \psi_{+-} = -A_m y^{-m} \left(\frac{l}{R} \right)^{m} - A_{m_1} y^{-m_1} \left(\frac{l}{R} \right)^{m_1}, \qquad (6.78)$$

where $m = 6$, $m_1 = 8$.

The correction to $\Delta \Phi_z$ due to the potential of Eq. (6.78) is

$$\Delta \Phi_z = -A_m z_m^{+-} y^{-m} - A_{m_1} z_{m_1}^{+-} y^{-m_1}, \qquad (6.79)$$

and the values of A_m, A_{m_1} are, according to Mayer's estimate [5],

$$A_6 = \frac{C}{z_3^{+-} l_0^6}, \quad A_8 = \frac{D}{z_3^{+-} l_0^8}, \quad C = 1.8 \cdot 10^{-58} \text{ erg} \cdot \text{cm}^6, \quad D = 1.8 \cdot 10^{-76} \text{ erg} \cdot \text{cm}^8. \qquad (6.80)$$

We shall use the quantities

$$\Delta \gamma_d = -\frac{m(m-1)}{x^2} \frac{z_{m+2}^{+-}}{z} A_m y^{-(m+2)} - \frac{m_1(m_1-1)}{x^2} \frac{z_{m_1+2}^{+-}}{z} A_{m_1} y^{-(m_1+2)},$$

$$\Delta \gamma_n = -\frac{(m+2)m(m-1) z_{m+2}^{+-}}{x^2} A_m y^{-(m+2)} - \frac{(m_1+2)m_1(m_1-1) z_{m_1+2}^{+-}}{x^2} \frac{}{z} A_{m_1} y^{-(m_1+2)}, \qquad (6.81)$$

where the subscripts "d" and "n" denote corrections in the denominator and numerator of the expression for γ.

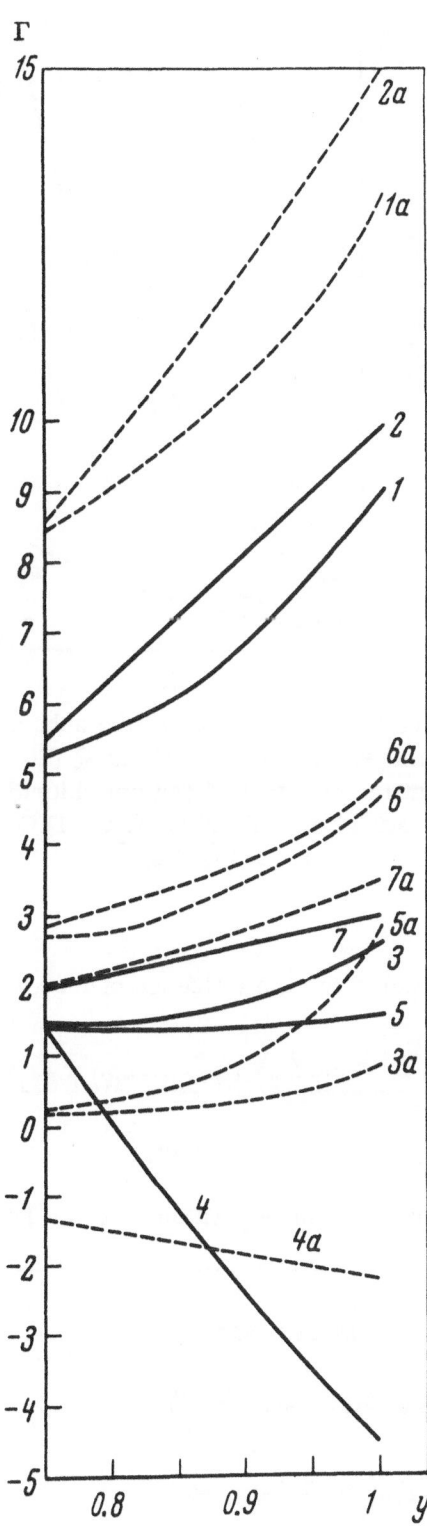

Fig. 6.11. Dependences of the quantities Γ [Eqs. (6.33a) and (6.33b)] on the lattice constant of sodium chloride with the NaCl structure. First variant: 1) Γ_{11}; 2) Γ'_{11}; 3) Γ_{12}; 4) Γ'_{12}; 5) Γ'_{44}; 6) Γ; 7) Γ'. Second variant: 1a) Γ_{11}; 2a) Γ'_{11}; 3a) Γ_{12}; 4a) Γ'_{12}; 5a) Γ'_{44}; 6a) Γ; 7) Γ'.

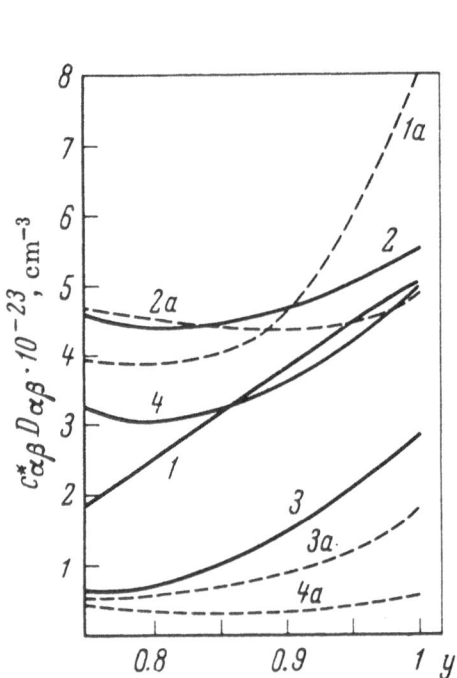 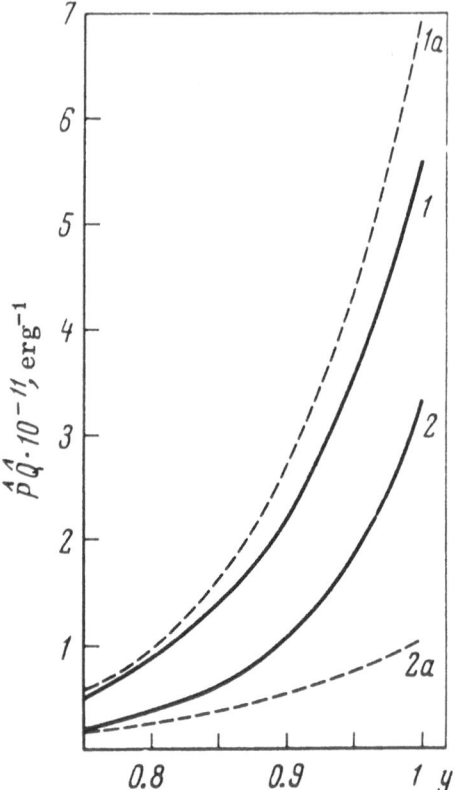

Fig. 6.12. Dependences of the temperature coefficients of the elastic moduli $c^*_{\alpha\beta}D_{\alpha\beta}$ [Eq. (6.31)] on the lattice constant of sodium chloride with the NaCl structure. First variant: 1) $c^*_{11}D_{11}$; 2) $c^*_{12}D_{12}$; 3) $c^*_{44}D_{44}$; 4) $D\chi/\chi^*$. Second variant: 1a) $c^*_{11}D_{11}$; 2a) $c^*_{12}D_{12}$; 3a) $c^*_{44}D_{44}$; 4a) $D\chi/\chi^*$.

Fig. 6.13. Dependences of the anharmonicity parameters $\hat{P}$ [Eq. (6.37)] and $\hat{Q}$ [Eq. (6.38)] on the lattice constant of sodium chloride with the NaCl structure. First variant: 1) $\hat{Q}$; 2) $\hat{P}$. Second variant: 1a) $\hat{Q}$; 2a) $\hat{P}$.

Then, the formula for the Grüneisen parameter corrected for the influence of the van der Waals forces, assumes the form

$$\gamma = \frac{1}{6}\left\{ xy\left(1 - \frac{2}{xy} - \frac{2}{x^2y^2}\right)A_{+-}\exp(-xy) + \frac{\bar{z}}{z}x_1y\left(1 - \frac{2}{x_1y} - \frac{2}{x_1^2y^2}\right)A_M\exp(-x_1y) + \Delta\gamma_n\right\} + \\
\times \left\{\left(1 - \frac{2}{xy}\right)A_{+-}\exp(-xy) + \frac{\bar{z}}{z}\left(1 - \frac{2}{x_1y}\right)A_M\exp(-x_1y) + \Delta\gamma_d\right\}^{-1}. \tag{6.82}$$

The rest of the calculation will be carried out ignoring the van der Waals forces. The quantities c^* of Eqs. (6.27a) and (6.24) become

$$c_{\alpha\beta}^* = \frac{1}{2V_{z0}^0}\left\{\frac{2}{3}x^2z^{\alpha\beta}A_{+-}y^{-1}\left(1 + \frac{1}{xy}\right)\exp(-xy) + \\
+ \frac{x_1^2\bar{z}^{\alpha\beta}}{3y}(A_{++}+A_{--})\left(1 + \frac{1}{x_1y}\right)\exp(-x_1y) - 2z_1^{\alpha\beta}\frac{e^2}{l_0}y^{-4}\right\}, \tag{6.83}$$

$$(\chi^*)^{-1} = \frac{1}{3}(c_{11}^* + 2c_{12}^*) = \frac{1}{18V_{z0}^0}\left\{2\frac{x^2z}{y}A_{+-}\left(1 + \frac{1}{xy}\right)\exp(-xy) + \\
+ \frac{x_1^2\bar{z}}{y}(A_{++}+A_{--})\left(1 + \frac{1}{x_1y}\right)\exp(-x_1y) - 6\frac{e^2}{l_0}z_1y^{-4}\right\}. \tag{6.84}$$

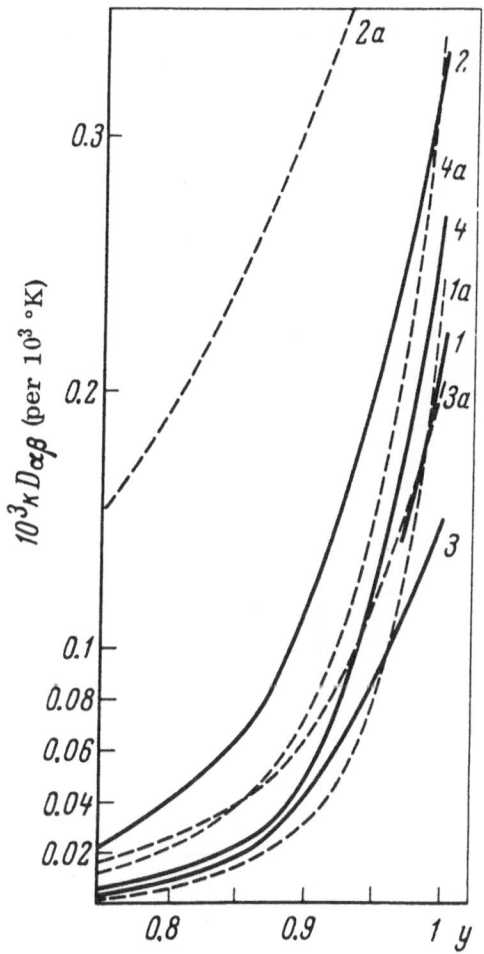

Fig. 6.14. Dependences of the relative temperature coefficients of the elastic moduli $10^3 k D_{\alpha\beta}$ [Eq. (6.31)] on the lattice constant of sodium chloride with the NaCl structure. First variant: 1) $10^3 k D_{11}$; 2) $10^3 k D_{12}$; 3) $10^3 k D_{44}$; 4) $10^3 k D_\chi$. Second variant: 1a) $10^3 k D_{11}$; 2a) $10^3 k D_{12}$; 3a) $10^3 k D_{44}$; 4a) $10^3 k D_\chi$.

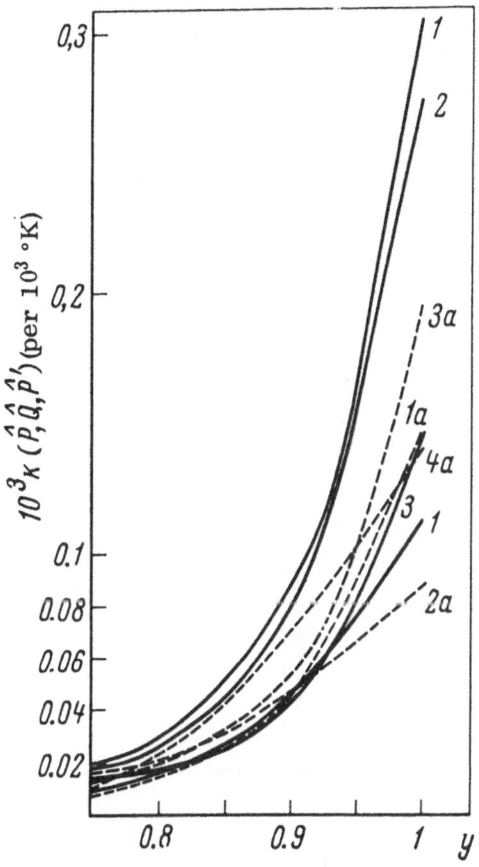

Fig. 6.15. Dependences of the anharmonicity parameters in cp [Eq. (6.42)] on the lattice constant of sodium chloride with the NaCl structure. First variant: 1) $1.5 \cdot 10^3 k \hat{Q}$; 2) $6 \cdot 10^3 k \hat{P}$; 3) $2 \cdot 10^3 k \hat{P}$; 4) $10^3 k (2\hat{P}' + 6\hat{P} - 1.5\hat{Q})$. Second variant: 1a) $1.5 \cdot 10^3 k \hat{Q}$; 2a) $6 \cdot 10^3 k \hat{P}$; 3a) $2 \cdot 10^3 k \hat{P}$; 4a) $10^3 k (2\hat{P}' + 6\hat{P} - 1.5\hat{Q})$.

The quantities with the superscript zero are related to the quantities with an asterisk by Eqs. (6.56) and (6.57). The quantities defined in Eq. (6.33) now become

$$
\left.
\begin{aligned}
\Gamma_{\alpha\beta} &= \frac{\chi^0}{18 V_{z0}{}^0} \left\{ 2\varkappa^3 z^{\alpha\beta} A_{+-} \exp(-\varkappa y)\left(1 + \frac{3}{\varkappa y} + \frac{3}{\varkappa^2 y^2}\right) + \right. \\
&\quad + \varkappa_1{}^3 \bar{z}^{\alpha\beta}(A_{++} + A_{--})\exp(-\varkappa_1 y)\left(1 + \frac{3}{\varkappa_1 y} + \frac{3}{\varkappa_1{}^2 y^2}\right) - 30 z_1{}^{\alpha\beta} \frac{e^2}{l_0} y^{-4}\right\}, \\
\Gamma_{11}' &= -\gamma^2 + \gamma\chi^0 \frac{c_{11}{}^*}{3} + I^{11} \\
\Gamma_{12}' &= \Gamma_{44}' - \gamma^2 = -\gamma^2 + \frac{1}{3}\gamma\chi^0 c_{12}{}^* + I^{12},
\end{aligned}
\right\}
\qquad (6.85)
$$

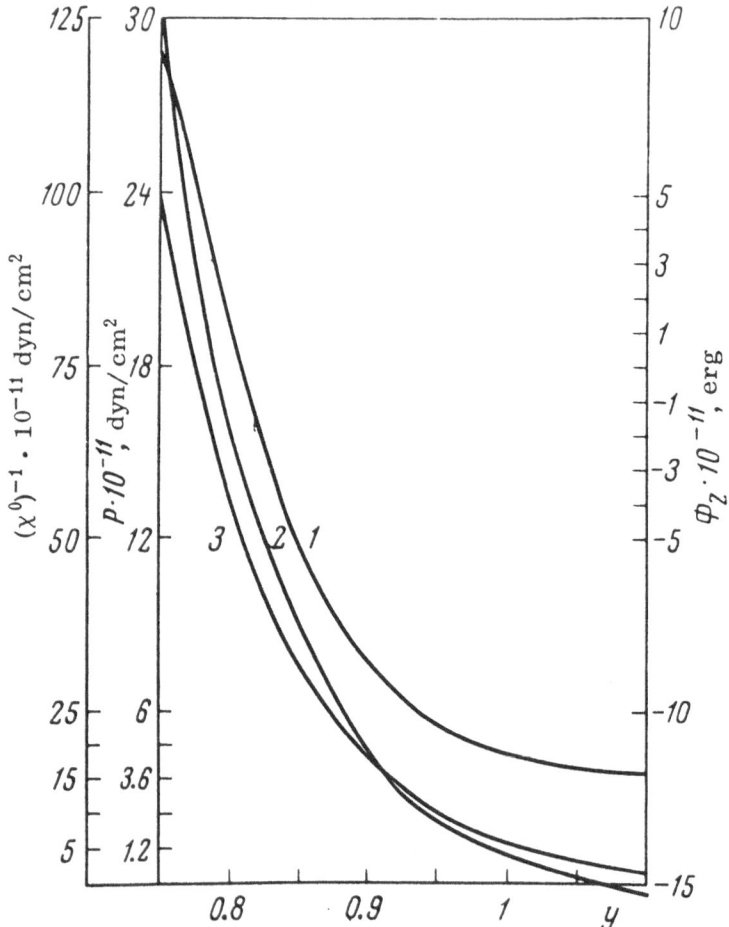

Fig. 6.16. Dependences of the energy, pressure, and the reciprocal of
compressibility on the lattice constant of sodium chloride with the
CsCl structure: 1) Φ_Z; 2) P; 3) $(\chi^0)^{-1}$.

$$I^{\alpha\beta} = \frac{z^{\alpha\beta}\left\{ x^2 y^2 A_{+-} \exp(-xy)\left(1 - \dfrac{1}{xy} - \dfrac{6}{x^2 y^2} - \dfrac{6}{x^3 y^3}\right)\right\}}{6z\left\{A_{+-}\left(1 - \dfrac{2}{xy}\right)\exp(-xy) + \dfrac{\bar{z}}{z}A_M\left(1 - \dfrac{2}{x_1 y}\right)\exp(-x_1 y)\right\}} +$$

$$+ \frac{\bar{z}^{\alpha\beta}x_1^2 y^2 A_M \exp(-x_1 y)\left(1 - \dfrac{1}{x_1 y} - \dfrac{6}{x_1^2 y^2} - \dfrac{6}{x_1^3 y^3}\right)}{6z\left\{A_{+-}\left(1 - \dfrac{2}{xy}\right)\exp(-xy) + \dfrac{\bar{z}}{z}A_M\left(1 - \dfrac{2}{x_1 y}\right)\exp(-x_1 y)\right\}} . \qquad (6.86)$$

The quantities Γ and Γ' are still given by the formulas in Eq. (6.33b). Knowing the dependences
of Γ_{ik} and Γ_{ik}^0 on the volume, we can use Eqs. (6.30), (6.31), and (6.56) to determine the temperature dependences of the elastic moduli at high pressures.

The quantities $\hat{P}$ and $\hat{Q}$ of Eqs. (6.36)–(6.39) are found using Eqs. (6.63)–(6.65), (6.75),
(6.2), (6.2a), and (6.2b).

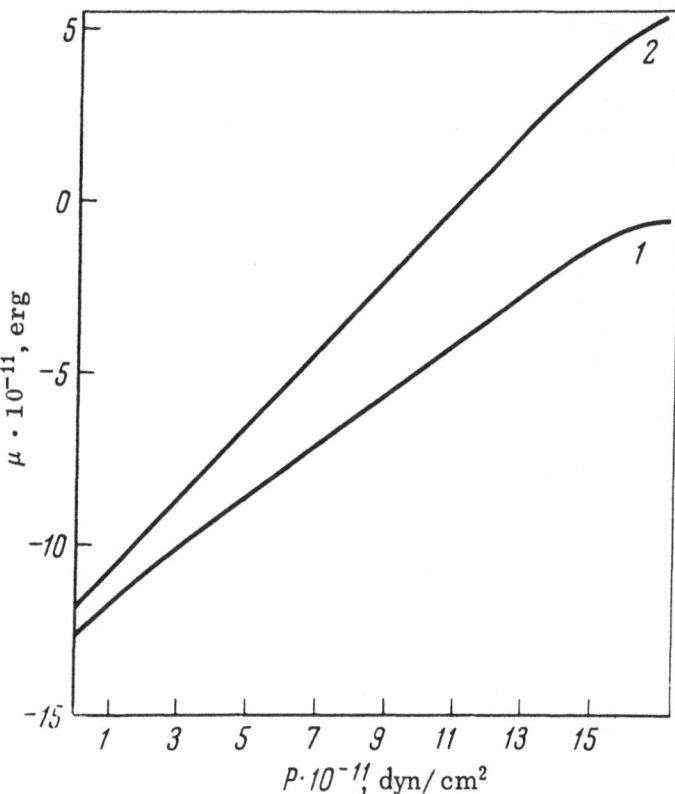

Fig. 6.17. Pressure dependences of the chemical potential μ of sodium chloride with the NaCl and CsCl structures: 1) μ for the NaCl structure; 2) μ for the CsCl structure.

We thus obtain

$$\hat{Q} = \frac{A_{+-}\exp(-\varkappa y)\left(1 - \dfrac{4}{\varkappa y}\right) + \dfrac{\bar{z}}{z}A_{2M}\exp(-\varkappa_1 y)\left(1 - \dfrac{4}{\varkappa_1 y}\right)}{z\left\{A_{+-}\left(1 - \dfrac{2}{\varkappa y}\right)\exp(-\varkappa y) + \dfrac{\bar{z}}{z}A_M\left(1 - \dfrac{2}{\varkappa_1 y}\right)\exp(-\varkappa_1 y)\right\}^2}, \tag{6.87}$$

where

$$A_{2M} = 2M^2\left(\frac{A_{++}}{M_+^2} + \frac{A_{--}}{M_-^2}\right); \tag{6.88}$$

$$\hat{P} = \frac{3}{z^2\Omega_1}\left\{\frac{M^2}{M_+M_-}A_{+-}^2\left[1 + \frac{6}{\varkappa^2 y^2}\left(1 + \frac{1}{\varkappa y}\right)^2\right]\exp(-2\varkappa y) + \frac{\bar{z}}{z}A_{3M}^3\left[1 + \frac{6}{\varkappa_1^2 y^2}\left(1 + \frac{1}{\varkappa_1 y}\right)^2\right]\exp(-2\varkappa_1 y) - \right.$$

$$- \frac{12M^2}{\varkappa^3 M_+M_-}A_{+-}\exp(-\varkappa y)\frac{e^2}{l_0}y^{-4}\left[1 + \frac{3}{\varkappa y}\left(1 + \frac{1}{\varkappa y}\right)\right] + \frac{12\bar{z}}{z\varkappa_1^3}\frac{e^2}{l_{10}}y^{-4}\left[1 + \frac{3}{\varkappa_1 y}\left(1 + \frac{1}{\varkappa_1 y}\right)\right]\bar{A}_{3M}\exp(-\varkappa_1 y) +$$

$$\left. + \frac{90}{\varkappa^8}\frac{e^4}{l_0^2}y^{-8}\left[\frac{z_8^{+-}}{z}\frac{M^2}{M_+M_-} + \frac{z_8^{++}}{z}\left(\frac{M^3}{M_+^3} + \frac{M^3}{M_-^3}\right)\right]\right\}, \tag{6.89}$$

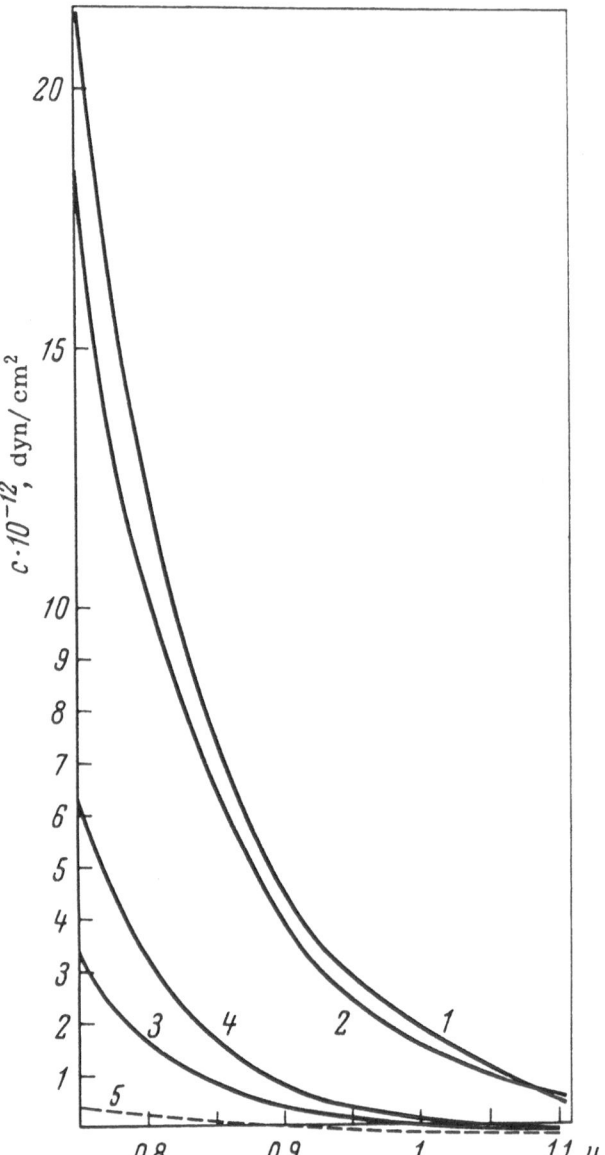

Fig. 6.18. Dependences of the elastic moduli on the lattice constant of sodium chloride with the CsCl structure: 1) c_{11}^{*}; 2) c_{11}^{0}; 3) c_{12}^{*}; 4) c_{12}^{0}; 5) c_{44}^{0}.

where

$$\Omega_1 = \left\{ \left[A_{+-}\left(1 - \frac{2}{\varkappa y}\right)\exp(-\varkappa y) + \frac{\bar{z}}{z}A_M\left(1 - \frac{2}{\varkappa_1 y}\right)\exp(-\varkappa_1 y)\right]^3 - \frac{\bar{z}}{z^3}\bar{A}_M{}^3\left(1 - \frac{2}{\varkappa_1 y}\right)^3\exp(-3\varkappa_1 y) \right\}, \quad (6.90)$$

$$A_{3M}^{2} = M^3\left(\frac{A_{++}^{2}}{M_+{}^3} + \frac{A_{--}^{2}}{M_-{}^3}\right), \quad \bar{A}_{3M} = M^3\left(\frac{A_{++}}{M_+{}^3} + \frac{A_{--}}{M_-{}^3}\right). \quad (6.91)$$

The actual calculations for sodium chloride with the NaCl structure were carried out in two ways. In the first variant of the calculations, the parameters were determined using Eqs.

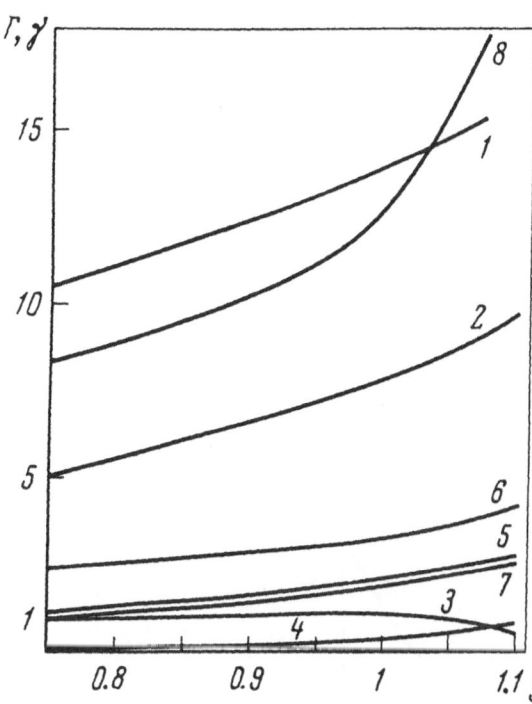

Fig. 6.19. Dependences of the quantities Γ [Eqs. (6.33a) and (6.33b)] and γ [Eq. (6.10b)] on the lattice constant of sodium chloride with the CsCl structure: 1) Γ_{11}; 2) Γ'_{11}; 3) Γ_{12}; 4) Γ'_{12}; 5) Γ'_{44}; 6) Γ; 7) Γ'; 8) 10γ.

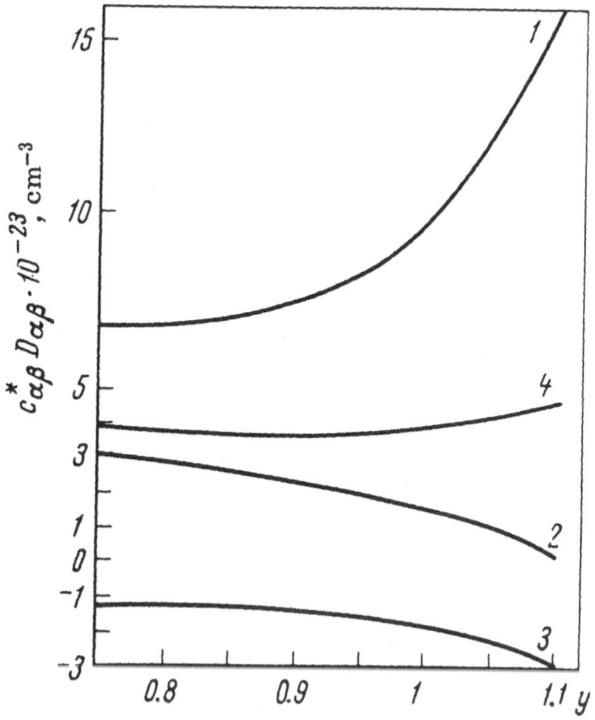

Fig. 6.20. Dependences of the temperature coefficients of the elastic moduli $c^*_{\alpha\beta}\,D_{\alpha\beta}$ [Eq. (6.31)] on the lattice constant of sodium chloride with the CsCl structure: 1) $c^*_{11}D_{11}$; 2) $c^*_{12}D_{12}$; 3) $c^*_{44}D_{44}$; 4) D_χ/χ^*.

(6.66)–(6.67). We took into account the repulsion between the nearest neighbors and between the nearest negative ions. The van der Waals attraction was ignored. The quantity ξ in Eq. (6.65) was assumed to be 0.9633. The potential used in this variant was the same as that employed in [6]. The results of the calculations are represented by the continuous curves in Figs. 6.8–6.15.

In the second variant, we included only the repulsion between unlike ions. The parameters for this repulsion were taken from [7]. In our notation, these parameters are

$$\varkappa = 9.69, \qquad A_{+-} = 0.805 \cdot 10^{14}\,V_{z0}. \tag{6.92}$$

The results of the second-variant calculations are represented by the dashed curves in Figs. 6.8–6.15. The Grüneisen parameter was considered separately (it will be discussed at the end of the section). Then, we calculated the corresponding values for sodium chloride having the CsCl structure. The results of these calculations are presented in Figs. 6.16–6.23.

It follows from Fig. 6.8 that both potentials give similar values of the energy Φ_z, the pressure P, and the compressibility χ^0 in a very wide range of changes of the lattice constant. This is not accidental.

The results are similar because Φ_z is expressed in terms of an integral of P and because $(\chi^0)^{-1}$ is expressed in terms of a derivative of P. The pressures calculated using the two potentials are similar [6] and, as shown in [8], they are in satisfactory agreement with the experimental data on the shock compression of sodium chloride up to $\sim 7 \cdot 10^5$ bar.

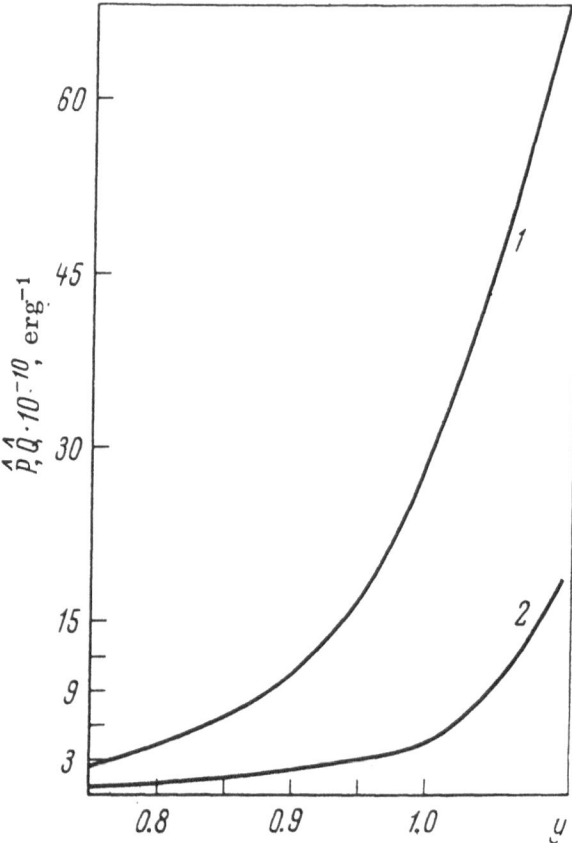

Fig. 6.21. Dependences of the anharmonicity parameters
$\hat{P}$ [Eq. (6.37)] and $\hat{Q}$ [Eq. (6.38)] on the lattice constant of
sodium chloride with the CsCl structure: 1) $\hat{Q}$; 2) $\hat{P}$.

The dependences of the elastic moduli on the lattice constant are presented in Figs. 6.9–6.10. We can see that c_{11}^0 of Eq. (6.65) depends weakly on the type of the potential but c_{44}^0 and c_{12}^0 are appreciably different for the two variants. In spite of the fact that the dashed curves for y = 1 satisfy best the experimental data, we shall conclude that the potential used in the first variant (two exponential functions) is more correct. This is because the first and second variants predict quite different behavior of the elastic modulus c_{44}^0. In agreement with the experimental data reported in [9], the first variant predicts a weak rise of c_{44}^0 with decreasing y, whereas the second variant predicts a fall in c_{44}^0. Moreover, it seems to us that the small deviation from the experimental data for y = 1 in the first variant can be used to refine the potential. It follows from Figs. 6.11 and 6.12 that the selected potential affects the values of $\Gamma_{\alpha\beta}$ [Eqs. (6.33a) and (6.33b)] and $c_{\alpha\beta}^* D_{\alpha\beta}$ [Eq. (6.31)]. Figure 6.13 shows the anharmonicity parameters $\hat{Q}$ [Eq. (6.38)] and $\hat{P}$ [Eq. (6.37)]. A very important conclusion can be drawn from this figure: the parameters $\hat{Q}$ and $\hat{P}$ decrease exponentially with decreasing y.

Figure 6.14 shows the dependences of the relative temperature coefficients of the elastic moduli on the lattice constant, per 10^3 °K. It is evident from this figure that these coefficients decrease very rapidly with increasing pressure.

Figure 6.15 shows the parameters $\hat{P}$, $\hat{Q}$, and $\hat{P}'$, per 10^3 °K, which determine the temperature dependence of the specific heat c_P of Eq. (6.42) at high pressures. We can see that the temperature dependence of c_P weakens exponentially with increasing pressure.

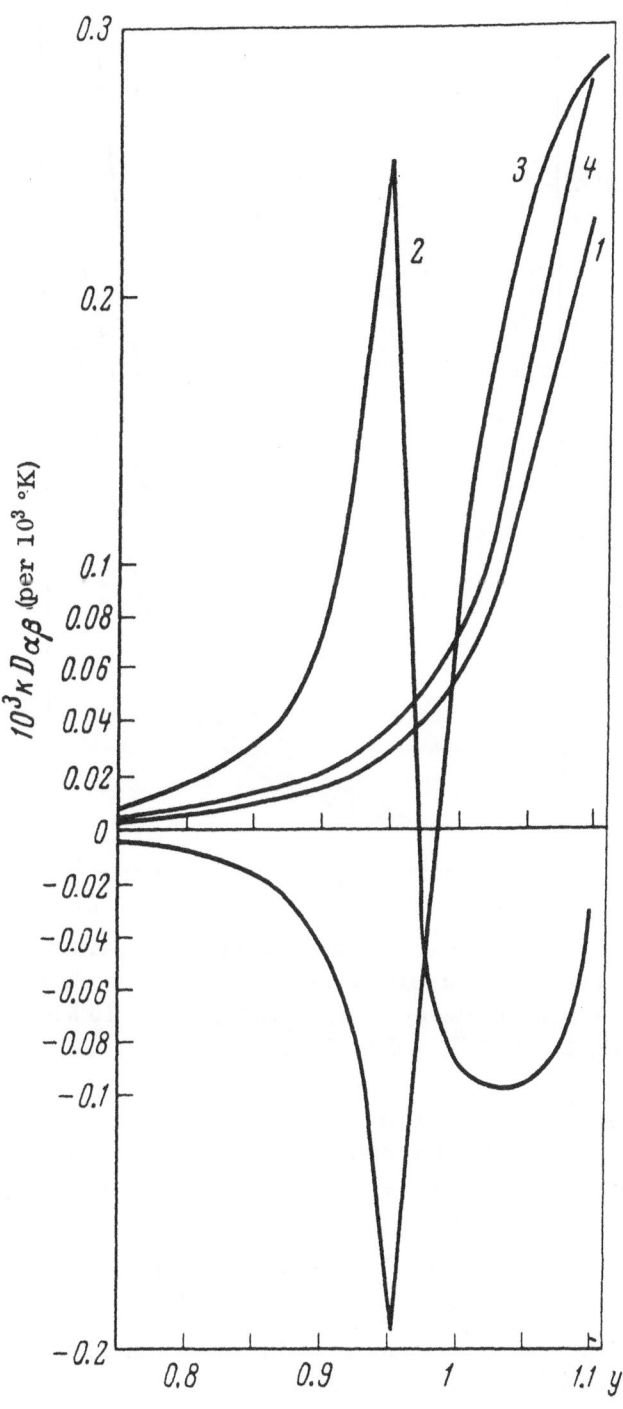

Fig. 6.22. Dependence of the relative temperature coefficients of the elastic moduli $10^3kD_{\alpha\beta}$ [Eq. (6.31)] on the lattice constant of sodium chloride with the CsCl structure: 1) 10^3kD_{11}; 2) 10^3kD_{12}; 3) 10^3kD_{44}; 4) 10^3kD_{χ}.

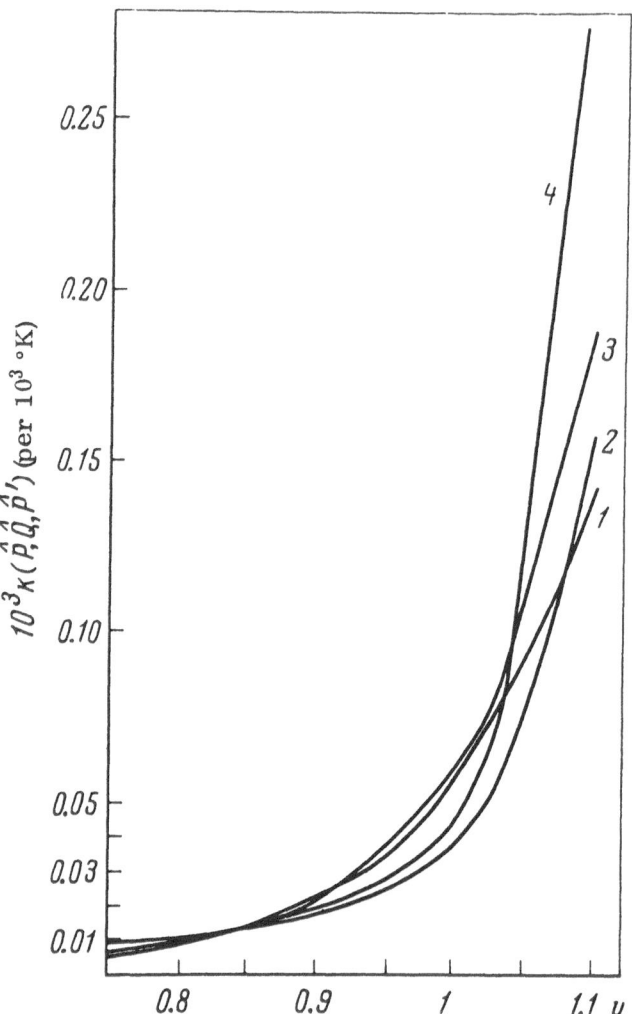

Fig. 6.23. Dependences of the anharmonicity parameters in c p [Eq. (6.42)] on the lattice constant of sodium chloride with the CsCl structure: 1) $1.5 \cdot 10^3 k\hat{Q}$; 2) $6 \cdot 10^3 k\hat{P}$; 3) $2 \cdot 10^3 k\hat{P}$; 4) $10^3 k(2\hat{P}' + 6\hat{P} - 1.5\hat{Q})$.

The calculations for sodium chloride with the CsCl structure (Figs. 6.16-6.22) were carried out using the parameters of the first variant described earlier in the section. The value $y = 1$ was taken to represent the shortest distance between Cl^- and Na^+ in the CsCl structure, which was $l_0 = 2.81 \cdot 10^{-8}$ cm.

The ratio of the volumes per ion pair V_z for the NaCl- and CsCl-type structures was found to be 1.3 (for the same value of l_0 in both structures).

Figure 6.17 shows the pressure dependence of the chemical potential of sodium chloride with the NaCl- and CsCl-type structures at $T = 0°K$:

$$\mu = \Phi_z(P) + PV_{z0}y^3.$$

The parameters of the first variant were used in the calculations for the NaCl-type structure.

It is evident from Fig. 6.17 that, throughout the investigated range of pressures, the NaCl-type structure is thermodynamically stable. However, this result is very sensitive to the

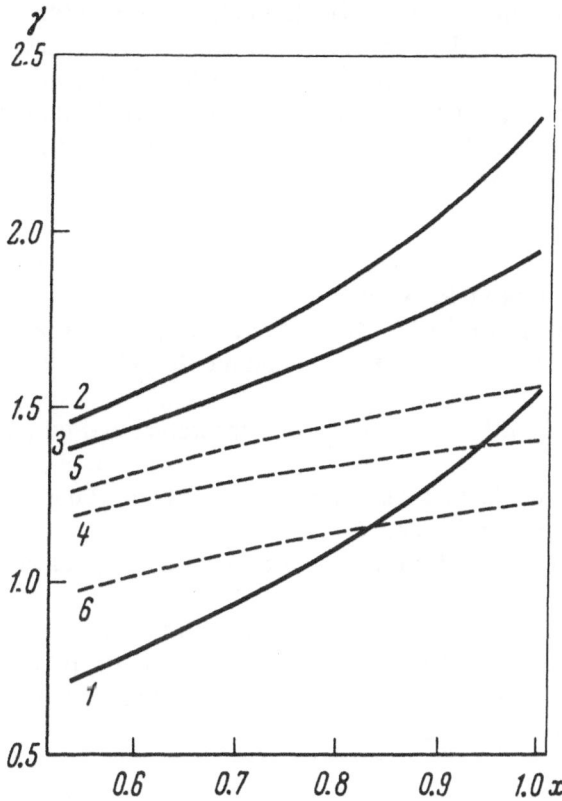

Fig. 6.24. Dependence of the Grüneisen parameter of sodium chloride with the NaCl structure on the relative volume. 1) Experimentally determined Grüneisen parameter [8]; 2) Grüneisen parameter calculated using the Landau — Slater formula [8]; 3) Grüneisen parameter calculated using the Dugdale — MacDonald formula [8]; 4) Grüneisen parameter calculated using Eq. (6.76) and the parameters of the first variant; 5) Grüneisen parameter calculated using Eq. (6.76) and the parameters of the second variant; 6) Grüneisen parameter calculated using Eq. (6.82) and allowing for the van der Waals forces.

selection of the potential and it cannot be regarded as a firm theoretical proof. We must point out that μ_{NaCl} and μ_{CsCl} are represented by nearly parallel curves right up to P ~ 10^6 bar. In order to calculate the phase diagram of sodium chloride, we would have to determine the potential very accurately and to include all other interactions, particularly the van der Waals attraction between the Na^+ and Cl^- ions.

At present, it is not yet possible to select the parameters of these interactions in such a way as to obtain theoretical phase diagrams of practical interest. We shall make no comments on the results of the calculations presented in Figs. 6.16 and 6.18-6.23. They are basically similar to the results obtained for the NaCl-type structure. We shall only point out that the selected potential yields negative values of c_{44}^0 in a wide range of pressures (up to ~9 · 10^5 bar). This means that the structure of CsCl is mechanically unstable. We may find that a more accurate form of the potential will show that this structure is stable. However, it is interesting to stress that at high pressures we may meet an interesting situation in which a new phase does not form because of its mechanical instability.

Figure 6.24 shows the dependence of the Grüneisen parameter γ on the relative volume $x = y^3$. The same figure includes the results of our experimental investigation of this parameter [8]. Curve 2 can be regarded as representing the low-frequency γ and curves 4-6 as representing the high-frequency γ. Figure 6.24 shows convincingly that the low-frequency and high-frequency Grüneisen parameters differ quite appreciably and neither agrees with the experimental values of γ (curve 1).

Literature Cited

1.   V. N. Zharkov, in: Solids at Pressures and Temperatures in Earth's Interior [in Russian], Nauka, Moscow (1964), p. 41; Dokl. Akad. Nauk SSSR, 154:302 (1964).
2.   G. Leibfried and W. Ludwig, "Theory of anharmonic effects in crystals," Solid State Phys., 12:276 (1961).
3.   L. D. Landau and E. M. Lifshitz, Theory of Elasticity, 2nd ed., Pergamon Press, Oxford (1969).
4.   J. E. Lennard-Jones and A. E. Ingham, Proc. Roy. Soc. (London), A107:66 (1925); R. D. Misra, Proc. Cambridge Phil. Soc., 36:173 (1940).
5.   M. Born and K. Huang, Dynamical Theory of Crystal Lattices, Oxford University Press (1954).
6.   V. N. Zharkov, Trudy Inst. Fiziki Zemli Akad. Nauk SSSR, No. 11(178), p. 14 (1960).
7.   V. I. Davydov, Izv. Akad. Nauk SSSR, Ser. Geofiz., No. 12, p. 1411 (1956).
8.   V. N. Zharkov and V. A. Kalinin, Dokl. Akad. Nauk SSSR, 145:551 (1962).
9.   D. Lazarus, Phys. Rev., 76:545 (1949).

CHAPTER 7

APPLICATIONS OF EQUATIONS OF STATE TO THE PHYSICS OF THE EARTH'S INTERIOR

The methods described so far have been used in investigations of the earth's interior [1-7]. The earth is a natural high-pressure laboratory and provides a convenient object for the application of the methods used in high-pressure physics, which yield very illuminating information about the earth's interior. Such applications to the physics of the earth were begun in 1952 by Magnitskii [8-10] and Birch [11].

7.1. Internal Structure of the Earth

Current ideas about internal structure of our planet are given in Magnitskii's book [8]. Our treatment is based on Zharkov's paper [6].

Geophysical Data

According to the seismological data the earth can be divided into three main regions: the crust, the mantle, and the core. The thickness of the earth's crust is irregular: it varies from ~10 km in oceanic regions to many tens of kilometers in mountainous regions of the continents. The contribution of the crust to the total mass of the earth and to its moment of inertia, relative to its rotation axis, is small. Therefore, for convenience, the terrestrial crust is usually

TABLE 7.1

Layer	Depth, km	Longit. wave velocity, km/sec	Trans. wave velocity, km/sec	Fraction of vol.	Mass, 10^{25} g	Fraction of mass
			Crust			
A	0—33	Erratic	Erratic	0.0155	5	0.008
			Mantle			
B	33—410	7.8—9.0	4.4—5.0	0.1667	62	0.194
C	410—1000	9.0—6.4	5.0—6.4	0.2131	98	0.164
D	1000—2700	11.4—13.6	6.4—7.3	0.4428	245	0.410
D'	2700—2900	13.6	7.3			
			Core			
E	2900—4980	8.1—10.4	Not observed	0.1516		
F	4980—5120	Acc. to Jeffreys 10.4—9.5 Acc. to Gutenberg 10.1—11.23	The same	0.0028	188	0.315
G	5120—6373	11.2—11.3	»	0.0076		

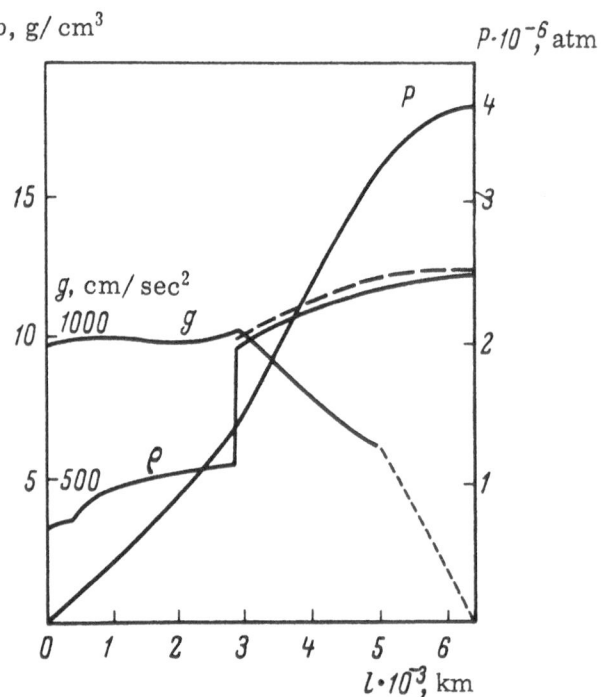

Fig. 7.1. Distribution of the density ρ, the pressure P, and the acceleration due to gravity g, in the earth's interior. The dashed curve shows the upper limit of the density for a uniform core, calculated by Molodenskii [16].

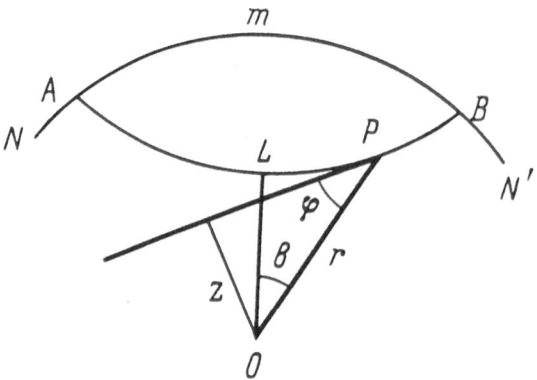

Fig. 7.2. Propagation of a seismic ray in the earth.

represented by a uniform layer with an effective depth of ~33 km. The mantle, which lies under the crust, extends from 33 to 2900 km and is of silicate composition. Finally, the central part of the earth, extending from depths of 2900 to 6373 km, is the core. Seismological data indicate that the mantle and the core have a definite structure. Table 7.1 (based on Bullen's data [12]) and Fig. 7.1 give some characteristics of the various regions and layers of the earth.

The problem of the dependence of the density $\rho(l)$, the pressure $P(l)$, and the adiabatic bulk modulus $K_S(l)$ on the depth l is basically solved. The information has been obtained by seismological measurements of the velocities of the longitudinal $v_P = v_P(l)$ and transverse $v_S = v_S(l)$ waves in the mantle and the core.* These measurements are discussed in the geophysical literature (for example, see [12, 13]) and, therefore, we shall simply present here the results relevant to our discussion.

An earthquake generates elastic vibrations, known as seismic waves, which originate from the focus of the earthquake, which is a finite region usually located under the surface of the earth. The time interval during which most of the earthquake energy is released is of the order of 1 sec (with the exception of extremely strong earthquakes). The linear dimensions of the focal region are of the order of several kilometers. Seismic waves are also generated by man-made explosions.

Bearing in mind that the distances from the source of seismic waves amount to hundreds or even thousands of kilometers, we shall assume that the focus is a point and, for simplicity, we shall take the focus to be located on the surface of the earth. The travel times of the seismic waves are of the order of minutes or tens of minutes and, therefore, we may assume that the energy release at the focus is instantaneous. Records of seismic waves, obtained by special instruments known as seismographs, are described as seismograms and the beginning of each trace in a seismogram is called the arrival time. A seismogram contains the primary experimental material from which the majority of conclusions about the internal structure of the earth is drawn.

In the interpretations of seismograms, the mantle is regarded, to a good approximation, as an ideal elastic isotropic body, and the core as an ideal liquid.

The wavelengths of seismic waves are of the order of 10 km and, therefore, the polycrystalline mantle behaves as an elastic isotropic body. Consequently, elastic properties of the mantle can be characterized by only two parameters and its density. These two parameters are usually the bulk modulus K_S and the shear modulus μ, or K_S and Poisson's ratio σ.

The values of K_S, μ, and σ increase with depth mainly because of the pressure of the upper layers. Variations of the elastic moduli over distances of the order of the seismic wavelengths can be ignored. This gives rise to two remarkable phenomena.

1) The volume (bodily) waves in the mantle can be divided easily into the longitudinal and transverse waves, whose velocities v_P and v_S are

$$v_P^2 = \frac{K_S + \frac{4}{3}\mu}{\rho},$$
$$v_S^2 = \frac{\mu}{\rho}.$$

---

*The core of the earth does not transmit the transverse waves. Therefore, at least in the outer part of the core (layer E), we have $v_S(l) = 0$.

The longitudinal waves have a higher velocity, they are recorded first on the seismograms, and are called primary (the P waves); the transverse waves are called secondary (the S waves). In liquid regions, for example, in the outer core (layer E), we have

$$v_P{}^2 = \frac{K_S}{\rho}, \qquad v_S = 0.$$

2) The propagation of seismic waves in the earth obeys very well the laws of geometrical optics. Moreover, the earth exhibits spherical symmetry in relation to the propagation of seismic rays. Deviations from the spherical symmetry are very small.

A seismogram consists of several phases (groups) of seismic waves, corresponding to the different paths of a ray traveling from the earthquake focus to the recording station; a large number of such paths is due to the reflection from the surface of the earth and the reflection or refraction in the internal regions. The seismic wave phases differ also in another respect: over certain paths a seismic wave is either longitudinal (P) or transverse (S). Each phase is usually represented by a group of oscillations in a seismogram. Seismograms contain information on the arrival times, amplitudes, and periods of these oscillations. The most accurate information on the internal structure of the earth is obtained from the arrival times deduced from seismograms. We shall consider these arrival times in more detail.

Since the earth can be regarded as spherically symmetrical, the travel time t between a pair of points on the surface depends only on the angular epicentral distance between them, Δ, measured from the center of the earth, but it is independent of the position of the earthquake focus and the seismic station. In Fig. 7.2 the curve NAmBN' is the surface of the earth, the point O is the center of the earth, ALPB is a seismic ray whose epicentral distance Δ is equal to the length of an arc AmB, expressed in degrees. Measurements of the travel times of seismic rays t along different trajectories make it possible to determine t for the whole earth as a function of the epicentral distance Δ: $t = t(\Delta)$. The experimental curve drawn in this way is known as the seismic hodograph. The accuracy of seismic hodographs is 1-2%. The experimentally determined function $t = t(\Delta)$ makes it possible to determine the distribution of the velocities of the P and S waves within the earth. Let us consider Fig. 7.2 where ALPB is a seismic ray, P is a running point on the ray, and L is the vertex of the ray. The convex part of the ray is facing the center of the earth because the seismic wave velocities increase with the depth. The polar coordinates of the point P, relative to the center of the earth O, will be denoted by r and θ; z is the normal from O onto the tangent to the ray at the point P; v is the wave velocity of a P or S wave at the point P. As in geometrical optics, the travel time t along the path ALPB has its extremum value (Fermat's principle). This yields an equation for the seismic ray:

$$\frac{r \sin \varphi}{v} = \frac{z}{v} = \lambda,$$

where the parameter $\lambda = \lambda(r)$ is constant along each ray but varies from one ray to another. Considering a pair of neighboring rays, we can easily establish that $\lambda = dt/d\Delta$. For the ray ALPB, we have $\Delta = 2\int d\theta$, where the integral is evaluated between the limits L and B. Moreover,

$$\frac{d\theta}{dr} = \frac{z}{r\sqrt{r^2 - z^2}}.$$

After the substitution of the variables $\eta = r/v$, the expression for Δ assumes the form

$$\Delta = 2 \int_{\lambda}^{\eta_0} -\frac{\lambda}{r\sqrt{\eta^2 - \lambda^2}} \frac{dr}{d\eta} d\eta. \tag{7.1}$$

Here, the zero subscript is used for the values on the surface of the earth. The lower limit of integration is λ, because $\eta = r/v = z/v$ at the point L, which is equal to λ. If we know v as a function of r, then r is a known function of η. In this case, the integral equation just given defines $\Delta = \Delta(\lambda)$ and the differential equation $\lambda = dt/d\Delta$ makes it possible to determine the hodograph theoretically. In fact, we are faced with the reverse problem: knowing the experimental function $t(\Delta)$ we have to determine $v(r)$.

The shortest path is found by applying the operator

$$\int_{\eta_1}^{\eta_0} \frac{d\lambda}{\sqrt{\lambda^2 - \eta^2}}$$

to both parts of Eq. (7.1). Here, η_1 is any convenient value. After simplifications [12], we obtain the required formula

$$\int_0^{\Delta_1} \cosh^{-1}\left(\frac{\lambda}{\eta_1}\right) d\Delta = \pi \ln \frac{r_0}{r_1}, \qquad (7.2)$$

where r_1 corresponds to η_1. Let us assume that we are given the value of η_1. Then, since

$$\eta_1 = \lambda_{\Delta=\Delta_1} = \left(\frac{dt}{d\Delta}\right)_{\Delta_1},$$

the corresponding value of Δ_1 can be obtained from the hodograph. Moreover, the hodograph gives λ as a function of Δ. Consequently, the value of r_1 can be found from Eq. (7.2) for a known value of η_1. Numerical integrations of this type yield r as a function of η; consequently, $v = r/\eta$ is now given in the form of a function of r. This method is valid only if the quantity $d\eta/dr$ does not change its sign over the integration contour of Eq. (7.1). A change in the sign makes the integral diverge. In the mantle and the core of the earth, the wave velocity increases with depth,* so that the following condition is satisfied:

$$\frac{dv}{dr} - \frac{v}{r} \leqslant 0.$$

Great difficulties are encountered in regions where the velocity increases very rapidly with depth. There are two such regions: layer C in the mantle and layer F in the core (Table 7.1 and Fig. 7.3). In particular, these difficulties have forced Jeffreys to define the distribution of velocities in the core layer F by a formal limiting condition

$$\frac{dv}{dr} - \frac{v}{r} = 0, \qquad v = cr \text{ (layer F)}.$$

Here, c is a constant.

Since the "difficult" layer F is relatively thin, it does not affect appreciably the accuracy of the determination of the wave velocities elsewhere in the core (layers E and G). The velocity of the P waves changes suddenly at the mantle — core boundary and the velocity of the S waves vanishes at this point. Consequently, the distribution of velocities in the core of the earth is determined as follows: the seismic hodograph does not refer to the surface of the earth but to

---

*It is rare for the condition $d\eta/dr > 0$ not to be satisfied in the mantle and the core of the earth, and even when this inequality is not satisfied, the region involved is not very large. Consequently, the method described gives reliable information on the velocity distribution in the earth.

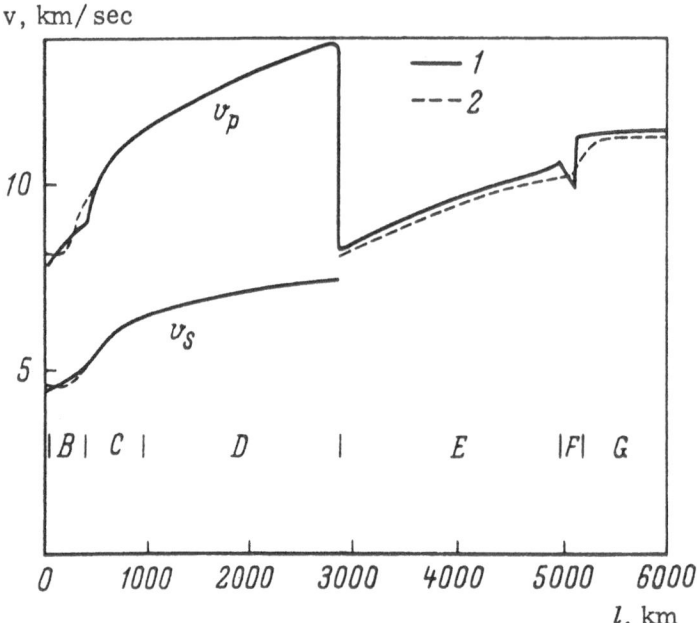

Fig. 7.3. Distribution of the seismic wave velocities in the earth according to Jeffreys (1) and Gutenberg (2).

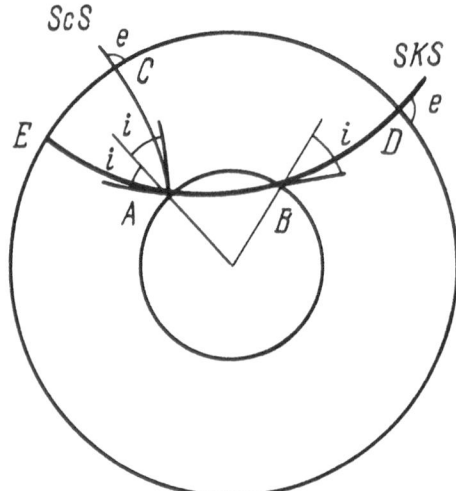

Fig. 7.4. Propagation of a seismic ray in the earth's core.

the surface of the core. This is explained in Fig. 7.4, where ECD is the surface of the earth, AB is the surface of the core, C is the point of incidence of a reflected transverse wave (EAC), denoted by ScS. The angle of incidence of this wave on the surface of the earth is equal to the angle of emergence of a wave SKS at a point D on the earth's surface. SKS is part of a ray EABD, which is propagated in the form of a transverse wave S in the mantle (arcs EA and BD) and in the form of a longitudinal wave P in the core (denoted by K). We can easily see that the epicentral distance, Δ_{AB}, corresponding to the path traveled in the core, is

$$\Delta_{AB} = \Delta_{ED} - \Delta_{EC},$$

and the corresponding travel time is

$$t_{AB} = t_{ED} - t_{EAC}.$$

We shall now apply our method to this hodograph since the velocity of the longitudinal waves increases uniformly in the core.

Determinations of the distribution of the P and S wave velocities in the earth are based on work carried out by the leading seismologists and geophysicists. The currently accepted distribution of velocities has been derived by Jeffreys, Bullen, Gutenberg, and Richter (Fig. 7.3). The distributions of Jeffreys and Gutenberg have been obtained by various methods for averaging the experimental data. These two distributions differ appreciably in layer F, which separates the outer from the inner core. However, these distributions are identical in the main parts of the core (layers E and G). The differences between these distributions in layer F are due to the indeterminacy of the solutions obtained for this layer. Gutenberg's solution seems to be more realistic. Therefore, we shall use his results. We shall not use the Jeffreys distribution because it does not correspond to any reasonable physical interpretation. This is justified by the following considerations.

All the results indicate that the core of the earth is chemically uniform as a whole. The higher velocities of the longitudinal seismic wave in the inner core v_{Pi}, compared with the velocities in the outer core v_{Pe}, is due to the fact that the outer core behaves effectively as a liquid while the inner core behaves effectively as a solid with respect to vibrations whose periods are of the order of seconds, i.e.,

$$v_{Pe}^2 = \frac{K_S}{\rho}, \qquad v_{Pi}^2 = \frac{K_S + 4/3\mu}{\rho}.$$

Such behavior of the longitudinal wave velocities can be expected for an amorphous core in which the viscosity increases from the periphery to the center in such a way that the transition from the effective liquid to the effective solid state takes place in layer F. If this is correct, the core can be regarded as a rheological body in the Maxwellian sense.

The value of K_S does not exhibit relaxation but μ does. At a given frequency the shear modulus μ varies, with the viscosity, from values close to zero (at low frequencies) to the true value μ_i (at high frequencies).

Since the value of K_S is constant, there are no grounds for assuming a fall of the velocity v_{Pe} in layer F, as postulated in the Jeffreys distribution of velocities. In general, we must say that in any model of the core in which viscosity (diffusion) relaxation processes are important, it is reasonable to assume that the relaxing quantity is μ (the law of relaxation of μ need not be Maxwellian) and that K_S is constant. Therefore, a velocity dip in layer F cannot be explained on the basis of any model of a uniform core with relaxation properties.

Conversely, a relatively smooth rise of the velocity in layer F (the Gutenberg model) can be explained in a natural way by the model of a uniform core with relaxation properties, provided the parameters are suitably selected.

A different assumption, frequently discussed in connection with the problem of the structure of the core, is that put forward by Birch and supported by Bullen and others: according to this hypothesis the inner core is crystalline and the outer core is liquid; layer F is a transition region and it should consist of a mixture of crystalline and liquid phases.

The Gutenberg distribution of velocities differs from the Jeffreys distribution not only in layer F but also in the subcrustal region (layer B). In this region the Gutenberg distribution of velocities predicts a small minimum at depths of 5–250 km. This is very important in many branches of geophysics, which we shall not consider here. Until recently, the Gutenberg and Jeffreys velocity distributions had been regarded as equiprobable but some preference was given to the Jeffreys distribution.

Lately, the situation has changed in view of successful investigations of natural oscillations of the earth [14]. These studies show that the Gutenberg distribution is qualitatively more correct than the Jeffreys distribution.

Variations of the Density, Pressure, Acceleration Due to Gravity, and Compressibility in the Interior of the Earth

We shall assume that the earth can be divided into shell-like layers, each of which is of constant chemical composition (Table 7.1). Then, the derivative of the density with respect to depth is

$$\frac{d\rho}{dl} = \frac{\rho}{K_T}\frac{dP}{dl} - \rho\varkappa\frac{dT}{dl}, \qquad (7.3)$$

where K_T is the isothermal bulk modulus; T is the absolute temperature; $\varkappa$ is the thermal expansion coefficient. The distribution of the longitudinal and transverse wave velocities in the mantle and of the longitudinal wave velocities in the core can be used to determine experimentally the function

$$\Phi = K_S/\rho,$$

which assumes the following form for the mantle:

$$\Phi = K_S/\rho = v_P{}^2 - \frac{4}{3}v_S{}^2$$

but in the core it becomes *

$$\Phi = K_S/\rho = v_P{}^2,$$

where K_S is the adiabatic bulk modulus, otherwise known as the incompressibility.

The quantity Φ can be introduced into Eq. (7.3) using the thermodynamic equation

$$\left(\frac{\partial V}{\partial P}\right)_T = \left(\frac{\partial V}{\partial P}\right)_S - \frac{T}{c_P}\left(\frac{\partial V}{\partial T}\right)_P^2,$$

where c_P is the specific heat at constant pressure. Separating the adiabatic component $(\partial T/\partial l)_{ad}$ of the total temperature gradient,

$$\frac{dT}{dl} = \tau + \left(\frac{\partial T}{\partial l}\right)_{ad}, \qquad \left(\frac{\partial T}{\partial l}\right)_{ad} = \frac{Tg\varkappa}{c_P} \qquad (7.4)$$

---

* $\Phi = v_P^2$ only for the outer core (layer E). All the available results show that the core is uniform. Since the pressure and the density vary weakly in layer F and in the inner core (layer G), extrapolation of the experimental value of Φ for the outer core should give the approximately correct value of Φ for the whole core.

and using the formulas of the gravitational potential theory

$$\frac{dP}{dl} = g\rho, \qquad g = \frac{Gm}{r^2}, \qquad m = 4\pi \int\limits_{0}^{r} \rho y^2 \, dy, \qquad (7.5)$$

we obtain the basic equation for the variation of the density with the depth solely because of the pressure of the upper layers:

$$\frac{d\rho}{dl} = \frac{Gm\rho}{r^2\Phi} - \varkappa\rho\tau, \qquad (7.6)$$

where r is the distance from the center of the earth; G is the gravitational constant. "Corrections" to the density gradient of Eq. (7.6) are usually omitted because the actual temperature distribution differs little from the adiabatic case and the equation for the density gradient becomes

$$\frac{d\rho}{dl} = \frac{Gm\rho}{r^2\Phi}. \qquad (7.6a)$$

Equation (7.6a), together with the relationship

$$\frac{dm}{dr} = 4\pi r^2 \rho \qquad (7.7)$$

and the well-known boundary conditions for ρ and m, can be used to determine the distribution of the density throughout the earth's interior on the assumption that the earth consists of chemically uniform layers. Equation (7.6a) was first deduced by Williamson and Adams [15] in an investigation of the variation of the density in the mantle and was later used by Bullen [12], Molodenskii [16], and others [17, 18] to determine $\rho(l)$ for the earth's interior.

The nonadiabatic gradient τ in Eq. (7.6) may differ from zero only above layer D (i.e., in layers C and B), i.e., in those layers where stable nonzero values of τ may be due to inhomogeneities. It is reasonable to assume that $\tau = 0$ in the D and deeper layers:

$$\tau \sim 0 \quad \text{at} \quad l \gtrsim 900 \text{ (km)}. \qquad (7.8)$$

Consequently, at depths greater than ~900 km the accuracy of Eq. (7.6a) is governed by the value of Φ. The quantity Φ has been determined by seismologists with an accuracy of ~2%.

An analysis shows that if $\tau \sim 1$ deg/km in the outer layers of the mantle (layers C and B), $d\rho/dl$ is about 10% less compared with the values deduced from Eq. (7.6a); consequently, the use of Eq. (7.6a) in the determination of $\rho(l)$ in layers B and C can give rise to errors ~10%. Below layer C the accuracy of this equation is ~2% provided the chemical composition is uniform. In the actual solution of Eq. (7.6a) the density is represented by a quadratic function of the radius (or depth) for each layer (B, C, D, and E). Since the seismic wave velocities in the mantle and the core are continuous functions of the depth, it follows that the density ρ is also a continuous function of the depth in the mantle and the core. However, the density has a discontinuity at the mantle — core boundary.

There are two well-established numerical parameters which can be used to represent the density of the earth. One of them is the average density of the earth, which is $\rho_m = 5.517$ g/cm³. The corresponding mass of the earth is M = $5.977 \cdot 10^{27}$ g. The other parameter is the moment of inertia of the earth I = $0.334Ma^2$, where a is the average radius of the earth. These two quantities are known with a higher accuracy than Eq. (7.6a). Using the known value of M and calculating the correction for the mass of the crust, we can obtain a fairly accurate boundary value for m in the upper part of layer B (Table 7.1). The boundary value of ρ, for example

ρ', cannot be found without additional data. The value of ρ' is usually selected on the assumption that rocks for which the values of K_S/ρ and μ/ρ are approximately the same as in the upper part of layer B, are of density close to 3.3–3.4 g/cm^3. In an actual calculation, Bullen assumed $\rho' = 3.32$ g/cm^3. In this way, we can calculate the theoretical distributions of ρ and m for the whole mantle. Then, using the known moment of inertia of the earth and the relationship

$$\frac{dI}{dr} = \frac{8\pi}{3} r^4\rho,$$

Bullen found that the moment of inertia of the core is $0.57(mr^2)_1$, where the subscript "1" denotes the core boundary. However, this result is unlikely to be correct in view of the stability conditions, since the coefficient 0.57 exceeds the value 0.40, which applies to a sphere of constant density.

It follows that one of the assumptions made so far is incorrect. It is found that the incorrect assumption must be that which postulates that the earth's mantle can be divided into chemically uniform layers. It follows that there is an appreciable chemical inhomogeneity somewhere between the crust and the core. Further numerical calculations led Bullen to the conclusion that this chemically inhomogeneous zone should lie far from the core, probably at depths lower than 1000 km. He assumed that this chemical inhomogeneity is located in layer C (Table 7.1). Later developments in geophysics have confirmed the correctness of this hypothesis. It has been shown that the derivative $d\Phi/dr$ in layer C behaves irregularly, in contrast to its behavior in the neighboring layers B and D, where the relevant derivative is approximately constant. It has been shown experimentally that the electrical conductivity increases by about three orders of magnitude in layer C. Moreover, the deepest earthquake focuses have been found in layer C.

Bullen calculated the dependence of the density on the depth in layer C in such a way as to obtain an acceptable value for the polar moment of inertia of the core: $\leq 0.4(mr^2)_1$. Since the core is assumed to be uniform, the density distribution in the core is defined completely by the boundary value of the density at the core surface and by Eq. (7.6a). The density distribution satisfying all these requirements is obtained by assuming that $\rho = 9.7$ g/cm^3 at the boundary of the core. The distributions of the density, the pressure, and the acceleration due to gravity, obtained in this way, are presented in Fig. 7.1. The distribution of K_S is obtained by multiplying the experimental values of Φ by the values of ρ. Bullen's distribution of the density is probably close to the true distribution. Deviations of the density calculated by Bullen from the true values probably do not exceed 10% at any depth.

In contrast to the distributions of the density, the pressure, and the acceleration due to gravity (which are known relatively accurately), the distribution of the temperature in the earth's interior is not yet known exactly. Some ideas of the temperature in the earth's interior can be obtained on the basis of the following observations. The average geothermal gradient at the surface of the earth is about 20 deg/km. This rapid rise of temperature should gradually slow down and at depths of the order of 100 km the temperature is about 1200–1800°C.

The "thermometers," used to deduce the temperature at these depths, are the molten primary volcanic focuses (the melting point of the lavas is known to be ~1200°C). Seismic wave studies show that the mantle behaves as a solid. Consequently, the highest temperature in the mantle is set by the melting curve. Laboratory data on the melting point at a depth of 100 km indicate a value of ~1500°C (or ~1800°K). This value, combined with the geophysical data and the semiempirical formulas for the melting curve (Chapter 3), can be used to find the distribution of the melting points in the mantle of the earth. It is found that at the boundary between the mantle and the core T is ~5000–6500°K. The distribution along the adiabat is regarded as the lower limit of the temperatures in the mantle. This gives T ~ 2500°K in the lowest part of the mantle.

The core is in the molten state. Consequently, the lowest temperatures in the core should be set by the melting curve. Assuming that the core consists of iron, we find that the laboratory data on the melting point of iron yield $\leq 4600°K$ for $P \sim 1.4 \cdot 10^6$ bar (this is the pressure at the mantle — core boundary). It is evident that the core does not consist of pure iron but it contains lighter elements which reduce the melting point of the molten mixture. On the basis of these data we may assume that the temperature at the mantle — core boundary lies within the limits ~4000-5000°K. In the liquid core the temperatures cannot be higher than the adiabatic temperatures with the initial value $T_{2900 km}$ = 4000-5000°K and, as indicated by the theory of the terrestrial magnetism, the core temperatures cannot be appreciably lower than the adiabatic values. This gives ~6000°K for the center of the earth but this value is accurate only to within 1000°K.

7.2. Determination of the Composition of Layer D in the Earth's Mantle

One of the most important problems in geophysics is the determination of the composition of the earth's interior. This problem can be solved by combining the laboratory data for a given substance with the geophysical results. A solution of this problem, obtained by a special method [5] will be described in the present section for an important mineral known as gabbro. It is shown in [5] that if we know the equation of state for a mineral, a comparison of its laboratory and geophysical data can give unambiguous results.

The equation of state for gabbro is deduced in [5] from the dynamic data using methods described in Chapter 5. Gabbro undergoes a phase transition at high pressures (Fig. 7.5). We are interested in the equation of state for the high-pressure phase of gabbro. This equation of state is of the form

$$\left.\begin{array}{c} P = P_p(x) + \dfrac{3R\rho_0}{\mu} \dfrac{\gamma}{x} T, \qquad \rho_0 = 2.993 \text{ g/cm}^3, \\[2mm] P_p(x) = Ax^{-7/3} \exp[b\,(1 - x^{1/3})] - Kx^{-4/3} \\[2mm] A = 2.0252 \cdot 10^6 \text{ bar}, \qquad K = 2.5036 \cdot 10^6 \text{ bar}, \qquad b = 4.322. \end{array}\right\} \quad (7.9)$$

The Grüneisen parameter is found using the unnormalized Dugdale — MacDonald formula

$$\gamma(x) = \frac{1}{6} \frac{Ab^2 x^{2/3} \exp[b(1 - x^{1/3})] - 6Kx^{-2/3}}{Abx^{1/3} \exp[b(1 - x^{1/3})] - 2Kx^{-2/3}}. \qquad (7.9a)$$

Let us now compare the results which follow from the equation of state for gabbro with the geophysical data on layer D in order to find whether this layer may consist of gabbro. We shall use the results of Bullen: the pressure at the mantle — core boundary will be assumed to be practically independent of the model of the earth used in the calculations and to lie on the Hugoniot curve $P_H = (1.35-1.4) \cdot 10^6$ bar. Consequently, we know the values of P_H and of $\Phi = K_S/\rho$ at the mantle — core boundary:

$$\Phi = \frac{K_S}{\rho} = v_P^2 - \frac{4}{3} v_S^2, \qquad (7.10)$$

where K_S is the adiabatic bulk modulus; ρ is the density; v_P and v_S are the velocities of the longitudinal and transverse seismic waves in the mantle. The values of Φ are listed in Table 7.2. Since P_H and Φ are known, we shall show that the problem of comparison of the laboratory data in Eq. (7.9) with the geophysical data for layer D can be made in such a way as to obtain unambiguous results. The value Φ can be calculated using the equation of state (7.9)

$$\Phi = \frac{A}{3\rho_0} (bx^{2/3} + 2x^{1/3}) \exp[b(1 - x^{1/3})] - \frac{4}{3\rho_0} Kx^{-1/3} + \frac{3RT}{\mu} \gamma \left(1 + \gamma - \frac{d\ln\gamma}{d\ln x}\right). \qquad (7.11)$$

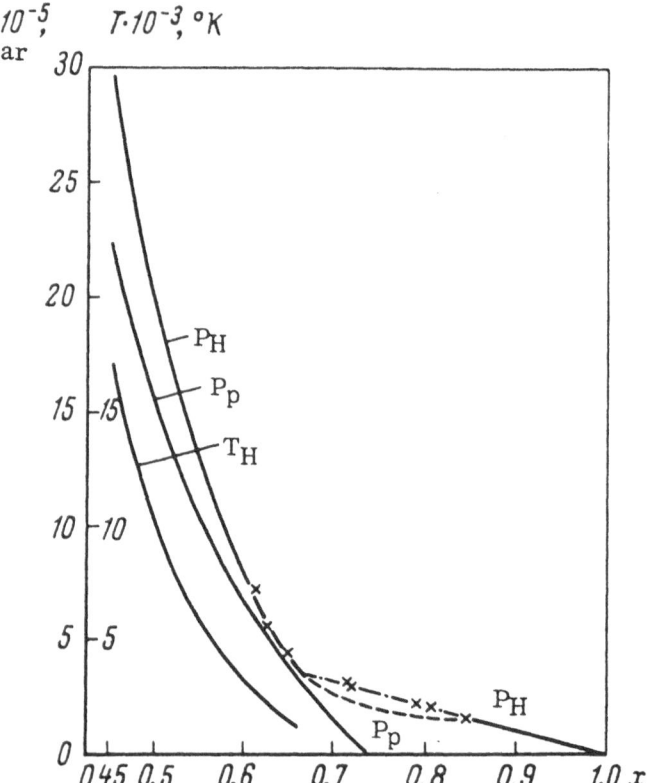

Fig. 7.5. Experimental Hugoniot curve of gabbro: theoretical P_H curve calculated using Eq. (5.75a) and the parameters given in Eq. (7.9); P_p is the potential pressure calculated from Eq. (7.9); T_H is the temperature on the theoretical Hugoniot curve given by Eq. (5.77).

Table 7.2*

h	Φ, km²/sec²	h	Φ, km²/sec²	h	Φ, km²/sec²
0	116.404	0.7	104.039	1.5	85.481
0 1	115.066	0.9	99.588	1.7	80.817
0.3	111.903	1.1	94.944	1.9	76.278
0 5	108.186	1.3	90.213		

* The value of Φ_H is obtained by extrapolating Φ_h from $h \geq 0.1$ to $h = 0$.

Eliminating the temperature from Eqs. (7.9) and (7.11), we obtain:

$$\Phi = \frac{A}{3\rho_0}(bx^{2/3} + 2x^{1/3}) \exp[b(1-x^{1/3})] - \frac{4}{3\rho_0}Kx^{-1/3} + \frac{x}{\rho_0}[P - P_p(x)]\left(1 + \gamma - \frac{d\ln\gamma}{d\ln x}\right). \qquad (7.12)$$

Replacing P in Eq. (7.12) with $P_H \sim (1.35-1.4) \cdot 10^6$ bar $= (1.35-1.4) \cdot 10^{12}$ dyn/cm² and using the known value of $\Phi_H = 1.164 \cdot 10^{12}$ cm²/sec², as well as the known values of A, K, b, and ρ_0, we can determine the limits of x_H and, consequently the density $\rho_H = \rho_0/x_H$. Knowing ρ_H, we can find the dependences of the density and pressure on the depth in layer D using Eqs. (7.5) and (7.6):

$$\frac{dP}{dh} = -g\rho, \qquad \frac{d\rho}{dh} = -\frac{g\rho}{\Phi}, \qquad dh = dr = -dl, \tag{7.13}$$

where l is the depth measured from the surface of the earth (at the mantle — core boundary $l = 2900$ km); r is the radius measured from the center of the earth (at the mantle — core boundary $r = 3473$ km); h is the thickness of the mantle measured from the mantle — core boundary along the radius vector: $h = 2900 - l$ km.

The acceleration due to gravity g is approximately constant throughout the mantle and is $\sim(0.98-1) \cdot 10^3$ cm^2/sec. We shall represent the experimental values of Φ_h^{-1} (Table 7.2) in the form of a polynomial

$$\Phi_h^{-1} = a + bh + ch^2. \tag{7.14}$$

Integrating Eq. (7.13) in layer D ($900 \le l \le 2900$ km) for $g = \bar{g} = 10^3$ cm^2/sec and using Eq. (7.14), where $a = 8.59077 \cdot 10^{-3}$, $b = 9.21920 \cdot 10^{-4}$, $c = 7.66619 \cdot 10^{-4}$,* we obtain the distribution of the density in layer D

$$\ln \frac{\rho_h}{\rho_H} = \ln \frac{x_H}{x_h} = -\bar{g}\left(ah + \frac{b}{2}h^2 + \frac{c}{3}h^3 \right),$$

or

$$\rho_h = \rho_H \exp\left[-\bar{g}\left(ah + \frac{b}{2}h^2 + \frac{c}{3}h^3 \right)\right],$$

$$x_h = x_H \exp\left[\bar{g}\left(ah + \frac{b}{2}h^2 + \frac{c}{3}h^3 \right)\right]. \tag{7.15}$$

In order to determine the distribution of the pressure, we shall represent first the exponential factor in Eq. (7.15) as a quadratic polynomial in h:

$$\exp\left[-\bar{g}\left(ah + \frac{b}{2}h^2 + \frac{c}{3}h^3 \right)\right] = 1 + n_1 h + n_2 h^2, \tag{7.16}$$

where $n_1 = -8.35355 \cdot 10^{-2}$, $n_2 = -5.52783 \cdot 10^{-3}$, and h is in thousands of kilometers.

Using Eqs. (7.13) and (7.16), we find the distribution of the pressure in layer D:

$$P_h = P_H - g\rho_H\left(h + \frac{n_1}{2}h^2 + \frac{n_2}{3}h^3 \right) \cdot 10^2. \tag{7.17}$$

Agreement with the geophysical data is obtained when the substitution of P_h of Eq. (7.17) for P in Eq. (7.12) and of x_h of Eq. (7.15) for x in Eq. (7.12) yields values of Φ_h which are equal to the experimental values listed in Table 7.2.

Let us apply this method to the case of gabbro. Substituting into Eq. (7.12) the values $P = (1.35-1.4) \cdot 10^{12}$ dyn/cm^2, $\Phi = 1.164 \cdot 10^{12}$ cm^2/sec^2, and using the values of A, K, b, and ρ_0 given in Eq. (7.9), we find that there is no value of x for which this equation can have a solution. This means that gabbro cannot satisfy the geophysical data characterizing layer D. We shall now consider how far the geophysical data for gabbro fail to satisfy Eq. (7.12) at the mantle — core boundary. For $0.55 \ge x \ge 0.5$, the pressure on the cold-compression isotherm

---

*For these values of a, b, c, and for h expressed in thousands of kilometers, we obtain $\Phi = K_S/\rho$ in km^2/sec^2.

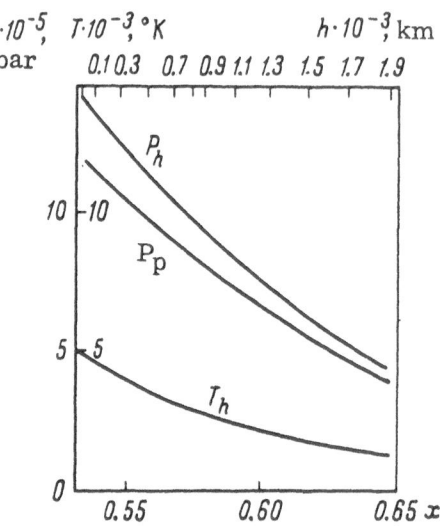

Fig. 7.6. Pressure and temperature distributions in layer D of the earth's mantle, calculated on the assumption that this layer consists of gabbro: P_h is the pressure in the layer; P_p is the cold-compression isotherm of gabbro; T_h is the temperature in the layer; h is the depth measured along the radius from the mantle — core boundary.

is $1.03 \cdot 10^6$ bar $\leq P_p(x) \leq 1.53 \cdot 10^6$ bar. The corresponding values of $\Phi(x, T = 0)$ are $8.74 \cdot 10^{11}$ cm^2/sec^2 $\leq \Phi_p \leq 9.66 \cdot 10^{11}$ cm^2/sec^2. Thus, even at x = 0.5, at which the pressure $P_p \sim 1.53 \cdot 10^6$ bar exceeds the pressure at the mantle — core boundary ($P_H \sim 1.4 \cdot 10^6$ bar), the value of Φ_p is still ~20% smaller than the value at the boundary $\Phi_H \sim 11.64 \cdot 10^{11}$ cm^2/sec^2. This is a very large discrepancy, particularly if we bear in mind that the thermal correction to Φ at T ~ 5000°K amounts to ~4%. For x = 0.55, $P_p \sim 1.03 \cdot 10^6$ bar, and the thermal pressure, which is $P_H \sim 1.4 \cdot 10^6$ bar, corresponds to $T_H \sim 8900$°K. The corresponding value of Φ is about 22% smaller than $\Phi_H \sim 11.64 \cdot 10^{11}$ cm^2/sec^2. A solution of Eq. (7.12) with $\Phi_H \sim 11.64 \cdot 10^{11}$ cm^2/sec^2 exists for $P_H > 2 \cdot 10^6$ bar.

We must draw attention to the following point. Let us select reasonable values of P_H and ρ_H (for example the values used in Bullen's model) and, using Eq. (7.17), let us calculate the distribution of P_h in layer D. Then, assuming that this layer consists of gabbro, let us determine the distribution of x_h from Eq. (7.15) and the corresponding distribution $P_p(x_h)$ from Eq. (7.9). Let us also assume that $x_H = 0.5325$ and $P_H = 1.4 \cdot 10^6$ bar. Let us now select for each point x_h such a temperature T_h that the value of P of Eq. (7.9) becomes equal to p_h. The obtained distribution of T_h, as well as the distributions P_h and $P_p(x_h)$, are shown in Fig. 7.6. The distribution of T_h in layer D is now found to be quite reasonable. Thus, we have demonstrated that a check based only on the pressure data can give agreement between the laboratory and geophysical data but a fuller comparison of these data, by the method just described, may give the opposite result.

7.3. Determination of the Composition

of the Earth's Core

There is an extensive literature on this subject, going back some 50 years. We shall consider two hypotheses: the hypothesis of an iron core and the hypothesis of a core consisting of metallized silicates. According to the first hypothesis the core consists mainly of molten iron with admixtures of some widely encountered elements (for example, Ni, Si, etc.). The Lodochnikov — Ramsey hypothesis on the metallization of silicates is based on the following considerations. At pressures ruling the boundary of the earth's core, $P \sim 1.4 \cdot 10^6$ bar, silicates undergo a phase transition to the metallic state. This transition is accompanied by a considerable rise of the density from ~ 5.7 to ~ 10 g/cm^3.

The current evidence shows that the Lodochnikov — Ramsey hypothesis must be rejected. The decisive data on this point were provided by the experiments of Al'tshuler and his colleagues [19], who carried out dynamic compression of the more important minerals using pressures of a few megabars. Their experiments extended well beyond the range of $\sim 1.4 \cdot 10^6$ bar and $\sim 4000-5000°$K at the core — mantle boundary but they found no evidence of the expected transition to the metallic state. In spite of the very short durations of the dynamic experiments, there is no doubt that this transition could not occur if the pressure were static or applied for a longer time. Three points must be stressed in this connection: 1) the metallization is an electronic transition and such transitions are practically instantaneous; 2) if we assume that the silicates subjected to a shock wave are in a metastable state, it follows that the "metastability" factor $P\Delta V$ is so large (because of the very large postulated change in volume ΔV) that it cancels out the requirement of a finite transition time; 3) the dislocation mechanism of phase transitions in shock waves (Chapter 5) should also give rise to an instantaneous transition at pressures exceeding the expected transition pressure by about 10^6 bar. In view of this, and for other reasons, the majority of investigators now prefers the hypothesis that the core consists of iron.

When the laboratory data for iron and its alloys or solutions are compared with the geophysical data, we must bear in mind that the core of the earth is liquid and the density of molten iron may be several percent lower than the density of iron in the solid state at the same pressure. This is very important since the difference between the laboratory and geophysical data lies within the limits 5-10%.

The first attempt to compare the density of iron at a pressure of $1.4 \cdot 10^6$ bar, using the laboratory and geophysical data for the density at the outer surface of the earth's core was carried out by Davydov in 1956 [20]. It was found that the "laboratory" density of iron (calculated ignoring the effects of melting and thermal expansion) exceeds the upper limit of the density in the core given by Molodenskii (Fig. 7.1). Davydov's paper appeared before the publication of the dynamic (shock-wave) data on the compression of iron. The first extensive reports of the Soviet and American investigations of the shock compression of metals were published in 1957 and 1958. It was found that a shock wave produced a phase transition in iron at a pressure of $\sim 1.3 \cdot 10^5$ bar and that this transition was accompanied by an increase in the density. The appearance of these dynamic data made it possible to derive the equation of state for the high-pressure phase of iron [21] and to develop a detailed method [4] for comparison of the laboratory and geophysical data for the earth's core. A considerable part of this section will deal with this comparison.

We know that the earth's core can be divided, according to the seismic data, into two parts: the outer core (layer E) extending from the boundary with the mantle at $l = 2900$ km to a depth $l \sim 5000$ km, and the inner core (layer G) extending from $l \sim 5000$ km to the center of the earth at $l \sim 6370$ km. The outer core is separated from the inner core by a transition region whose thickness is $\Delta l = 200$ km (layer F). The outer core is in the liquid state. Since it does not transmit transverse waves of periods $\tau \sim 1$ sec, its viscosity should be $\eta \lesssim 10^{12}$ P. The temperature distribution in the core should be nearly adiabatic since temperature gradients weaker than the adiabatic gradient could not produce convective currents responsible for the geomagnetic fields and gradients considerably higher than the adiabatic value would produce

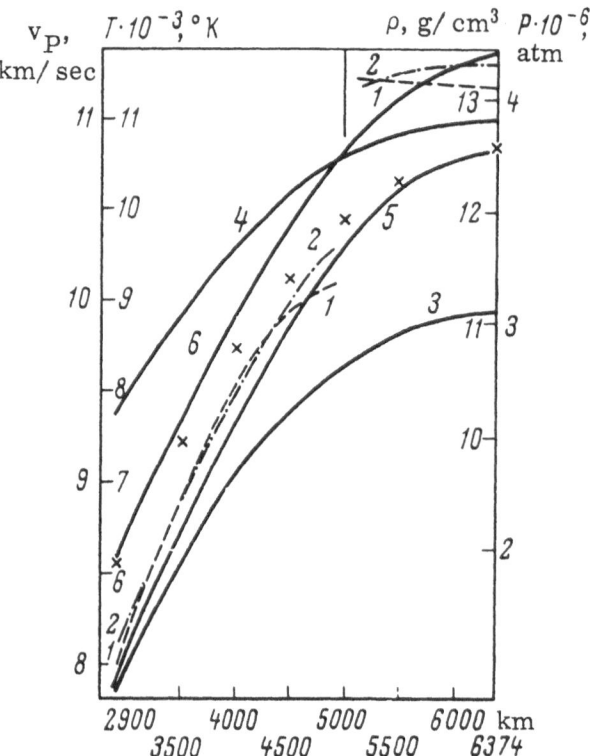

Fig. 7.7. Distributions of various parameters in the iron core of the earth. 1) Gutenberg distribution of the longitudinal wave velocity; 2) Jeffreys distribution of the longitudinal wave velocity; 3) distribution of velocities in a liquid iron core according to Eqs. (7.18) and (7.19); 4) distribution of the density; 5) distribution of the pressure; 6) distribution of the temperature in accordance with Eq. (7.23). The crosses represent adiabatic temperatures.

convective currents corresponding to a practically infinite thermal conductivity which would restore the temperature to the adiabatic value. There is less information about the state of the inner core and of the transition layer F. Two independent distributions have been proposed for the longitudinal wave velocity in the earth's core (Fig. 7.7). In layer F these two distributions are different and they are omitted in Fig. 7.7.

It is generally accepted that the substance of which the core is composed is in the metallic state. The chemical composition of the core has not yet been finally determined.

The equation of state for the high-pressure phase of iron at temperatures T, greater than the Debye temperature of iron θ, is ,

$$P = P_p(x) + \frac{3R}{\mu}\frac{\rho_0\gamma}{x} T + P_e, \quad P_p(x) = Ax^{-2/3}\exp[b(1-x^{1/3})] - Kx^{-1/3}$$

$$P_e = \frac{1}{2}\delta\rho_0 x^{-1/3}ag_e T^2, \quad \delta = \left(\frac{\pi}{3}\right)^{2/3}\frac{k^2 m_0 N_0}{\mu\hbar^2}\left(\frac{\mu}{\rho_0}\right)^{2/3}, \quad a = Z^{1/3}\frac{m^*}{m_0}, \tag{7.17a}$$

$$a = a_0\exp[B(x^{1/3}-1)], \quad g_e = \frac{2}{3} + \frac{d\ln a}{d\ln x} = \frac{2}{3} + \frac{B}{3}x^{1/3},$$

where P is the pressure; $x = V/V_0$ is the relative volume; $V_0 = 1/\rho_0$ is the volume under normal (atmospheric) conditions; k is the Boltzmann constant; R is the gas constant; $\hbar$ is the Planck constant, divided by 2π; m_0 is the free-electron mass; m^* is the effective mass of conduction electrons; N_0 is the Avogadro number; Z is the number of conduction electrons per atom; A, K, b, α_0, and B are constants determined using experimental data: $A = 9.4389 \cdot 10^5$ bar, $K = 1.0740 \cdot 10^6$ bar, $b = 7.7845$, $\alpha_0 = 17.2$, $B = 9.86$. The Grüneisen parameter γ depends only on the volume and it is determined from the unnormalized Dugdale — MacDonald formula using the cold-compression isotherm [Eq. (7.9a)].

Using Eq. (7.17a), we can easily find $\Phi = K_S/\rho$, where K_S is the adiabatic bulk modulus and ρ is the density:

$$\Phi = \frac{K_S}{\rho} = \frac{A}{3\rho_0}(bx^{1/3} + 2x^{1/3})\exp[b(1 - x^{1/3})] - \frac{4}{3\rho_0}Kx^{-1/3} +$$

$$+ \frac{3R}{\mu}\gamma\left(1 - \frac{d\ln\gamma}{d\ln x}\right)T + \frac{x}{\rho_0}\left(1 - g_e - \frac{d\ln g_e}{d\ln x}\right)P_e + \frac{(3R\gamma/\mu + g_e\,\delta x^{2/3}\alpha T)^2}{3R/\mu + \delta\alpha x^{2/3}T}T. \qquad (7.18)$$

The velocity of longitudinal waves in the liquid part of the core, v_P, is

$$v_P = \Phi^{1/2}. \qquad (7.19)$$

The adiabatic temperature gradient is defined in the usual manner and after elementary transformations it assumes the form:

$$\frac{dT}{dl} = \frac{\bar{g}\varkappa T}{c_P} = \frac{\bar{g}T}{\Phi}\frac{3R\gamma/\mu + g_e\,\alpha\delta x^{2/3}T}{3R/\mu + \delta\alpha x^{2/3}T}, \qquad (7.20)$$

where $\bar{g}$ is the acceleration due to gravity; c_P is the specific heat at constant pressure; $\varkappa$ is the thermal expansion coefficient

$$\varkappa = \frac{1}{x}\left(\frac{\partial x}{\partial T}\right)_P = \frac{1}{K_T}\left(\frac{\partial P}{\partial T}\right)_x = \frac{\dfrac{3R}{\mu}\dfrac{\rho_0\gamma}{x} + \delta\rho_0 x^{-1/3}\alpha g_e\,T}{K_T}, \qquad (7.21)$$

K_T is the isothermal bulk modulus

$$K_T = \frac{A}{3}(bx^{-1/3} + 2x^{-2/3})\exp[b(1 - x^{1/3})] - \frac{4}{3}Kx^{-1/3} + \frac{3R}{\mu}\frac{\rho_0\gamma}{x}\left(1 - \frac{d\ln\gamma}{d\ln x}\right)T + \left(1 - g_e - \frac{d\ln g_e}{d\ln x}\right)P_e.$$

The specific heat at constant volume is

$$c_V = \frac{3R}{u} + \delta\alpha x^{2/3}T. \qquad (7.22)$$

Let us now consider the data on the earth's core obtained by geophysical methods. We shall assume the core to be uniform and we shall postulate that the increase in the velocity of waves from the outer to the inner core is due to a transition from a liquid region where $v_P^2 = K_S/\rho$ (layer E) to a solid region where $v_P^2 = (K_S + 4\mu/3)/\rho$ (layer G). Here, μ is the high-frequency value of the shear modulus (for details see [6]). Next (Sec. 7.1), we shall use

$$\frac{dP}{dl} = \bar{g}\rho, \qquad \bar{g} = \frac{Gm}{r^2}, \qquad m = 4\pi\int_0^r \rho(y)y^2\,dy, \qquad (7.5)$$

and when the temperature varies adiabatically, the equation for the distribution of the density with depth becomes

$$\frac{d\rho}{dl} = \frac{Gm\rho}{r^2\Phi},$$ (7.6a)

where G is the gravitational constant and r is the distance from the center of the earth. We shall use the quantity $\Phi = v_p^2$ employed by seismologists. Selecting the initial value of the density on the surface of the core by means of Eqs. (7.5) and (7.6a), we can employ numerical integration to determine the distribution of the density ρ (and, therefore, of P and g) for the whole core. An analysis of various models of the density and pressure distributions show that the pressure at the mantle — core boundary is $P_{2900\ km} \sim (1.35\text{-}1.4) \cdot 10^6$ bar and the corresponding value of $\sqrt{\Phi}$ is $\Phi_{2900\ km}^{1/2} = v_P \sim 8$ km/sec. Using these initial conditions, we can employ Eqs. (7.17a) and (7.18) to determine $x_{2900} = \rho_0/\rho_{2900}$ and T_{2900} (we shall omit "km" from the subscripts). We find that $\rho_{2900} = 9.7$ g/cm^3, $T_{2900} \approx 6750°$K. If we use slightly different initial conditions, $P_{2900} \sim 1.4 \cdot 10^6$ atm, $\Phi_{2900}^{1/2} = (v_P)_{2900} \sim 7.85$ km/sec (these values are still within the limits of the accuracy of the determination of v_P by seismic methods), we find that $\rho_{2900} \approx 10.2$ g/cm^3 and $T_{2900} \approx 6080°$K. The second set of values of ρ_{2900} and T_{2900} will be used as the initial conditions.*

Next, we shall use Eqs. (7.5) and (7.6a) with $\rho_{2900} = 10.2$ g/cm^3 to find the density (ρ_c) and pressure (P_c) distributions in a uniform core (Fig. 7.7); we can also find the distribution of g. We note that the obtained density distribution is very close to the upper limit of the density distribution in the core deduced by Molodenskii. Knowing ρ_c (and consequently x) as well as P_c, we can easily determine the temperature in the core by solving Eq. (7.17) for T:

$$T = \frac{\left\{\left(\frac{3R}{\mu}\frac{\gamma\rho_0}{x}\right)^2 + 2\delta\rho_0 x^{-1/3} a g_e\ [P_c - P_p(x)]\right\}^{1/2} - \frac{3R}{\mu}\frac{\gamma\rho_0}{x}}{\delta\rho_0 x^{-1/3} a g_e}.$$ (7.23)

The values of the temperature calculated using Eq. (7.23) (Fig. 7.7) can be substituted, together with the corresponding values of x, into Eqs. (7.18) and (7.19). The velocity distribution for a uniform liquid-iron core, based on Eqs. (7.18) and (7.19), is represented by curve 3 in Fig. 7.7. Finally, we can integrate Eq. (7.20) for an initial temperature $T_{2900} = 6080°$K; the adiabatic temperatures obtained in this way are represented by crosses in Fig. 7.7. These temperatures are close to the temperatures calculated from Eq. (7.23). The calculated distribution of v_P is also close to the experimentally determined distribution (Fig. 7.7), provided we bear in mind that the comparison should be made only for the results obtained for the outer "liquid" core.

Summarizing, we conclude that the available experimental data do not contradict the hypothesis of an iron core of the earth.

However, we must stress the following difficulties: 1) the values of v_P, determined by seismic methods, increase toward the center of the earth more rapidly than the values of v_P for iron obtained on the basis of the laboratory observations using Eqs. (7.18) and (7.19), the greatest discrepancy between these values being ~5% near the inner core; 2) the temperature rise of about 6000°K between the mantle — core boundary and the center of the earth is fairly large: it is difficult to reconcile temperatures of 12,000°K at the center of the earth with the theory of cold formation of our planet.

---

*The use of the other set of the initial conditions, $\rho_{2900} \sim 9.7$ g/cm^3 and $T_{2900} \sim 6750°$K, gives qualitatively the same results, but the temperature distribution is shifted by 500-700°K.

These two difficulties are associated with the fact that the rise of the pressure with decreasing volume on the cold-compression isotherm of iron is slower than the corresponding pressure rise in the earth's core. Consequently, the difference $P_c - P_p(x)$ in Eq. (7.23) increases appreciably from the periphery to the center of the core. Compensation of this difference by the thermal pressure is responsible for the temperature rise.

Recently, Kormer and Funtikov [22] have determined the Hugoniot curve of an $Fe - Si$ alloy (Si $\sim$ 19%). They compared their results with the geophysical parameters of the core and reported good agreement between the two sets of data. However, more reliable conclusions can be drawn only by the application of the method described in the present section, which is still to be carried out.

7.4. Distribution of the Electrical Conductivity

in the Earth's Mantle

The electrical conductivity of the earth's mantle can be approached in two ways, which are complementary. The first method is purely geophysical. In this method the electrical conductivity profile of the mantle is selected so as to fit best the observed attenuation of long-period electromagnetic fluctuations. Several models have been proposed for the distribution of the electrical conductivity in the earth's mantle [23-26] (Fig. 7.8). A characteristic feature of all the proposed distributions of the electrical conductivity is an anomalous rise in the transition layer C at a depth of 400-900 km. It is extremely difficult to determine the electrical conductivity in layer D (below 1000 km) by purely geophysical methods. This is because even an analysis of six-month variations [26] gives little information on the detailed distribution of the electrical conductivity in layer D. An attempt to avoid these difficulties by a purely geophysical method is reported in [27]. In this method the electrical conductivity of the whole mantle is estimated from the attenuation of secular geomagnetic variations penetrating the mantle from the direction of the core. However, it is pointed out in [28] that the analysis given in [27] is not sufficiently convincing and the suggested distribution of the electrical conductivity is open to question.

Another way, which can be used to determine (at least in principle) the distribution of the electrical conductivity in the earth's mantle, is purely physical. If we know the substance or the type of substance of which the mantle is composed, then, by investigating the electrical conduction mechanism in the laboratory and using the dependence of the electrical conductivity $\sigma = \sigma(T, P)$ on the temperature T and the pressure P, we can estimate the conductivity of the mantle from the known distributions of P and T. This method is far from being fully developed although some results are already available [17, 28-30].

In this section we shall apply the methods described in Chapter 3 to the electrical conductivity of the earth's mantle. Our treatment is based on that described in [1, 7].

Electrical Conductivity of the Subcrustal Layer (Layer B)

The electrical conductivity of the subcrustal layer is relatively low and, therefore, it is masked (in magnetotelluric investigations) by the high conductivity of sedimentary layers near the surface. The main features of the distribution of the electrical conductivity in the subcrustal layer have been mentioned in the first physical investigations of this problem [1]. We shall base our treatment on the results reported in this early work as well as on the experimental data obtained under laboratory conditions by Volorovich, Bondarenko, and Parkhomenko [31]. According to the experimental data, at moderate temperatures (t < 1000-1200°C) the electrical conductivity of minerals is of the impurity type and charge can be carried by electrons or holes as well as by impurity ions. The intrinsic ionic conductivity begins to predominate in the temperature range $t \approx 1000-1200°C$ and at still higher temperatures this conductivity represents the main mechanism. Thus, Bondarenko [31] gives the following estimates for the electrical conductivity at $t \sim 1000-1200°C$:*

		$\sigma, \Omega^{-1} \cdot cm^{-1}$
Gabbro, basalt, peridotite, olivinite	. . . .	$2 \cdot 10^{-4}$-10^{-3}
Granite, dunite	. . . .	$7 \cdot 10^{-6}$-10^{-4}
Eclogite	. . . .	$2 \cdot 10^{-3}$-$7 \cdot 10^{-3}$

The pressure dependence of the electrical conductivity σ in the intrinsic ionic conductivity region has been investigated also by Zharkov [1] (see also Chapter 3). It is found that, like the self-diffusion coefficient, the electrical conductivity σ can be represented in the form

$$\sigma(P, T) = \sigma_0 \exp\left[-\xi T_m(P) / T\right] = \sigma_0 \exp(-\xi / \vartheta), \\ \vartheta = T/T_m(P), \xi \approx 15 \div 20, \quad\quad (7.24)$$

where $T_m(P)$ is the pressure dependence of the melting point. Equation (7.24) is obtained on the assumption that the concentration of thermal defects on the melting curve remains approximately constant (for details see Chapter 3). It follows that the activation energy of the electrical conduction process increases, with increasing pressure, proportionally to the melting point.

The formula (7.24) can be used to analyze the distribution of the electrical conductivity in the subcrustal layer. The electrical conductivity increases rapidly with increasing depth. Finally, we reach the region of the "first conducting layer,"† which is the region of intrinsic ionic conductivity characterized by

$$\sigma \sim 2 \cdot 10^{-4} - 5 \cdot 10^{-3} \, \Omega^{-1} \cdot cm^{-1} \quad\quad (7.25)$$

The depth of the "first conducting layer" can be regarded as a temperature calibration point ($t \sim 1100-1200°C$). We have mentioned already that the temperature increases rapidly with depth in the outermost 100-km deep layer of the earth and then the rise slows down since otherwise the mantle would be in the liquid state. Consequently, at depths of 100-200 km the temperature approaches closely the melting point of the constituent rocks. Deeper in the earth the temperature begins to fall below the melting point because of the dependence $T_m(P)$, i.e., the value of ϑ begins to fall. Such behavior of the temperature in layer B gives rise to an important singularity in the distribution of the electrical conductivity. According to Eq. (7.24), the value of σ increases as long as ϑ is increasing but it passes through a maximum at a point where $\vartheta = \vartheta_{max} < 1$. At greater depths the temperature T falls below the melting point $T_m(P)$ and the reduced temperature $\vartheta = T/T_m(P)$ also decreases, which is responsible for the fall of the electrical conductivity with increasing depth. This fall of the electrical conductivity continues up to the transition layer of Golitsyn (layer C) at depths of 300-400 km and it probably extends to the top of this layer. In the transition layer C the electrical conductivity again increases rapidly. In this layer the conduction mechanism changes to intrinsic electronic con-

---

*It must be pointed out that some samples of minerals tested by Bondarenko have suffered chemical changes at high temperatures and this may have given rise to considerable errors in the values of σ listed in the above table.

† The "second conducting layer" is located in layer D and is due to the intrinsic electronic conduction. Layer D is an n-type semiconductor. This "second conducting layer" will be described later.

ductivity in layer D (Table 7.1). Estimates show that at the electrical conductivity minimum the value of σ can decrease by a factor of 5 compared with its value in the "first conducting layer."

The rest of this section will be devoted to an estimate of the distribution of the electrical conductivity in layer D by purely physical methods. Knowing the value of the conductivity at a depth $l \approx 1000$ km and assuming the conduction mechanism to be the same as that in an intrinsic semiconductor, we shall estimate the distribution of σ in layer D from the known distributions of T and P. We shall use the results given in Chapter 3.

Basic Relationships

The general formula for the electrical conductivity of a semiconductor is [14]

$$\sigma = ne\mu_n + pe\mu_p,$$

where n and p are, respectively, the numbers of electrons and holes per unit volume (it follows from the electrical neutrality condition that n = p); μ_n and μ_p are, respectively, the mobilities of electrons and holes; e is the electronic charge. The mobility is defined as the velocity which an electron or a hole acquires in a unit electric field. The mobility μ is related by a simple expression to the average time interval between collisions τ and the mean free path λ of an electron or a hole:

$$\mu = \frac{e\tau}{m} = \frac{e\lambda}{mv},$$

where m and v are, respectively, the effective mass and the average velocity. In the case of semiconductors, the values of τ and λ, and consequently the mobility μ, are governed by the processes of scattering of electrons and holes by ionized impurities, thermal vibrations of the lattice (i.e., by acoustical and optical phonons), as well as by some processes associated with the complex band structure of semiconductors [27]. However, at the high pressures in layer D and at high temperatures we can reasonably assume that the dominant process is the scattering by the lattice vibrations. We shall assume that the minerals of which layer D is composed are closer in their nature to ionic crystals than to covalent semiconductors. In such crystals we must take into account not only the scattering of carriers by the acoustical phonons (as is the case in other semiconductors) but also the scattering by the longitudinal optical phonons, known also as the polarization waves. The mobility governed by the scattering on the acoustical phonons is described by the formula [32, 33]:

$$\mu_a = \frac{\sqrt{8\pi}}{3} \frac{e\hbar^4 \rho c_l^2}{(kT)^{3/2} m^{5/2} E_1^2}, \tag{7.26}$$

where $\hbar$ is the Planck constant divided by 2π; ρ is the density; k is the Boltzmann constant; E_1 is the deformation potential; c_l is the velocity of longitudinal elastic waves; m is the effective mass of an electron or a hole. Let us assume that ε_c and ε_v are, respectively, the energies of the conduction and valence band edges. Then, the deformation potential in the case of electron scattering is given by:

$$\frac{d\varepsilon_c}{d \ln V} = E_{1c}, \tag{7.27a}$$

and in the case of hole scattering, this potential is

$$\frac{d\varepsilon_v}{d \ln V} = E_{1v} \tag{7.27b}$$

TABLE 7.3

E, eV	Diamond	Silicon	Germanium	Tellurium
$\|E_{1c}\|$	8.8	6.5	1.7	2.4
$\|E_{1v}\|$	<30	11.3	2.4	2.4
$\|E_{1c}\|+\|E_{1v}\|$	<39	17.8	4.1	4.8

or

$$E_{1c} = \varepsilon_c \frac{d \ln \varepsilon_c}{d \ln V}, \qquad E_{1v} = \varepsilon_v \frac{d \ln \varepsilon_v}{d \ln V}. \tag{7.28}$$

The value of the deformation potential E_1 cannot be determined accurately and its order of magnitude is

$$E_1 \approx \ 5\text{-}10 \text{ eV}. \tag{7.29}$$

Table 7.3 gives the estimated values of E_1 for diamond, silicon, germanium, and tellurium, as well as the sum $E_{1c} + E_{1v}$, determined by a different method [32]. In order to estimate the dependence of E_1 on the volume, we can reasonably assume that the logarithmic derivatives in Eq. (7.28) are constant (at least in the first approximation). Then

$$E_1 = E_{10}(\varepsilon/\varepsilon_0).$$

We shall also assume that the energies at the conduction and valence band edges (ε_c and ε_v) vary similarly to the forbidden band width (energy gap) ε_G. Consequently,

$$E_1 = E_{10}(\varepsilon_G/\varepsilon_{G0}). \tag{7.30}$$

By measuring the mobility μ_a relative to some reference value, which we shall denote by the subscript zero, we obtain the following expression:

$$\mu_a = \mu_{a0}\left(\frac{\rho}{\rho_0}\right)\left(\frac{c_l}{c_{l0}}\right)^2\left(\frac{m_0}{m}\right)^{5/2}\left(\frac{E_{10}}{E_1}\right)^2\left(\frac{T_0}{T}\right)^{3/2}, \tag{7.31}$$

where m_0 is the free-electron mass. The mobility governed by the scattering by the polarization waves is given by the formula taken from [34]:

$$\mu_p = \frac{1}{3}\left(\frac{2}{\pi}\right)^{3/2} \frac{eMa^3(\hbar\omega_p)^2}{m^{3/2}(Ze^2)^2(kT)^{1/2}}. \tag{7.32}$$

Equation (7.32) is derived for a diatomic ionic crystal. In this equation, M is the reduced mass of two ions; a is the lattice constant; Ze is the ionic charge; ω_p is the limiting frequency of the longitudinal optical mode. We shall introduce an optical Debye temperature $k\theta_p = \hbar\omega_p$ and we shall assume that $\theta_p(\omega_p)$ depends on the volume in the same way as the conventional Debye temperature. Then by measuring the mobility μ_p relative to some reference value, which we shall denote by the subscript zero, we find that

$$\mu_p = \mu_{p_0}\left(\frac{\theta_p}{\theta_{p_0}}\right)^2\left(\frac{m_0}{m}\right)^{3/2}\left(\frac{\rho_0}{\rho}\right)\left(\frac{T_0}{T}\right)^{1/2}. \tag{7.33}$$

If several scattering processes are active simultaneously and if these processes can be regarded as independent, the effective relaxation time τ is found from the formula

$$\frac{1}{\tau} = \frac{1}{\tau_a} + \frac{1}{\tau_p},$$

where τ_a is the relaxation time for the scattering by the acoustical phonons and τ_p is the relaxation time for the scattering by the optical phonons. Since $\mu \propto \tau$, the effective mobility μ is given by a similar equation

$$\frac{1}{\mu} = \frac{1}{\mu_a} + \frac{1}{\mu_p}. \qquad (7.34)$$

We shall assume that our reference depth is $l = 1000$ km. According to the data reported in [35], it is reasonable to assume that the Debye temperature of silicates under normal conditions is $\theta_p \approx 1200°$K and that the Debye temperature at a depth $l \approx 1000$ km is $\theta_{p0} \approx 1900°$K. The average atomic mass of silicates is 21. Therefore, we can substitute $M \approx 10 M_p$ in Eq. (7.32) (M_p is the mass of the hydrogen atom). The ionic charge of silicates will be assumed to be 2 ($Z = 2$). We shall assume that the deformation potential is $E_{10} = 7.5$ eV. We shall now use these values to estimate μ_{a0} and μ_{p0}. We shall assume that $m = m_0$, $\rho = 4.7$ g/cm^3, $c_l = 1.16 \cdot 10^6$ cm/sec. In order to obtain the mobility in units of cm$^2 \cdot$ V$^{-1} \cdot$ sec^{-1}, we must divide μ_0 by 300. We then obtain

$$\mu_{a0} = 38.5 \text{ cm}^2 \cdot \text{V}^{-1} \cdot \text{sec}^{-1}, \quad \mu_{p0} = 10.9 \text{ cm}^2 \cdot \text{V}^{-1} \cdot \text{sec}^{-1}. \qquad (7.35)$$

Substituting Eqs. (7.31) and (7.33) into Eq. (7.34), we obtain

$$\frac{1}{\mu} = \frac{1}{\mu_{a0}} \left(\frac{\rho_0}{\rho}\right) \left(\frac{c_{l0}}{c_l}\right)^2 \left(\frac{m}{m_0}\right)^{5/2} \left(\frac{E_1}{E_{10}}\right)^2 \left(\frac{T}{T_0}\right)^{3/2} \left\{ 1 + \frac{\mu_{a0}}{\mu_{p0}} \left(\frac{\rho}{\rho_0}\right)^2 \left(\frac{\theta_{p0} c_l}{\theta_p c_{l0}}\right)^2 \left(\frac{m_0}{m}\right) \left(\frac{T_0}{T}\right) \left(\frac{E_{10}}{E_1}\right)^2 \right\}. \qquad (7.36)$$

We shall calculate μ_{a0} and μ_{p0} assuming that $T_0 = 2.7 \cdot 10^3 °$K at depth $l \approx 1000$ km. If we use any other initial temperature T_{01}, we then find that

$$\left. \begin{array}{c} \mu_{a01} = \mu_{a0} \left(\dfrac{T_0}{T_{01}}\right)^{3/2}, \qquad \mu_{p01} = \mu_{p0} \left(\dfrac{T_0}{T_{01}}\right)^{1/2}, \\[3mm] \dfrac{\mu_{a01}}{\mu_{p01}} = \dfrac{\mu_{a0}}{\mu_{p0}} \left(\dfrac{T_0}{T_{01}}\right). \end{array} \right\} \qquad (7.37)$$

For simplicity we shall assume that the charge is transported by one species of carrier, for example, by electrons. The number of electrons per unit volume in an intrinsic semiconductor is given by the formula [32]

$$n = [A_c(T) A_v(T)]^{1/2} \exp(-\varepsilon_G/2kT), \qquad (7.38)$$

where

$$A_c(T) = 2 \left(\frac{m_e kT}{2\pi\hbar^2}\right)^{3/2}, \quad A_v(T) = 2 \left(\frac{m_h kT}{2\pi\hbar^2}\right)^{3/2}, \quad \varepsilon_G = \varepsilon_c - \varepsilon_v.$$

Here, m_e is the effective mass of electrons and m_h is the effective mass of holes. Substituting the numerical values in $A(T)$, we obtain

$$[A_c(T) A_v(T)]^{1/2} = 4.82 \cdot 10^{15} \left(\frac{m_e m_h}{m_0^2}\right)^{3/4} T^{3/2} \text{ cm}^{-3}. \qquad (7.39)$$

Finally, we shall consider the problem of the dependence of the forbidden band width on the temperature and pressure. Assuming that the value of ε_G depends weakly on temperature, we can expand ε_G in terms of temperature and retain only the first term

TABLE 7.4

Substance	Mp, °C	ε_{G_0}, eV	ε_0 (300° K), eV	$(\partial\varepsilon_0/\partial T)_P \cdot 10^4$, eV/°K
PbS	1110	0.37	0.30	+4
PbSe	1065	—	0.22	+4
PbFe	904	—	0.27	+4
AlSb	1060	1.6	1.5	—3
GaP	—	—	2.4	—5.5
GaAs	1240	—	1.1—1.35	—5
GaSb	720	0.80	0.7	—3.5
InAs	940	0.47	0.35	—4
InAs	535	0.27	0.18	—3

TABLE 7.5

Substance	ε_{G_0}, eV	$(\partial\varepsilon_G/\partial T)_P \cdot 10^4$, eV/°K	ε_G (300° K), eV	ε_0 (300° K), eV	$(\partial\varepsilon_0/\partial T)_P \cdot 10^4$, eV/°K
Silicon	1.21	—4.2	1.09	1.05	—4
Germanium	0.785	—4.0	0.65	0.62	—4.4

TABLE 7.6

Substance	$\left(\dfrac{\partial\varepsilon_G}{\partial T}\right)_P \cdot 10^4$, eV/°K	$-\dfrac{\alpha}{\beta_T}\left(\dfrac{\partial\varepsilon_G}{\partial P}\right)_T \cdot 10^4$, eV/°K
Germanium		
gap relative to [111] conduction band minimum	—4	—0.7
gap relative to [100] conduction band minimum	—4	—1.8
Silicon		
gap relative to [100] conduction band minimum	—2.5	0.19

$$\varepsilon_G(P, T) = \varepsilon_{G_0}(P) + \left(\frac{\partial\varepsilon_G}{\partial T}\right)_{,P} T = \varepsilon_{G_0}(P) + \left[\left(\frac{\partial\varepsilon_G}{\partial T}\right)_V + \right.$$

$$+ \alpha\left(\frac{\partial\ln\varepsilon_G}{\partial\ln V}\right)_T \varepsilon_G\left.\right] T = \varepsilon_{G_0}(P) + \left[\left(\frac{\partial\varepsilon_G}{\partial T}\right)_V - \frac{\alpha}{\beta_T}\left(\frac{\partial\varepsilon_G}{\partial P}\right)_T\right] T, \qquad (7.40)$$

where ε_{G_0} is the forbidden band width at T = 0°K; α and β_T are, respectively, the thermal expansion coefficient and the isothermal compressibility. The value of $(\partial\varepsilon_G/\partial T)_V \neq 0$ is governed by the interaction of electrons and holes with phonons. Experimental data shows that

$$\left(\frac{\partial\varepsilon_G}{\partial T}\right)_V \gtrless \alpha\left(\frac{\partial\ln\varepsilon_G}{\partial\ln V}\right)_T \varepsilon_G. \qquad (7.41)$$

Finally, $\varepsilon_G(P)$ can be expressed in terms of a logarithmic derivative $L = (\partial \ln \varepsilon_G / \partial \ln V)_T$ and the compressibility:

$$\varepsilon_G(P) = \varepsilon_G(P_0) \exp\left\{ - \int_{P_0}^{P} \left(\frac{\partial \ln \varepsilon_G}{\partial \ln V} \right)_T \beta_T \, dP \right\}. \qquad (7.42)$$

Equation (7.42) is obtained by direct integration of the identity

$$\left(\frac{\partial \varepsilon_G}{\partial P} \right)_T = - \frac{\varepsilon_G}{K_T} \left(\frac{\partial \ln \varepsilon_G}{\partial \ln V} \right)_T, \qquad K_T = \frac{1}{\beta_T}. \qquad (7.43)$$

The logarithmic derivative depends much more weakly on the volume than do all the other quantities. Assuming that this derivative is constant and equal to L, we can reduce Eq. (7.42) to a very simple expression

$$\varepsilon_G(P) = \varepsilon_G(P_0) \left(\frac{\rho_0}{\rho} \right)^L. \qquad (7.44)$$

We shall now consider the experimental data. Table 7.4 gives the data on the temperature dependence of the forbidden band width of semiconductors. The results presented in that table are taken from Chapter 3. Table 7.5 gives the values of $(\partial \varepsilon_G / \partial T)_P$ and $(\partial \varepsilon_0 / \partial T)_P$ for germanium and silicon (ε_0 is the optical value of the forbidden band width which, in general, differs from the thermal value ε_G, which governs the electrical conductivity). We can see from that table that the temperature dependences of the optical and thermal values of the forbidden band width are similar. More detailed data are given in Table 7.6. (Germanium and silicon have a complex energy band structure, with several energy gaps. When the pressure is increased only one of those gaps becomes important, and that is the gap which governs the electrical conductivity. In this sense properties of solid dielectrics become simpler with increasing pressure.)

It follows from these data that

$$\left(\frac{\partial \varepsilon}{\partial T} \right)_P \approx - \ (3\text{-}5) \ \cdot 10^{-4} \ \text{eV/deg.} \qquad (7.45)$$

It is interesting to note that the value of the derivative of Eq. (7.45) depends on the volume. The dependence of the component $(\partial \varepsilon / \partial T)_P$, associated with the thermal expansion, can be expressed in terms of ε_G, $(\partial \ln \varepsilon_G / \partial \ln V)_T$, and α, i.e., in terms of quantities whose volume dependences can be determined. The more important is the volume dependence of the derivative $(\partial \varepsilon_G / \partial T)_V$, which determines, to a great extent, the temperature dependence of the forbidden band width. The formula which expresses $(\partial \varepsilon_G / \partial T)_V$ in terms of the quantities referred to earlier, is given in [36]:

$$\left(\frac{\partial \varepsilon}{\partial T} \right)_V = - \frac{1}{\pi^2} \frac{k q_m}{\rho (\hbar c_l)^2} (m_e E_{1c}^2 + m_h E_{1v}^2),$$

where q_m is the maximum value of the reduced wave number of the lattice. Expressing q_m in terms of the Debye temperature, $q_m = k\theta / \hbar c_l$), we obtain

$$\left(\frac{\partial \varepsilon}{\partial T} \right)_V = - \frac{1}{\pi^2} \frac{k^2 \theta}{\rho (\hbar c_l)^3} (m_e E_{1c}^2 + m_h E_{1v}^2). \qquad (7.46)$$

Using the subscript zero to denote the derivative obtained at some initial value, and omitting — for the sake of simplicity — one of the terms in parentheses of Eq. (7.46), we find that

$$\left(\frac{\partial \varepsilon}{\partial T} \right)_V = \left(\frac{\partial \varepsilon}{\partial T} \right)_{V_0} \left(\frac{\theta}{\theta_0} \right) \left(\frac{\rho_0}{\rho} \right) \left(\frac{c_{l_0}}{c_l} \right)^3 \left(\frac{m_e}{m_{e0}} \right) \left(\frac{E_{1c}}{E_{1c0}} \right)^2. \qquad (7.47)$$

TABLE 7.7

Substance	ν_0, cm^{-1}	ε_0, eV	$\Delta\nu/\nu_0$	$\dfrac{\Delta V}{V_s}$	$\dfrac{d\ln\varepsilon_0}{d\ln V}=\dfrac{\Delta\nu}{\nu_0}\dfrac{V_0}{\Delta V}$
TlI	21840	0.902	0.64	0.70	0.9
TlBr	23950	0.989	0.63	0.375	2.8
TlCl	37300	1.540	0.38	0.385	1.3
S	29300	1.210	0.61	0.32	1.9
	23250	0.960	0.09	0.5	0.19
Se	17750	0.733	1.05	0.33	3.3
	13600	0.562	0.15	0.35	0.43
As	5600	0.213	3.4	0.88	3.86
	7250	0.299	0.27	0.23	1.17
Red	4750	0.196	0.76	0.16	4.75
phosphorus	10960	0.453	0.18	0.23	0.78
	8460	0.349	0.425	0.16	2.66

TABLE 7.8

t	ρ	$c_l \cdot 10^{-5}$	T_{ad}	T_{mp1}	T_{mp2}	θ/θ_0
1000	4.63	1.16	1950	2700	3700	1
1800	5.13	1.27	2160	3300	4300	1.14
2900	5.63	1.37	2400	4100	5100	1.28

We shall now consider the experimental data on the influence of pressure (volume) on the forbidden band width (energy gap). The relevant data for semiconductors are given in Chapter 3. As already mentioned, these data are very complex because the energy band structure is complex at normal pressures. Thus, the forbidden band width of some substances increases with increasing pressure while that of others decreases. However, if the pressure is sufficiently high, an increase of the forbidden band width is always followed by its decrease. The falling part of the dependence of the forbidden band width of germanium is observed at pressures P > 50 kbar.

In order to determine the logarithmic derivative L we must use the falling part of the dependence $\varepsilon_G = \varepsilon_G(P)$. We shall consider the results of Drickamer (Chapter 3) on the influence of pressure on the optical value of the forbidden band width of semiconductors and insulators obtained at pressures up to 400 kbar. These data are presented in graphical form. We use these graphs to calculate the values of the logarithmic derivatives. The results of such calculations are given in Table 7.7. In those cases where the curves $\varepsilon = \varepsilon(V)$ had a considerable curvature, Table 7.7 lists the values at the beginning (the upper row) and the end (the lower row) of the curve. The general conclusion from Table 7.7 is

$$L \approx 1\text{-}3. \tag{7.48}$$

We may assume that for ionic compounds, such as minerals in layer D, the logarithmic derivative is $L \approx 2\text{-}3$.

In order to obtain the value of the electrical conductivity σ in $\Omega^{-1} \cdot$ cm^{-1}, we must multiply the electronic charge expressed in coulombs, e = 1.6 $\cdot$ 10^{-19}, by the mobility μ expressed in cm$^2 \cdot$ V$^{-1} \cdot$ sec^{-1}, and by the number of carriers n per cm^3.

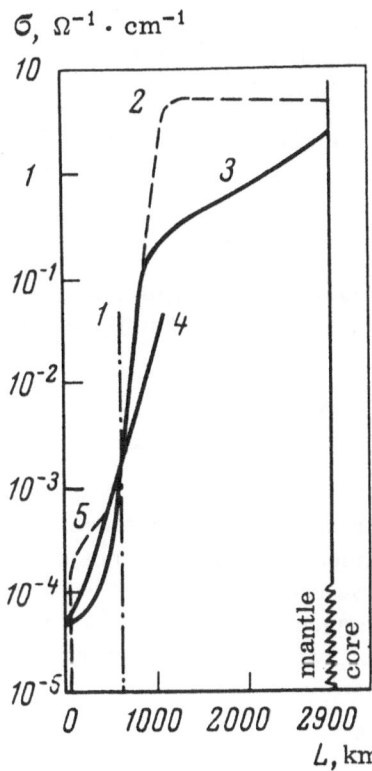

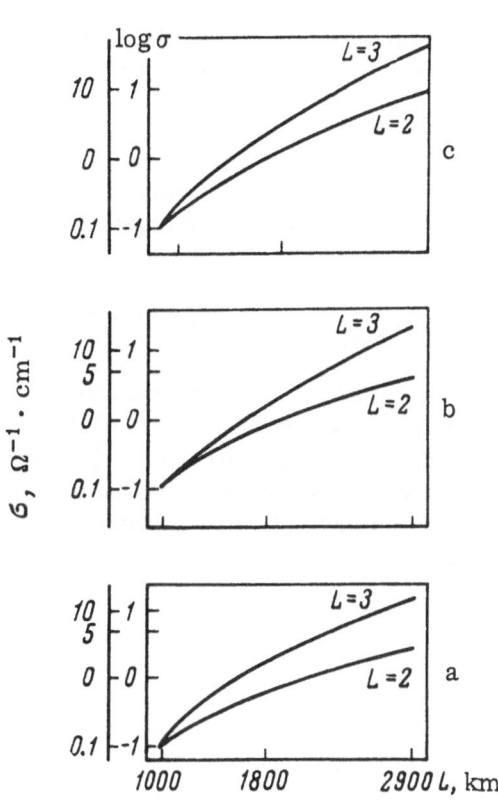

Fig. 7.8. Distributions of the electrical conductivity σ in the earth's mantle [26] according to various authors. 1) Lahiri and Price
[23]; 2) Eckhardt's [26] modification of
McDonald's curve [27]; 3) McDonald [27]; 4)
Lahiri and Price, different model [23]; 5)
combination of McDonald's curve [27] with
that of T. Cantwell ["Detection and analysis of
l.f. magnetotelluric signals," Ph.D. Thesis,
Massachusetts Inst. of Technology (1960)].

Fig. 7.9. Distributions of the electrical conductivity σ in layer D, obtained by physical
methods. a) $T = T_{ad}$; b) $T = T_{mp1}$; c) $T =
T_{mp2}$.

Electrical Conductivity in Layer D

From the above, if we know the value of the electrical conductivity at a depth $l = 1000$ km, the
formulas just given and the geophysical data can be used to determine the electrical conductivity elsewhere in layer D. The greatest indeterminacy in such determination is due to the following factors: a) our lack of knowledge of the temperature distribution in layer D; b) the inaccurate value of the logarithmic derivative L; c) the unreliable value of the initial electrical conductivity σ_{1000} at a depth of 1000 km.

We shall assume that σ_{1000} has the value

$$\sigma_{1000} \approx 10^{-1} \ \Omega^{-1} \cdot cm^{-1}. \tag{7.49}$$

This value is close to that which is found in the distribution of σ given by McDonald (Fig. 7.8,
curve 3). We shall carry out calculations for three variants of the temperature distribution:
1) the adiabatic temperatures T_{ad} [37]; 2) the temperatures along the melting curve, determined by the method of critical concentration of thermal defects T_{mp1} [2]; 3) the temperatures

along the melting curve, determined by the Lindemann method (corrected Uffen's curve) T_{mp2} [2]. The geophysical data using these calculations are taken from [37, 35, 2] and are given in Table 7.8. For each temperature distribution we carried out a calculation for two values of the logarithmic derivative L: 2 and 3. The results are presented in Fig. 7.9.

We shall now describe in detail the way in which these calculations were carried out. For simplicity, we assumed that only one carrier species (electrons) was present and we used

$$\sigma \approx e\mu n,$$

where μ is given by Eq. (7.36) and the carrier density n (for $m_e \approx m_h \approx m_0$) is given by the following formula, which is obtained after including the dependences on the density ρ and on the temperature:

$$n \approx 4.82 \cdot 10^{15} T_0^{3/2} (T/T_0)^{3/2} e^A \exp\left[-\frac{\varepsilon_{G_0}}{2kT}\left(\frac{\rho_0}{\rho}\right)^L\right], \tag{7.50}$$

where the zero subscript indicates that the value refers to the initial depth $l = 1000$ km. The quantity A is due to the dependence of the forbidden band width ε on the temperature. According to Eq. (7.47), the quantity $(\partial\varepsilon/\partial T)_V$, which governs the temperature dependence of the forbidden band width, decreases with increasing density. When we used the geophysical data given in [37, 35, 2] and Eq. (7.47), we found that the derivative decreased by a factor of 3–5 when l increased from 0 to 1000 km. Therefore, instead of Eq. (7.45), we used the following estimate:

$$\left[\left(\frac{\partial\varepsilon}{\partial T}\right)_P\right]_{l=1000} \approx -10^{-4} \text{ eV/deg.} \tag{7.51}$$

We thus found

$$A \approx 0.58\left(\frac{\theta}{\theta_0}\right)\left(\frac{c_{l0}}{c_l}\right)\left(\frac{\rho_0}{\rho}\right)^{2L+1}. \tag{7.52}$$

The last unknown quantity in Eq. (7.49), ε_{G_0}, was determined from the condition $\sigma_{l=1000} \approx 10^{-1} \Omega^{-1} \cdot \text{cm}^{-1}$ and then the data given in Table 7.8 were used to calculate the distributions of σ in layer D shown in Fig. 7.9. We shall now quote some values. The value of the forbidden band width (energy gap) ε_{G_0}, at a depth $l = 1000$ km was found to be 3 eV ($T_{ad_0} = 1950°$K), 4.3 eV ($T_{mp10} = 2700°$K), and 6 eV ($T_{mp20} = 3700°$K).

According to our calculations, the electrical conductivity in layer D increases by a factor of 33 (L = 2) or 167 (L = 3) if we use T = T_{ad}; it increases by a factor of 48 (L = 2) or 245 (L = 3) if we use T = T_{mp1}; finally, the increase is a factor of 93 (L = 2) or 435 (L = 3) if we use T = T_{mp2}.

On the basis of these calculations we may conclude that the electrical conductivity increases by approximately two orders of magnitude in the depth represented by layer D:

$$\sigma_{2900}/\sigma_{1000} \approx 10^2. \tag{7.53}$$

This is an order of magnitude higher than the value predicted by McDonald (Fig. 7.8, curve 3).

Literature Cited

1.   V. N. Zharkov, Izv. Akad. Nauk SSSR, Ser. Geofiz., No. 4, p. 458 (1958).
2.   V. N. Zharkov, Izv. Akad. Nauk SSSR, Ser. Geofiz., No. 3, p. 465 (1959).
3.   V. N. Zharkov, Trudy Inst. Fiziki Zemli Akad. Nauk SSSR, No. 11(178), p. 36 (1960).
4.   V. N. Zharkov, Dokl. Akad. Nauk SSSR, 135 : 1378 (1960).

5.   V. N. Zharkov and V. A. Kalinin, Izv. Akad. Nauk SSSR, Ser. Geofiz., No. 3, p. 298 (1962).

6.   V. N. Zharkov, Trudy Inst. Fiziki Zemli Akad. Nauk SSSR, No. 20(187), p. 3 (1962).

7.   V. N. Zharkov, Izv. Akad. Nauk, Fiziki Zemli, No. 9, p. 3 (1966).

8.   V. A. Magnitskii, Internal Structure and Physics of the Earth [in Russian], "Nedra," Moscow (1965).

9.   V. A. Magnitskii, in: Problems in Cosmogony [in Russian], Izd. AN SSSR, Moscow (1952).

10. V. A. Magnitskii, Trudy Geofiz. Inst. Akad. Nauk SSSR, No. 26(153), p. 61 (1955).

11. F. Birch, J. Geophys. Res., 57:227 (1952).

12. K. E. Bullen, An Introduction to the Theory of Seismology, 3rd ed., Cambridge University Press (1963).

13. E. F. Savarenskii and D. P. Kirnos, Elements of Seismology and Seismometry [in Russian], Gostekhizdat, Moscow (1955).

14. Natural Oscillations of the Earth [Russian translation], "Mir," Moscow (1964).

15. E. D. Williamson and L. H. Adams, J. Wash. Acad. Sci., 13:413 (1923).

16. M. S. Molodenskii, Trudy Geofiz. Inst. Akad. Nauk SSSR, No. 26(153), p. 121 (1955).

17. E. C. Bullard, Verhandel. Ned. Geol.-Mijnbouwk. Genoot., 18:23 (1957).

18. V. L. Pan'kov and V. N. Zharkov, in: Earth Tides and Internal Structure of the Earth [in Russian], "Nauka," Moscow (1967).

19. L. V. Al'tshuler, Usp. Fiz. Nauk, 85:197 (1965).

20. B. I. Davydov, Izv. Akad. Nauk SSSR, Ser. Geofiz., No. 12, p. 1411 (1956).

21. V. N. Zharkov and V. A. Kalinin, Dokl. Akad. Nauk SSSR, 135:811 (1960).

22. S. B. Kormer and A. I. Funtikov, Izv. Akad. Nauk, Fizika Zemli, No. 5, p. 1 (1965).

23. V. N. Lahiri and A. T. Price, Phil. Trans. Roy. Soc. London, A237:509 (1939).

24. K. Terada, Geophys. Mag. (Tokyo), 16:5 (1948).

25. T. Rikitake, Bull. Earthquake Res. Inst., Tokyo Univ., 30:13 (1952).

26. D. Eckhardt, K. Larner, and T. Madden, J. Geophys. Res., 68:6279 (1963).

27. K. L. McDonald, J. Geophys. Res., 62:117 (1957).

28. D. C. Tozer, Phys. and Chem. Earth, 3:414 (1959).

29. H. P. Coster, Monthly Notices Roy. Astron. Soc., Geophys. Suppl., 5:193 (1948).

30. H. Hughes, J. Geophys. Res., 30:187 (1955).

31. Electrical and Mechanical Properties of Rocks under High Pressures [in Russian], "Nauka," Moscow (1966).

32. W. Shockley, Electrons and Holes in Semiconductors, van Nostrand, New York (1950).

33. F. J. Blatt, "Theory of mobility of electrons in solids," Solid State Phys., 4:200 (1957).

34. A. I. Ansel'm, Introduction to the Theory of Semiconductors [in Russian], Fizmatgiz, Moscow (1962).

35. V. N. Zharkov, Izv. Akad. Nauk SSSR, Ser. Geofiz., No. 11, p. 1342 (1958).

36. H. Y. Fan, Rep. Progr. Phys., 19:107 (1956).

37. V. N. Zharkov, Izv. Akad. Nauk SSSR, Ser. Geofiz., No. 9, p. 1414 (1959).

APPENDICES

We shall begin from the well-known thermodynamic identity for the internal energy

$$dE = TdS - PdV, \qquad E = E(S, V), \tag{A.1}$$

where the energy is regarded as a function of two independent variables: the entropy S and the volume V. It follows directly from Eq. (A.1) that

$$T = \left(\frac{\partial E}{\partial S} \right)_V, \qquad P = -\left(\frac{\partial E}{\partial V} \right)_S, \tag{A.2}$$

and we also obtain an expression for the specific heat at constant volume

$$c_V = \left(\frac{\partial E}{\partial T} \right)_V = T \left(\frac{\partial S}{\partial T} \right)_V. \tag{A.3}$$

If the independent variables are the entropy and pressure, we use a thermal function W, known as the enthalpy or heat content

$$W = E + PV \tag{A.4}$$

and

$$dW = TdS + VdP, \qquad W = W(S, P), \tag{A.5}$$

$$T = \left(\frac{\partial W}{\partial S} \right)_P, \qquad V = \left(\frac{\partial W}{\partial P} \right)_S. \tag{A.6}$$

An expression for the specific heat at constant pressure follows from Eq. (A.5):

$$c_P = \left(\frac{\partial W}{\partial T} \right)_P = T \left(\frac{\partial S}{\partial T} \right)_P. \tag{A.7}$$

If the independent variables are the volume and temperature, we use the function

$$F = E - TS, \tag{A.8}$$

known as the free energy and obeying the following relationships:

$$dF = -SdT - PdV, \qquad F = F(T, V), \tag{A.9}$$

$$S = -\left(\frac{\partial F}{\partial T}\right)_V, \qquad P = -\left(\frac{\partial F}{\partial V}\right)_T. \tag{A.10}$$

If the independent variables are the pressure and temperature, the characteristic function of the system is the thermodynamic potential of Gibbs

$$\Phi = E - TS + PV = F + PV = W - TS, \tag{A.11}$$

$$d\Phi = -S\,dT + V\,dP, \qquad \Phi = \Phi(T, P), \tag{A.12}$$

$$S = -\left(\frac{\partial \Phi}{\partial T}\right)_P, \qquad V = \left(\frac{\partial \Phi}{\partial P}\right)_T. \tag{A.13}$$

The quantities E, W, F, and Φ are known, in general, as the thermodynamic potentials, each of them associated with its own set of independent variables.

The internal energy E and the enthalpy W can be easily expressed in terms of F and Φ by the relationships

$$E = F - T\left(\frac{\partial F}{\partial T}\right)_V = -T^2\left(\frac{\partial}{\partial T}\frac{F}{T}\right)_V, \tag{A.14}$$

$$W = \Phi - T\left(\frac{\partial \Phi}{\partial T}\right)_P = -T^2\left(\frac{\partial}{\partial T}\frac{\Phi}{T}\right)_P. \tag{A.15}$$

If we know the equation of state

$$P = P(V, T),$$

we can easily derive various useful relationships between the derivatives of the thermodynamic quantities. It is convenient to carry out calculations using the Jacobians

$$\frac{\partial(u, v)}{\partial(x, y)} = \begin{vmatrix} \dfrac{\partial u}{\partial x} & \dfrac{\partial u}{\partial y} \\ \\ \dfrac{\partial v}{\partial x} & \dfrac{\partial v}{\partial y} \end{vmatrix}$$

The Jacobians have the following properties:

$$\frac{\partial(u, v)}{\partial(x, y)} = -\frac{\partial(v, u)}{\partial(x, y)}, \qquad \frac{\partial(u, y)}{\partial(x, y)} = \left(\frac{\partial u}{\partial x}\right)_y,$$

$$\frac{\partial(u, v)}{\partial(x, y)} = \frac{\partial(u, v)}{\partial(t, s)} \cdot \frac{\partial(t, s)}{\partial(x, y)},$$

$$\frac{d}{dt}\frac{\partial(u, v)}{\partial(x, y)} = \frac{\partial(du/dt, v)}{\partial(x, y)} + \frac{\partial(u, dv/dt)}{\partial(x, y)}.$$

We calculate the derivative $\left(\dfrac{\partial c_P}{\partial P}\right)_T$

$$\left.\begin{aligned}\left(\frac{\partial c_P}{\partial P}\right)_T &= T\frac{\partial^2 S}{\partial P\,\partial T} = -T\frac{\partial^3 \Phi}{\partial P\,\partial T^2} = -T\frac{\partial^2}{\partial T^2}\left(\frac{\partial \Phi}{\partial P}\right)_T = -T\left(\frac{\partial^2 V}{\partial T^2}\right)_P,\\[2mm]\left(\frac{\partial c_P}{\partial P}\right)_T &= -T\left(\frac{\partial^2 V}{\partial T^2}\right)_P\end{aligned}\right\} \tag{A.16}$$

Similarly, we obtain the formula

$$\left(\frac{\partial c_V}{\partial V}\right)_T = T\left(\frac{\partial^2 P}{\partial T^2}\right)_V. \tag{A.17}$$

Since F of Eq. (A.9) and Φ of Eq. (A.12) are total differentials, it follows that:

$$\left(\frac{\partial S}{\partial V}\right)_T = \left(\frac{\partial P}{\partial T}\right)_V = \frac{\alpha}{\beta_T}, \tag{A.18}$$

$$\left(\frac{\partial S}{\partial P}\right)_T = -\left(\frac{\partial V}{\partial T}\right)_P = -V\alpha, \tag{A.19}$$

where α is the volume thermal expansion coefficient. Using thermodynamic identities and the established relationships between the derivatives, we easily obtain the following formulas:

$$\left(\frac{\partial E}{\partial V}\right)_V = T\left(\frac{\partial P}{\partial T}\right)_V - P, \qquad \left(\frac{\partial E}{\partial P}\right)_T = -TV\alpha + PV\beta_T, \tag{A.20}$$

$$\left(\frac{\partial W}{\partial V}\right)_T = T\left(\frac{\partial P}{\partial T}\right)_V - K_T, \qquad \left(\frac{\partial W}{\partial P}\right)_T = V - TV\alpha, \tag{A.21}$$

$$\left(\frac{\partial E}{\partial T}\right)_P = c_P - PV\alpha, \qquad \left(\frac{\partial W}{\partial V}\right)_V = c_V + V\left(\frac{\partial P}{\partial T}\right)_V, \tag{A.22}$$

where the compressibility β_T and the bulk modulus K_T are defined by the relationship

$$\beta_T = \frac{1}{K_T} = -\frac{1}{V}\left(\frac{\partial V}{\partial P}\right)_T. \tag{A.23}$$

Using the properties of the Jacobians, we can derive formulas relating the difference between the specific heats $(c_P - c_V)$ to other thermodynamic derivatives:

$$c_P = T\left(\frac{\partial S}{\partial T}\right)_P = T\frac{\partial(S,P)}{\partial(T,P)} = \frac{\partial(S,P)}{\partial(T,V)}\,\frac{\partial(T,V)}{\partial(T,P)} = T\frac{\left(\frac{\partial S}{\partial T}\right)_V\left(\frac{\partial P}{\partial V}\right)_T - \left(\frac{\partial S}{\partial V}\right)_T\left(\frac{\partial P}{\partial T}\right)_V}{\left(\frac{\partial P}{\partial V}\right)_T}.$$

Using Eq. (A.18), we obtain

$$c_P - c_V = TV\frac{\left(\frac{\partial P}{\partial T}\right)_V^2}{K_T}. \tag{A.24}$$

Similarly,

$$c_V = T\left(\frac{\partial S}{\partial T}\right)_V = T\frac{\partial(S,V)}{\partial(T,V)} = T\frac{\partial(S,V)}{\partial(T,P)}\cdot\frac{\partial(T,P)}{\partial(T,V)} = T\frac{\left(\frac{\partial S}{\partial T}\right)_P\left(\frac{\partial V}{\partial P}\right)_T - \left(\frac{\partial S}{\partial P}\right)_T\left(\frac{\partial V}{\partial T}\right)_P}{\left(\frac{\partial V}{\partial P}\right)_T}.$$

Using Eq. (A.19), we find that

$$c_P - c_V = TVK_T\alpha^2. \tag{A.25}$$

We shall now calculate the derivatives of the temperature with respect to the pressure and volume at constant entropy

$$\left(\frac{\partial T}{\partial P}\right)_S = \frac{\partial(T,S)}{\partial(P,S)} = \frac{\frac{\partial(T,S)}{\partial(P,T)}}{\frac{\partial(P,S)}{\partial(P,T)}} = -\frac{\left(\frac{\partial S}{\partial P}\right)_T}{\left(\frac{\partial S}{\partial T}\right)_P} = -\frac{T}{c_P}\left(\frac{\partial S}{\partial P}\right)_T$$

and, substituting Eq. (A.19), we obtain

$$\left(\frac{\partial T}{\partial P}\right)_S = \frac{TV\alpha}{c_P}. \tag{A.26}$$

Equally simply we can obtain the formulas

$$\left(\frac{\partial T}{\partial V}\right)_S = -\frac{T}{c_V}\left(\frac{\partial P}{\partial T}\right)_V, \qquad \left(\frac{\partial V}{\partial S}\right)_P = \frac{T}{c_P}\left(\frac{\partial V}{\partial T}\right)_P \tag{A.27}$$

We shall now derive a formula relating the ratio of the specific heats to the ratio of the compressibilities

$$\left(\frac{\partial V}{\partial P}\right)_T = \frac{\partial(V,T)}{\partial(P,T)} = \frac{\partial(V,S)}{\partial(P,S)}\frac{\frac{\partial(P,S)}{\partial(P,T)}}{\frac{\partial(V,S)}{\partial(V,T)}} = \left(\frac{\partial V}{\partial P}\right)_S\frac{\left(\frac{\partial S}{\partial T}\right)_P}{\left(\frac{\partial S}{\partial T}\right)_V} = \frac{c_P}{c_V}\left(\frac{\partial V}{\partial P}\right)_S$$

or

$$\frac{\beta_T}{\beta_S} = \frac{K_S}{K_T} = \frac{c_P}{c_V}. \tag{A.28}$$

Combining Eq. (A.28) with Eqs. (A.24) and (A.25), we obtain

$$\beta_S = \beta_T - \frac{TV\alpha^2}{c_P}, \tag{A.29}$$

$$K_S = K_T + \frac{TV}{c_V}\left(\frac{\partial P}{\partial T}\right)_V^2. \tag{A.30}$$

Using the same technique and the second law of thermodynamics, we easily obtain the following inequalities:

$$c_V > 0, \qquad (\partial P/\partial V)_T < 0. \tag{A.31}$$

The states which do not satisfy these conditions are unstable and cannot exist in nature.

All the thermodynamic potentials E, F, W, and Φ have the property of additivity. This allows us to represent them as functions of the number of particles N in the following form:

$$E = Nf(S/N, V/N), \tag{A.32}$$

$$F = Nf(V/N, T), \tag{A.33}$$

$$W = Nf(S/N, P), \tag{A.34}$$

$$\Phi = Nf(P, T). \tag{A.35}$$

If we regard N as an additional thermodynamic variable, we can expand thermodynamic identities by including terms proportional to dN:

$$dE = TdS - PdV + \mu dN, \quad \mu = (\partial E/\partial N)_{S, V}; \tag{A.36}$$

$$dW = TdS + VdP + \mu dN, \quad \mu = (\partial W/\partial N)_{S, P}; \tag{A.37}$$

$$dF = -SdT - PdV + \mu dN, \quad \mu = (\partial F/\partial N)_{T, V}; \tag{A.38}$$

$$d\Phi = -SdT + VdP + \mu dN, \quad \mu = (\partial \Phi/\partial N)_{P, T}. \tag{A.39}$$

The quantity μ is known as the chemical potential.

Comparing Eqs. (A.35) and (A.39), we can see that the chemical potential is equal to the Gibbs thermodynamic potential per single particle:

$$\Phi = N\mu \tag{A.40}$$

and

$$d\mu = -sdT + vdP, \tag{A.41}$$

where s and v are the entropy and volume per single particle.

So far we have assumed that we are dealing with a fixed amount of matter, i.e., that the number of particles is constant but the volume is variable. We shall now assume the volume to be given and the number of particles to be variable. Then Eq. (A.38) assumes the form

$$dF = -SdT + \mu dN,$$

i.e., F = F(T, N). Changing from the independent variable N to μ, we obtain

$$d(F - \mu N) = -SdT - Nd\mu.$$

Denoting this new potential by Ω and using $\Phi = \mu N$, as well as $F - \Phi = -PV$, we obtain

$$\Omega = -PV, \tag{A.42}$$

$$d\Omega = -SdT - Nd\mu, \quad N = -(\partial \Omega/\partial \mu)_{T, V} = V(\partial P/\partial \mu)_{T, V}. \tag{A.43}$$

The potential Ω will be useful in the derivation of the isentropes of the electron gas in Appendix C.

B. Debye Approximation

B. 1. Calculation of the Debye Function

The Debye function

$$D\left(\frac{\theta}{T}\right) = \frac{3T^3}{\theta^3} \int\limits_0^{\theta/T} \frac{y^3\,dy}{e^y - 1}$$

cannot be represented by a single analytic expression, which could be used over the whole range of variation of the argument θ/T. However, in the limiting cases of low and high values of θ/T, we can obtain formulas which overlap the whole range of possible values of the argument. For brevity we shall use $y_m = \theta/T$ and we shall consider the two limiting cases.

High Temperatures $(y_m < 2\pi)$. We shall calculate by parts the integral in the Debye function $D(y_m)$

$$\int\limits_0^{y_m} \frac{y^3\,dy}{e^y - 1} = \frac{1}{4}\frac{y_m^4}{e^{y_m} - 1} + 2 \int\limits_0^{y_m/2} \frac{z^4\,dz}{\sinh^2 z},$$

where $z = y/2$. This integral can be expressed as a series converging when $z < \pi$:

$$\int \frac{z^4 dz}{\sinh^2 z} = -z^4 \coth z + 4 \sum_{k=0}^{\infty} \frac{2^{2k} B_{2k}}{(2k+3)(2k)!} z^{2k+3},$$

where B_{2k} are the Bernoulli numbers. The first few Bernoulli numbers are: $B_0 = 1$, $B_2 = \frac{1}{6}$, $B_4 = -\frac{1}{30}$, etc. We finally obtain

$$D\left(\frac{\theta}{T}\right) = -\frac{3}{8}\frac{\theta}{T} + 3 \sum_{k=0}^{\infty} \frac{B_{2k}}{(2k+3)(2k)!}\left(\frac{\theta}{T}\right)^{2k} \tag{B.1}$$

for $(\theta/T) < 2\pi$.

Low Temperatures $(y_m > 0)$. In this case the integral in $D(y_m)$ can be represented conveniently by a sum of two integrals:

$$\int\limits_0^{y_m} \frac{y^3\,dy}{e^y - 1} = \int\limits_0^{\infty} \frac{y^3\,dy}{e^y - 1} - \int\limits_{y_m}^{\infty} \frac{y^3\,dy}{e^y - 1}.$$

The first of these integrals is known and it is equal to $\pi^4/15$, whereas the integrand in the second integral can be represented as a series

$$\frac{y^3}{e^y - 1} = y^3 \sum_{n=0}^{\infty} e^{-(n+1)y},$$

which converges for $y > 0$. The second integral can thus be written in the form

$$\int\limits_{y_m}^{\infty} \frac{y^3\,dy}{e^y - 1} = \sum_{n=0}^{\infty} \left(\frac{y_m^3}{n+1} + \frac{3y_m^2}{(n+1)^2} + \frac{6y_m}{(n+1)^3} + \frac{6}{(n+1)^4} \right) e^{-(n+1)y_m}.$$

We finally obtain

$$D\left(\frac{\theta}{T}\right) = \frac{\pi^4}{5}\left(\frac{T}{\theta}\right)^3 - 3 \sum_{k=1}^{\infty} \frac{1}{k}\left[1 + \frac{3}{k}\left(\frac{T}{\theta}\right) + \frac{6}{k^2}\left(\frac{T}{\theta}\right)^2 + \frac{6}{k^3}\left(\frac{T}{\theta}\right)^3\right] e^{-k\theta/T} \tag{B.2}$$

for $(T/\theta) > 0$. The last expression for $D(\theta/T)$ is valid over the full range of values of θ/T and, therefore, the separation into high and low temperatures in calculations of the Debye function is simply made for convenience because the series in Eqs. (B.1) and (B.2) converge rapidly.

B.2. Thermodynamic Functions in the Debye Approximation

Using the expression for the free energy, given by Eq. (2.23),

$$F = E_p(x) + \frac{RT}{\mu}\left[\frac{9}{8}\frac{\theta}{T} + 3\ln(1 - e^{-\theta/T}) - D\left(\frac{\theta}{T}\right)\right],$$

we can easily obtain all the principal thermodynamic functions of a crystal.

We shall first calculate the derivative of the Debye function with respect to $\theta/T \equiv y_m$, which is used quite frequently:

$$\frac{\partial D(y_m)}{\partial y_m} = \frac{3}{e^{y_m} - 1} - \frac{3}{y_m}D(y_m). \tag{B.3}$$

The internal energy is now given by

$$E = F - T\left(\frac{\partial F}{\partial T}\right)_V = E_p(x) + \frac{R}{\mu}\left[\frac{9}{8}\theta + 3TD\left(\frac{\theta}{T}\right)\right]. \tag{B.4}$$

The entropy is

$$S = -\left(\frac{\partial F}{\partial T}\right)_V = \frac{R}{\mu}\left[4D\left(\frac{\theta}{T}\right) - 3\ln(1 - e^{-\theta/T})\right]. \tag{B.5}$$

The specific heat at constant volume is

$$c_V = T\left(\frac{\partial S}{\partial T}\right)_V = \frac{3R}{\mu}\left[4D\left(\frac{\theta}{T}\right) - \frac{3\theta/T}{e^{\theta/T} - 1}\right]. \tag{B.6}$$

The pressure is

$$P = P_p(x) + \frac{R}{\mu}\frac{\rho_0\gamma}{x}\left[\frac{9}{8}\theta + 3TD\left(\frac{\theta}{T}\right)\right]. \tag{B.7}$$

The isothermal bulk modulus is

$$K_T = K_p(x) + \frac{R}{\mu}\frac{\rho_0\gamma}{x}\left\{\left[\frac{9}{8}\theta + 3TD\left(\frac{\theta}{T}\right)\right]\left(1 + \gamma - \frac{\partial\ln\gamma}{\partial\ln x}\right) - 12\gamma TD\left(\frac{\theta}{T}\right) + \frac{9\theta\gamma}{e^{\theta/T} - 1}\right\}. \tag{B.8}$$

The specific heat at constant pressure c_P is related to c_V by

$$c_P = c_V + \frac{x\alpha T}{\rho_0}\left(\frac{\partial P}{\partial T}\right)_x$$

The derivative $(\partial P/\partial T)_x$ can be found from Eqs. (B.3) and (B.7):

$$\left(\frac{\partial P}{\partial T}\right)_x = \frac{3R}{\mu}\frac{\rho_0\gamma}{x}\left[4D\left(\frac{\theta}{T}\right) - \frac{3\theta/T}{e^{\theta/T} - 1}\right]. \tag{B.9}$$

Finally, we find c_P

$$c_P = \frac{3R}{\mu}\left[4D\left(\frac{\theta}{T}\right) - \frac{3\theta/T}{e^{\theta/T} - 1}\right](1 + \alpha\gamma T). \tag{B.10}$$

The thermal expansion coefficient is

$$\alpha = \frac{1}{x}\left(\frac{\partial x}{\partial T}\right)_P = -\frac{(\partial P/\partial T)_x}{x(\partial P/\partial x)_T} = \frac{3R}{\mu}\frac{\rho_0\gamma}{x}\frac{1}{K_T}\left[4D\left(\frac{\theta}{T}\right) - \frac{3\theta/T}{e^{\theta/T}-1}\right]. \qquad (B.11)$$

C.   Ideal Gas

This subject covers a wide range of topics which can be found in books on statistical thermodynamics. We shall consider briefly the information on a classical ideal gas, a degenerate Fermi gas, and a photon (radiation) gas.

C.1. Boltzmann Ideal Gas

The equation of state for an ideal gas can be calculated theoretically using a general formula of the statistical thermodynamics for the free energy

$$F = -kT\ln\sum_n\exp(-E_n/kT). \qquad (C.1)$$

The quantity under the logarithm sign is known as the partition function. In this quantity summation extends over all the energy levels of a system being considered. The partition function of the ideal gas simplifies considerably since the energy of the system can be written in the form of sums of energies of identical particles and all the terms of the sum are of the same form:

$$\varepsilon_i = p^2/2m + \varepsilon_i'. \qquad (C.2)$$

Here, the first term is the kinetic energy of the translational motion of a particle and ε_i' represents the energy levels corresponding to the internal state of the particle; m and $\mathbf{p}$ (p_x, p_y, p_z) are the mass and momentum of the particle.

Reducing the summation in Eq. (C.1) to the summation over the states of a single particle, we must bear in mind that each state should be taken into account only once. The number of possible permutations of N identical particles is N!. Consequently, the value of the partition function in (C.1) can be represented in the form

$$\sum_n\exp(-E_n/kT) = \frac{1}{N!}\left[\sum_i\exp(-\varepsilon_i/kT)\right]^N. \qquad (C.3)$$

The division by N! in Eq. (C.3) is due to the fact that in the new summation each state would be taken into account N! times. N is a large number and, therefore, the following relationship applies:

$$\ln N! \simeq N\ln(N/e).$$

We can thus reduce Eq. (C.1) to the form

$$F = -NkT\ln\left[\frac{e}{N}\sum_i\exp(-\varepsilon_i/kT)\right]. \qquad (C.4)$$

The summation of the translational motion of the particles can be replaced by the integration over a phase space ($\mathbf{P}$, $\mathbf{r}$) with the statistical weight

$$d\tau = \frac{d\mathbf{p}\cdot d\mathbf{r}}{(2\pi\hbar)^3}, \qquad (C.5)$$

where $\hbar$ is the Planck constant, divided by 2π. We thus find that

$$\sum_i \frac{1}{(2\pi\hbar)^3} \exp(-\varepsilon_i'/kT) \int\int_V \int\int_{-\infty}^{+\infty}\int \exp\left[-\frac{1}{2mkT}(p_x^2+p_y^2+p_z^2)\right]\times$$

$$\times dp_x\,dp_y\,dp_z\,dV = V\left(\frac{mkT}{2\pi\hbar^2}\right)^{3/2} \sum_i \exp(-\varepsilon_i'/kT) \quad . \tag{C.6}$$

This expression for the partition function yields the following formula for the free energy:

$$F = -NkT \ln\left[\frac{eV}{N}\left(\frac{mkT}{2\pi\hbar^2}\right)^{3/2} \sum_i \exp(-\varepsilon_i'/kT)\right]. \tag{C.7}$$

The quantity under the summation sign is independent of the volume and is only a function of the temperature. Separating this function within the expression for the free energy, we obtain

$$F = -NkT \ln (eV/N) + Nf(T). \tag{C.8}$$

This expression for F can be used, together with Eq. (A.10), to obtain the equation of state for an ideal gas

$$P = -\left(\frac{\partial F}{\partial V}\right)_T = \frac{NkT}{V}, \tag{C.9}$$

or

$$PV = NkT,$$

which is known as the Clapeyron—Clausius equation. The Clapeyron—Clausius equation can be written in any of the following equivalent forms:

$$P = N_0 kT = N_g \rho kT = \frac{NkT}{V} = \frac{R}{\mu}\rho T = R_0 \rho T, \tag{C.9a}$$

where N_0 is the number of particles in 1 cm^3; N_g is the number of particles in 1 g; R is the universal gas constant; R_0 is the gas constant per 1 g; μ is the average molecular weight. Using the free energy of Eq. (C.8), the equation of state of Eq. (C.9). and the formulas given in Appendix A, we can calculate all the other thermodynamic functions of the ideal gas.

The thermodynamic potential is

$$\Phi = NkT \ln P + N\chi(T), \quad \chi(T) = f(T) - kT \ln kT. \tag{C.10}$$

The entropy is given by

$$\left.\begin{aligned} S(V,T) &= -\left(\frac{\partial F}{\partial T}\right)_V = Nk \ln\left(\frac{eV}{N}\right) - Nf'(T), \\[2mm] S(P,T) &= -\left(\frac{\partial \Phi}{\partial T}\right)_P = -Nk \ln P - N\chi'(T). \end{aligned}\right\} \tag{C.11}$$

The internal energy is

$$E = Nf(T) - NTf'(T). \tag{C.12}$$

The enthalpy is

$$W = NkT + Nf(T) - NTf'(T).\qquad(C.13)$$

We can see that the internal energy and the enthalpy of the ideal gas are functions only of the temperature. Since $W = E + NkT$, it follows that

$$c_P - c_V = Nk.\qquad(C.14)$$

C.2. Ideal Gas with Constant Specific Heats

The assumption of the constancy of specific heats

$$C_P = Nc_P,\quad C_V = Nc_V\qquad(C.15)$$

(i.e., the assumption that they are independent of temperature), allows us to calculate the unknown function $f(T)$ introduced in preceding equations and thus obtain a complete thermodynamic description of the ideal gas.

Differentiating the internal energy of Eq. (C.12), we find the relationship between the second derivative $f''(T)$ and the specific heat c_V:

$$c_V = -Tf''(T).$$

Integration of this relationship yields the function $f(T)$:

$$f(T) = -c_V T \ln kT - k\xi T + \varepsilon_0.\qquad(C.16)$$

The constant of integration ξ is known as the chemical constant and ε_0 is the zero-point (zero-temperature) energy. Using Eq. (C.16), we can easily find the various thermodynamic functions:

$$F = N\varepsilon_0 - NkT \ln(eV/N) - Nc_V T \ln kT - Nk\xi T,\qquad(C.17)$$

$$\Phi = N\varepsilon_0 + NkT \ln P - Nc_P T \ln kT - Nk\xi T,\qquad(C.18)$$

$$E = N\varepsilon_0 + Nc_V T,\qquad(C.19a)$$

$$W = N\varepsilon_0 + Nc_P T,\qquad(C.19b)$$

$$S(V, T) = Nk \ln(eV/N) + Nc_V \ln kT + N(k\xi + c_V),\qquad(C.20a)$$

$$S(P, T) = -Nk \ln P + Nc_P \ln kT + N(k\xi + c_P).\qquad(C.20b)$$

Using Eqs. (C.20a) and (C.20b), we obtain the following equations for the isentropes of the ideal gas with constant specific heats:

$$\left.\begin{array}{ll}\rho = \rho_0(T/T_0)^{1/(\gamma-1)}, & P = P_0(\rho/\rho_0)^{\gamma},\\ P = P_0(T/T_0)^{\gamma/(\gamma-1)}.\end{array}\right\}\qquad(C.21)$$

The relationships given in Eq. (C.21) are known as Poisson's adiabats. The exponent γ is equal to the ratio of the specific heats:

$$\gamma = c_P/c_V.\qquad(C.22)$$

The quantity γ is always greater than unity; for a monatomic gas, $\gamma = 5/3$, and for a diatomic gas, $\gamma = 7/5$.

Using Eq. (C.21) we can determine the velocity of sound in the ideal gas under consideration:

$$c^2 = \left(\frac{\partial P}{\partial \rho}\right)_s = \gamma \frac{P}{\rho} = \gamma \frac{RT}{\mu}. \tag{C.23}$$

We shall use ε and w for the internal energy and enthalpy of a single particle. Then, by measuring the energy from the unimportant constant ε_0, we easily obtain the following formulas:

$$\varepsilon = c_V T = \frac{Pv}{\gamma - 1} = \frac{c^2}{\gamma(\gamma - 1)}, \tag{C.24}$$

$$w = c_P T = \frac{\gamma Pv}{\gamma - 1} = \frac{c^2}{\gamma - 1}, \tag{C.25}$$

where v is the volume per single particle, and

$$S = c_V \ln \frac{P}{\rho^\gamma} = c_P \ln \frac{P^{1/\gamma}}{\rho}. \tag{C.26}$$

C.3. Degenerate Electron Gas at T = 0

All properties of a degenerate electron gas at T = 0°K can be established using Pauli's principle. According to this principle, a given quantum state cannot be occupied by more than two electrons. The number of quantum states of an electron with the absolute momentum in the interval between p and p + dp is given by

$$2\frac{4\pi p^2 V \, dp}{(2\pi\hbar)^3} = \frac{V p^2 \, dp}{\pi^2 \hbar^3}. \tag{C.27}$$

This number is twice as large as the number of states given by (C.5) for classical spinless particles. At T = 0°K all the lowest states between p = 0 and p = p_0 are occupied. The number of occupied states is equal to the total number of electrons N:

$$N = \frac{V}{\pi^2 \hbar^3} \int_0^{p_0} p^2 \, dp = \frac{V p_0^3}{3\pi^2 \hbar^3},$$

and, therefore,

$$p_0 = (3\pi^2)^{1/3} (N/V)^{1/3} \hbar. \tag{C.28}$$

The energy of an electron is equal to the kinetic energy of its translational motion

$$\varepsilon_p = p^2 / 2m, \tag{C.29}$$

where m is the mass of the electron. The minimum energy of electrons at absolute zero

$$\varepsilon_0 = \frac{p_0^2}{2m} = (3\pi^2)^{2/3} \frac{\hbar^2}{2m} \left(\frac{N}{V}\right)^{2/3} \tag{C.30}$$

is known as the Fermi energy. We shall show later that this energy has the physical meaning of the chemical potential and is denoted by μ_0. The total energy of the gas is found by integrating the energy given by Eq. (C.29) over all the filled states of Eq. (C.27):

$$E_0 = \frac{V}{2m\pi^2 \hbar^3} \int_0^{p_0} p^4 \, dp = \frac{V p_0^5}{10m\pi^2 \hbar^3} = \frac{3}{10}(3\pi^2)^{2/3} \frac{\hbar^2}{m} \left(\frac{N}{V}\right)^{2/3} N. \tag{C.31}$$

According to the Nernst theorem, the entropy of the electron gas at T = 0°K is zero:

$$S_0 = 0. \tag{C.32}$$

The free energy F_0 at T = 0°K is equal to the total energy

$$F_0 = E_0 = \frac{3}{10} (3\pi^2)^{2/3} \frac{\hbar^2}{m} \left(\frac{N}{V} \right)^{2/3} N. \tag{C.33}$$

The pressure is

$$P_0 = -\frac{\partial F_0}{\partial V} = \frac{1}{5} (3\pi^2)^{2/3} \frac{\hbar^2}{m} \left(\frac{N}{V} \right)^{5/3}. \tag{C.34}$$

The thermodynamic potential of Eq. (A.11) becomes

$$\Phi_0 = F_0 + P_0 V = (3\pi^2)^{2/3} \frac{\hbar^2}{2m} \left(\frac{N}{V} \right)^{2/3} N = N\varepsilon_0. \tag{C.35}$$

Comparing Eqs. (C.35) and (A.40), we find that ε_0 is equal to the chemical potential of electrons

$$\mu_0 = \varepsilon_0 = (3\pi^2)^{2/3} \frac{\hbar^2}{2m} \left(\frac{N}{V} \right)^{2/3} = \mu_{00} x^{-2/3}, \qquad x = \frac{V}{V_0}. \tag{C.36}$$

These formulas can be used, as a first approximation, at temperatures fairly close to absolute zero. The criterion of their validity (the "strong degeneracy" condition) is the smallness of T compared with the degeneracy temperature T_0 ($T \ll T_0$), which is defined by

$$T_0 = \frac{\varepsilon_0}{k} = (3\pi^2)^{2/3} \frac{\hbar^3}{2mk} \left(\frac{N}{V} \right)^{2/3}. \tag{C.37}$$

We shall illustrate the use of these relationships by considering the electron gas in a metal. We shall substitute in Eqs. (C.35) and (C.30) the electron mass m; we shall assume that N is equal to Avogadro's number and that V is the molecular volume of the electron gas (in cm^3), which is equal to the atomic volume of the metal divided by the number of valence electrons. Then,

$$\mu_0 = 4.166 \cdot 10^{-11} V^{-2/3} \text{ erg/atom} = 26.0 V^{-2/3} \text{ eV},$$

$$T_0 = \mu_0 / k = 301{,}810 V^{-2/3} \text{ °K}, \tag{C.38}$$

$$E_0 = \frac{3}{5} N_0 \mu_0 = 359.7 \quad \text{kcal/mole} \tag{C.39}$$

$$P_0 = \frac{2}{3} \frac{E_0}{V} = 1.004 \cdot 10^{13} V^{-5/3} \text{ dyn/cm}^2 = P_{00} x^{-5/3} = 1.004 \cdot 10^7 V^{-5/3} \text{ bar}. \tag{C.40}$$

For the majority of metals the value of V in Eqs. (C.38) and (C.40) is approximately 10.

C.4. Degenerate Electron Gas at T ≠ 0

An electron gas obeys the Fermi — Dirac statistics and, at sufficiently low temperatures $T \ll T_0$, it is described by the Fermi distribution

$$\bar{n}_k = \frac{1}{1 + \exp[(\varepsilon_k - \mu)/kT]}, \tag{C.41}$$

which gives the average occupation numbers $\bar{n}_k$ in terms of the temperature, the chemical potential μ, and the electron energy ε_k in k-th state. The Fermi distribution is normalized by the condition

$$\sum_{k} \bar{n}_k = N. \tag{C.42}$$

This condition implicitly defines μ as a function of T and N.

The summation in Eq. (C.42) can be replaced by integration. Expressing the number of quantum states of Eq. (C.27) in terms of the electron energy (C.29)

$$\frac{\sqrt{2}\, V m^{3/2} \varepsilon^{1/2}}{\pi^2 \hbar^3}\, d\varepsilon, \tag{C.43}$$

we obtain the number of electrons $N(\varepsilon)d\varepsilon$ in the interval of energy values between ε and $\varepsilon + d\varepsilon$:

$$N(\varepsilon)d\varepsilon = \frac{\sqrt{2}}{\pi^2} V\left(\frac{m^{1/2}}{\hbar}\right)^3 \frac{\varepsilon^{1/2}}{1 + \exp[(\varepsilon - \mu)/kT]}\, d\varepsilon. \tag{C.44}$$

Integrating this expression over all energies and equating the result to the total number of particles N, we obtain

$$N = \frac{\sqrt{2}}{\pi^2} V\left(\frac{m^{1/2}}{\hbar}\right)^3 \int_0^{\infty} \frac{\varepsilon^{1/2}}{1 + \exp[(\varepsilon - \mu)/kT]}\, d\varepsilon. \tag{C.45}$$

Similarly, the energy of the system can be found from the formula

$$E = \int_0^{\infty} \varepsilon N(\varepsilon)d\varepsilon = \frac{\sqrt{2}}{\pi^2} V\left(\frac{m^{1/2}}{\hbar}\right)^3 \int_0^{\infty} \frac{\varepsilon^{3/2}}{1 + \exp[(\varepsilon - \mu)/kT]}\, d\varepsilon. \tag{C.46}$$

Consequently, our problem reduces to calculation of integrals of the type

$$I = \int_0^{\infty} \frac{f(\varepsilon)\, d\varepsilon}{1 + \exp[(\varepsilon - \mu)/kT]},$$

where $f(\varepsilon)$ is some function, which is equal to $\varepsilon^{1/2}$ in Eq. (C.45), $\varepsilon^{3/2}$ in Eq. (C.46), etc. The method for calculating these integrals is known and it reduces to the following procedure. First, we alter the variable so that $\varepsilon - \mu = kTz$, and then

$$I = \int_{-\mu/kT}^{\infty} \frac{f(\mu + kTz)}{1 + e^z}\, kT\, dz = kT \int_0^{\mu/kT} \frac{f(\mu - kTz)}{1 + e^{-z}}\, dz + kT \int_0^{\infty} \frac{f(\mu + kTz)\, dz}{1 + e^z}.$$

Using

$$\frac{1}{1 + e^{-z}} = 1 - \frac{1}{1 + e^z}$$

and separating the first integral into two, we obtain

$$I = \int_0^{\mu} f(\varepsilon)\, d\varepsilon - kT \int_0^{\mu/kT} \frac{f(\mu - kTz)}{1 + e^z}\, dz + kT \int_0^{\infty} \frac{f(\mu + kTz)}{1 + e^z}\, dz.$$

Since $\mu/kT \gg 1$ and the integrals under consideration converge rapidly, we shall replace the upper limit in the second integral by infinity; then,

$$I = \int_0^{\mu} f(\varepsilon)\, d\varepsilon + kT \int_0^{\infty} \frac{f(\mu + kTz) - f(\mu - kTz)}{1 + e^z}\, dz.$$

Expanding the numerator in the integrand of the second integral as a Taylor's series and integrating term by term, we obtain

$$I = \int_0^\mu f(\varepsilon)\,d\varepsilon + 2(kT)^2 f'(\mu) \int_0^\infty \frac{z\,dz}{1 + e^z} + \frac{1}{3}(kT)^4 f'''(\mu) \int_0^\infty \frac{z^3\,dz}{1 + e^z}.$$

Substituting the known values of the definite integrals

$$\int_0^\infty \frac{z^{2n-1}\,dz}{1 + e^z} = \frac{2^{2n-1} - 1}{2n}\,\pi^{2n} B_n,$$

where B_n are the Bernoulli numbers $(B_1 = \tfrac{1}{6}, B_2 = \tfrac{1}{30}, \ldots)$, we obtain the required formula

$$I = \int_0^\infty \frac{f(\varepsilon)\,d\varepsilon}{1 + \exp[(\varepsilon - \mu)/kT]} = \int_0^\mu f(\varepsilon)\,d\varepsilon + \frac{\pi^2}{6}(kT)^2 f'(\mu) + \frac{7\pi^4}{360}(kT)^4 f'''(\mu) + \ldots \qquad \text{(C.47)}$$

The Bernoulli numbers used here are defined differently (and have different values) than those in Eq. (C.1) of Section C.1.

We shall use Eq. (C.47) to determine μ. Employing Eqs. (C.36) and (C.30), we can transform Eq. (C.45) into

$$N = \int_0^\infty N(\varepsilon)\,d\varepsilon = \frac{3N}{2\mu_0^{3/2}} \int_0^\infty \frac{\varepsilon^{1/2}\,d\varepsilon}{1 + \exp[(\varepsilon - \mu)/kT]}, \qquad \text{(C.48)}$$

where μ_0 is the chemical potential of the electron gas at T = 0°K,

$$f(\varepsilon) = \varepsilon^{1/2}, \quad \int_0^\mu f(\varepsilon)\,d\varepsilon = \frac{2}{3}\mu^{3/2},$$

$$f'(\mu) = \frac{1}{2}\mu^{-1/2}, \quad f'''(\mu) = \frac{3}{8}\mu^{-5/2}.$$

These expressions and Eq. (C.47) reduce Eq. (C.48) to the form

$$1 = \left(\frac{\mu}{\mu_0}\right)^{3/2}\left[1 + \frac{\pi^2}{8}\left(\frac{kT}{\mu}\right)^2 + \frac{7\pi^4}{640}\left(\frac{kT}{\mu}\right)^4 + \ldots\right].$$

We shall raise the above expression to a power of $\tfrac{2}{3}$ and, using the expansion

$$\frac{1}{(1 + x)^{2/3}} = 1 - \frac{2}{3}x + \frac{5}{9}x^2 - \ldots,$$

we obtain

$$\mu = \mu_0\left[1 - \frac{\pi^2}{12}\left(\frac{kT}{\mu}\right)^2 + \frac{\pi^4}{720}\left(\frac{kT}{\mu}\right)^4 + \ldots\right].$$

Thus, the chemical potential of the electron gas (accurate to within T^2), is given by

$$\mu = \mu_0\left[1 - \frac{\pi^2}{12}\left(\frac{kT}{\mu_0}\right)^2\right]. \qquad \text{(C.49)}$$

In order to obtain μ with an accuracy to T^4, the value of μ in Eq. (C.49) should be substituted in the quadratic term of the preceding expression and in the last term we must substitute $\mu = \mu_0$.

The final expression for μ is of the form

$$\mu = \mu_0 \left[1 - \frac{\pi^2}{12} \left(\frac{kT}{\mu_0} \right)^2 - \frac{\pi^4}{80} \left(\frac{kT}{\mu_0} \right)^4 + \ldots \right], \tag{C.50}$$

where μ_0 is given by Eq. (C.36).

The internal energy

$$E = \int_0^\infty \varepsilon N(\varepsilon) \, d\varepsilon = \frac{3N}{2\mu_0^{3/2}} \int_0^\infty \frac{\varepsilon^{3/2} \, d\varepsilon}{1 + \exp[(\varepsilon - \mu)/kT]}$$

is calculated similarly by means of Eq. (C.47), which gives

$$f(\varepsilon) = \varepsilon^{3/2}, \qquad \int_0^\mu f(\varepsilon) \, d\varepsilon = \frac{2}{5} \mu^{5/2},$$

$$f'(\mu) = \frac{3}{2} \mu^{1/2}, \qquad f'''(\mu) = -\frac{3}{8} \mu^{-3/2},$$

and

$$E = \frac{3}{5} N \left(\frac{\mu}{\mu_0} \right)^{3/2} \mu \left[1 + \frac{5\pi^2}{8} \left(\frac{kT}{\mu} \right)^2 - \frac{7\pi^4}{384} \left(\frac{kT}{\mu} \right)^4 + \ldots \right].$$

Substituting in the above expression the value of μ from Eq. (C.50), we finally obtain

$$E = N\mu_0 \left[\frac{3}{5} + \frac{\pi^2}{4} \left(\frac{kT}{\mu_0} \right)^2 - \frac{3\pi^4}{80} \left(\frac{kT}{\mu_0} \right)^4 + \ldots \right]. \tag{C.51}$$

The specific heat at constant volume, given by Eq. (A.3), is

$$c_V = \left(\frac{\partial E}{\partial T} \right)_V = Nk \frac{\pi^2}{2} \frac{kT}{\mu_0} \left[1 - \frac{3\pi^2}{10} \left(\frac{kT}{\mu_0} \right)^2 + \ldots \right]. \tag{C.52}$$

Since at T = 0°K, the entropy S_0 of Eq. (C.32) is equal to zero, it follows that the integration of Eq. (A.3) yields

$$S = \int_0^T \frac{c_V}{T} \, dT = Nk \frac{\pi^2}{2} \frac{kT}{\mu_0} \left[1 - \frac{\pi^2}{10} \left(\frac{kT}{\mu_0} \right)^2 + \ldots \right] = \frac{N\mu_0}{T} \left[\frac{\pi^2}{2} \left(\frac{kT}{\mu_0} \right)^2 - \frac{\pi^4}{20} \left(\frac{kT}{\mu_0} \right)^4 + \ldots \right]. \tag{C.53}$$

The free energy F = E − TS becomes

$$F = \frac{3}{5} N\mu_0 \left[1 - \frac{5}{12} \pi^2 \left(\frac{kT}{\mu_0} \right)^2 + \frac{\pi^4}{48} \left(\frac{kT}{\mu_0} \right)^4 - \ldots \right]. \tag{C.54}$$

The pressure is now given by

$$P = -\left(\frac{\partial F}{\partial V} \right)_T = \frac{2}{5} \frac{N\mu_0}{V} \left[1 + \frac{5\pi^2}{12} \left(\frac{kT}{\mu_0} \right)^2 - \frac{\pi^4}{16} \left(\frac{kT}{\mu_0} \right)^4 + \ldots \right] = \frac{2}{3} \frac{E}{V}. \tag{C.55}$$

The relationship

$$PV = \frac{2}{3} E \tag{C.56}$$

for the electron gas is universal and independent of temperature.

The enthalpy W of Eq. (A.4) is

$$W = E + PV = \frac{5}{3} E = N\mu_0 \left[1 + \frac{5\pi^2}{12} \left(\frac{kT}{\mu_0} \right)^2 - \frac{\pi^4}{16} \left(\frac{kT}{\mu_0} \right)^4 + \cdots \right]. \tag{C.57}$$

The thermodynamic potential Φ of Eq. (A.11) now becomes

$$\Phi = W - TS = N\mu_0 \left[1 - \frac{\pi^2}{12} \left(\frac{kT}{\mu_0} \right)^2 - \frac{\pi^4}{80} \left(\frac{kT}{\mu_0} \right)^4 + \cdots \right] = N\mu. \tag{C.58}$$

Calculating the electronic analog of the Grüneisen parameter γ_e using Eq. (C.56), we obtain

$$\gamma_e = \frac{2}{3}. \tag{C.59}$$

Consequently, the value of γ_e for the Fermi electron gas is constant and equal to two-thirds irrespective of the temperature range. Using the general formula (C.56), the definition of the potential $\Omega = -PV$, and making the substitution $\varepsilon/kT = z$ in the energy integral of Eq. (C.46), we find that

$$\Omega = \Omega(T, \mu) = -PV = VT^{5/2} f(\mu/T), \tag{C.60}$$

where f is a function only of the argument (μ/T). Consequently, Ω/V is a homogeneous function of μ and T of the order of $5/2$. Using Eq. (A.43), we obtain

$$\frac{S}{V} = -\frac{1}{V} \left(\frac{\partial \Omega}{\partial T} \right)_{V,\mu} \qquad \frac{N}{V} = -\frac{1}{V} \left(\frac{\partial \Omega}{\partial \mu} \right)_{T,V}. \tag{C.61}$$

These quantities are functions of μ and T of the order of $3/2$ and their ratio, S/N, is a homogeneous function of zeroth order. Thus, in an adiabatic process (S = const) the ratio μ/T is constant and, consequently,

$$\rho = \rho_0 (T/T_0)^{3/2}, \qquad P = P_0 (T/T_0)^{5/2}, \tag{C.62}$$

which follows from Eqs. (C.61) and (C.60). Eliminating the temperature from Eq. (C.62), we obtain the equation for the electron gas isentrope

$$P = P_0 (\rho/\rho_0)^{5/3}. \tag{C.63}$$

Using Eq. (C.63), we can obtain the adiabatic bulk modulus K_S and the velocity of sound c:

$$K_S = \frac{5}{3} P, \qquad c^2 = \frac{5}{3} \frac{P}{\rho}. \tag{C.64}$$

Differentiating the pressure P of Eq. (C.55) at a constant temperature, we obtain the isothermal bulk modulus

$$K_T = -V \left(\frac{\partial P}{\partial V} \right)_T = \frac{2}{3} N \frac{\mu_0}{V} \left[1 + \frac{\pi^2}{12} \left(\frac{kT}{\mu_0} \right)^2 + \frac{3\pi^4}{80} \left(\frac{kT}{\mu_0} \right)^4 + \cdots \right]. \tag{C.65}$$

Using Eqs. (A.28), (C.64), (C.65), and (C.55), we can calculate the difference between the specific heats

$$c_P - c_V = c_V \frac{K_0}{K_T} \frac{\pi^2}{3} \left(\frac{kT}{\mu_0} \right)^2 \left[1 - \frac{3\pi^2}{10} \left(\frac{kT}{\mu_0} \right)^2 \right], \tag{C.66}$$

where

$$K_0 = \frac{2}{3} N \frac{\mu_0}{V} \qquad (C.67)$$

is the bulk modulus at T = 0°K. According to Eq. (C.66) the difference between the specific heats of the electron gas is proportional to a small quantity $(kT/\mu_0)^2$.

C.5. Photon Gas

We shall now determine the thermodynamic functions of a photon gas which is in thermal equilibrium with matter. As is known, the blackbody radiation can be regarded as a gas of non-interacting radiation quanta of energies

$$\varepsilon_k = \hbar\omega_k, \qquad (C.68)$$

where ω_k are the eigenfrequencies of the radiation in a given volume V. The distribution of photons over various quantum states of energies ε_k is given by the Planck formula

$$\bar{n}_k = \frac{1}{\exp(\hbar\omega_k/kT) - 1}. \qquad (C.69)$$

The photon momentum is defined by the formula

$$p_k = \varepsilon_k/c = \hbar\omega_k/c, \qquad (C.70)$$

where c is the velocity of light. The number of states in a photon gas with frequencies in the interval between ω and $\omega + d\omega$ is obtained by multiplying the general formula (C.5) by 2 (corresponding to two independent polarizations of a photon):

$$V\omega^2 d\omega/\pi^2 c^3. \qquad (C.71)$$

Having multiplied the Planck distribution of Eq. (C.69) by the above quantity, we obtain the number of photons for a given frequency interval:

$$dN_\omega = \frac{V}{\pi^2 c^3} \cdot \frac{\omega^2 d\omega}{\exp(\hbar\omega/kT) - 1}, \qquad (C.72)$$

and multiplying again by $\hbar\omega$, we obtain the energy of radiation in the part of the spectrum being considered

$$dE_\omega = \frac{V\hbar}{\pi^2 c^3} \cdot \frac{\omega^3 d\omega}{\exp(\hbar\omega/kT) - 1}. \qquad (C.73)$$

This is the well-known Planck formula for the spectral distribution of the energy of blackbody radiation. At low frequencies ($\hbar\omega \ll kT$) Eq. (C.73) reduces to the Rayleigh — Jeans formula

$$dE_\omega = V \frac{kT}{\pi^2 c^3} \omega^2 d\omega, \qquad (C.74)$$

and at high frequencies ($\hbar\omega \gg kT$), the Planck formula reduces to the Wien equation

$$dE_\omega = V \frac{\hbar}{\pi^2 c^2} \omega^3 \exp(-\hbar\omega/kT) d\omega. \qquad (C.75)$$

Making the substitution z = $\hbar\omega/kT$ in Eqs. (C.72) and (C.73) and integrating over all frequencies between zero and infinity, we obtain the internal energy and number of particles in a photon gas enclosed in a volume V:

$$E = \frac{V\hbar}{\pi^2 c^3}\left(\frac{kT}{\hbar}\right)^4 \int\limits_0^\infty \frac{z^3 dz}{e^z - 1},$$

$$N = \frac{V}{\pi^2 c^3}\left(\frac{kT}{\hbar}\right)^3 \int\limits_0^\infty \frac{z^2 dz}{e^z - 1}.$$

The values of the definite integrals in these formulas are

$$\int\limits_0^\infty \frac{z^3 dz}{e^z - 1} = \frac{\pi^4}{15}, \qquad \int\limits_0^\infty \frac{z^2 dz}{e^z - 1} = 2.404.$$

Using these values, we can write the internal energy E and the number of particles N in the final form

$$E = \frac{4\sigma}{c} V T^4, \tag{C.76}$$

$$N = 0.244\left(\frac{kT}{\hbar c}\right)^3 V, \tag{C.77}$$

where

$$\sigma = \frac{\pi^2 k^4}{60 \hbar^3 c^2} = 5.67 \cdot 10^{-5} \text{ g} \cdot \text{sec}^{-3} \cdot \text{deg}^{-4}. \tag{C.78}$$

is the Stefan — Boltzmann constant.

We shall now determine the other thermodynamic functions of the photon gas. The specific heat at constant volume is

$$c_V = \left(\frac{\partial E}{\partial T}\right)_V = \frac{16}{c}\sigma T^3 V. \tag{C.79}$$

The entropy is

$$S = \int\limits_0^T \frac{c_V}{T} dT = \frac{16\sigma}{3c} V T^3. \tag{C.80}$$

The free energy is given by

$$F = E - TS = -\frac{4\sigma}{3c} V T^4. \tag{C.81}$$

The pressure and the product of the pressure and the volume are

$$P = -\left(\frac{\partial F}{\partial V}\right)_T = \frac{4\sigma}{3c} T^4, \tag{C.82}$$

$$PV = \frac{1}{3} E. \tag{C.83}$$

The enthalpy is

$$W = E + PV = \frac{4}{3} E = \frac{16}{3}\frac{\sigma}{c} V T^4 = ST. \tag{C.84}$$

The thermodynamic potential is given by the expression

$$\Phi = W - TS = 0. \tag{C.85}$$

It follows that the chemical potential of the photon gas is zero.

Assuming that the entropy of Eq. (C.80) is constant and using Eq. (C.82), we obtain the equation for the isentrope of black-body radiation:

$$V = V_0(T/T_0)^{-3}, \qquad P = P_0(V/V_0)^{-4/3}. \tag{C.86}$$

D. Calculation of Parameters in the

Potential of a Solid

All the calculations will now be carried out for the potential of ionic crystals or metals given by Eq. (1.7):

$$E_p = \frac{3A}{b\rho_0} \exp[b(1 - x^{1/3})] - \frac{3K}{\rho_0} x^{-1/3}. \tag{D.1}$$

Similar treatments can be provided for other types of solid.

The phonon term will be described using the Debye theory. Using the condition $P = 0$ at $T = T_0$, $x = 1$ ($\rho = \rho_0$), as well as Eq. (B.7), we shall eliminate the parameter K, which can be expressed in terms of the parameter A, and the known quantities ρ_0, γ_0, θ_0, and T_0 [see Eq. (4.4)]

$$K = A + \rho_0\gamma_0 \frac{R}{\mu}\left[\frac{9}{8}\theta_0 + 3T_0 D\left(\frac{\theta_0}{T_0}\right)\right], \tag{D.2}$$

where R is the universal gas constant and μ is the average molecular weight per single atom.

The other two parameters, A and b, are found by minimizing the function

$$\zeta = \sum_{v=1}^{N} [P_{ev} - P(x_v)]^2 \frac{1}{\sigma_v^2},$$

which yields the following two expressions:

$$\left.\begin{array}{l}
\displaystyle\sum_{v=1}^{N} \frac{1}{\sigma_v^2}[P_{ev} - P(x_v)]\frac{\partial P(x_v)}{\partial A} = 0, \\[4mm]
\displaystyle\sum_{v=1}^{N} \frac{1}{\sigma_v^2}[P_{ev} - P(x_v)]\frac{\partial P(x_v)}{\partial b} = 0.
\end{array}\right\} \tag{D.3}$$

Here, P_{ev} is the experimental value of the pressure at a point x_v. This may be the static or the dynamic pressure, depending on the case considered. The quantity $P(x_v)$ is the corresponding theoretical expression for the pressure at the same point x_v; $1/\sigma_v^2$ is the statistical weight of the given point; σ_v is the standard deviation. If ξ_v is the relative experimental error, we may assume that $\sigma_v = \xi_v P_{ev}$.

The system of expressions given by Eq. (D.3) is complex and, therefore, we shall solve it by the method of successive approximations. We shall introduce small quantities α and β using the formulas

$$A = A_0(1 + \alpha), \qquad b = b_0(1 + \beta), \tag{D.4}$$

where A_0 and b_0 are the values of the parameters A and b in the zeroth approximation. We shall expand each term of the sums in Eq. (D.3) as a series in α and β up to linear terms and, assuming that the differences $P_{e\nu} - P(x_\mu)$ to be small, we shall neglect the terms $[P_{e\nu} - P(x_\mu)]\alpha$ and $[P_{e\nu} - P(x_\mu)]\beta$. We thus obtain the following linear system of equations for the small quantities α and β:

$$\left.\begin{array}{l}
A_0\alpha \sum\limits_{\nu=1}^{N} \dfrac{1}{\sigma_\nu{}^2}\left(\dfrac{\partial P(x_\nu)}{\partial A}\right)^2 + b_0\beta \sum\limits_{\nu=1}^{N} \dfrac{1}{\sigma_\nu{}^2}\dfrac{\partial P(x_\nu)}{\partial A}\dfrac{\partial P(x_\nu)}{\partial b} = \sum\limits_{\nu=1}^{N}\dfrac{1}{\sigma_\nu{}^2}[P_{e\nu} - P(x_\nu)]\dfrac{\partial P(x_\nu)}{\partial A}, \\[4mm]
A_0\alpha \sum\limits_{\nu=1}^{N} \dfrac{1}{\sigma_\nu{}^2}\dfrac{\partial P(x_\nu)}{\partial b}\dfrac{\partial P(x_\nu)}{\partial A} + b_0\beta \sum\limits_{\nu=1}^{N} \dfrac{1}{\sigma_\nu{}^2}\left(\dfrac{\partial P(x_\nu)}{\partial b}\right)^2 = \sum\limits_{\nu=1}^{N}\dfrac{1}{\sigma_\nu{}^2}[P_{e\nu} - P(x_\nu)]\dfrac{\partial P(x_\nu)}{\partial b},
\end{array}\right\} \tag{D.5}$$

where all the derivatives are calculated assuming that $A = A_0$ and $b = b_0$. The values of α and β obtained by solving the system (D.5) are then used to determine A_1 and b_1 using expressions similar to those in Eq. (D.4). These new values A_1 and b_1 are now used as A_0, b_0, and so on. The iteration is repeated as many times as is necessary to obtain α and β of the required smallness. This calculation process converges quite rapidly: more than ten approximations are rarely needed. Sometimes even a second or a third approximation yields α and β of the order of 10^{-5}–10^{-6}. The initial values of the parameters are usually taken to be $A_0 = 500$ kbar, $b_0 = 10$.

All our calculations were carried out using a BÉSM-type computer, employing a general-purpose program. Moreover, the results obtained were checked independently using the SP-123 program prepared by I. N. Silin at the Joint Institute for Nuclear Energy. The results obtained by these two programs were in good agreement.

D.1. Isotherm

We shall consider the case when the initial experimental data are in the form of an isotherm for $T = T_0$. In this case P(x) in Eq. (D.3) should be replaced with Eq. (4.1):

$$P(x) = P_p(x) + \frac{\rho_0\gamma}{x}\frac{R}{\mu}\left[\frac{9}{8}\theta + 3T_0 D\left(\frac{\theta}{T_0}\right)\right], \tag{D.6}$$

where $D(\theta/T_0)$ is the Debye function and the potential pressure $P_p(x)$ is found by differentiating the potential given by Eq. (D.1):

$$P_p(x) = Ax^{-1/3}\exp[b(1 - x)^{1/3}] - Kx^{-4/3}. \tag{D.7}$$

The Debye temperature θ is assumed to depend only on the volume x and, according to the definition of the Grüneisen parameter given by Eq. (2.26), it is calculated using the formula

$$\theta = \theta_0 \exp \int_x^1 \frac{\gamma}{x}\,dx, \tag{D.8}$$

where θ_0 is the Debye temperature at $x = 1$. The quantity γ depends on the parameters A and b and, therefore, the Debye temperature θ should also depend on these parameters. However, θ is very insensitive to changes in these parameters and, therefore, we can calculate the derivatives of P(x) with respect to A and b assuming that θ is independent of these parameters.

The derivatives $\partial P(x)/\partial A$ and $\partial P(x)/\partial b$ in Eq. (D.3) are then given by

$$\frac{\partial P(x)}{\partial A} = \frac{\partial P_p(x)}{\partial A} + \frac{\partial \gamma}{\partial A}\frac{\rho_0}{x}\frac{R}{\mu}\left[\frac{9}{8}\theta + 3T_0 D\left(\frac{\theta}{T_0}\right)\right], \tag{D.9}$$

and by a similar expression in the case of $\partial P(x)/\partial b$. The derivatives of γ with respect to A and b are given in Appendix D.3. It follows from Eqs. (D.7) and (D.2) that

$$\frac{\partial P_p(x)}{\partial A} = x^{-1/3}(\exp[b(1-x^{1/3})] - x^{-2/3}), \tag{D.10}$$

$$\frac{\partial P_p(x)}{\partial b} = Ax^{-2/3}(1-x^{1/3})\exp[b(1-x^{1/3})]. \tag{D.11}$$

Thus, we now have all the necessary formulas for the calculation of the parameters A and b in the case considered here.

D.2. Hugoniot Curve

If the initial experimental data are in the form of a Hugoniot curve, we must replace P(x) of Eq. (D.3) with Eq. (5.75):

$$P_H(x) = \frac{P_p(x) + \gamma\varphi(x)}{1 - \gamma(1-x)/2x}, \tag{D.12}$$

where

$$\varphi(x) = \frac{\rho_0}{x}[E_0 - E_p(x)]. \tag{D.13}$$

The energy of unit mass ahead of a shock-wave front

$$E_0 = E_p(1) + \frac{R}{\mu}\left[\frac{9}{8}\theta_0 + 3T_0D\left(\frac{\theta_0}{T_0}\right)\right] \tag{D.14}$$

depends on the parameters A and b through the function $E_p(1)$, i.e., through the potential energy of Eq. (D.1) for x = 1. The derivatives of $P_H(x)$ with respect to A and b, which occur in Eq. (D.3), are of the following form:

$$\frac{\partial P_H(x)}{\partial A} = \frac{1}{1-\gamma(1-x)/2x}\left\{\left[\frac{1-x}{2x}P_H(x) + \varphi(x)\right]\frac{\partial\gamma}{\partial A} + \frac{\partial P_p(x)}{\partial A} + \gamma\frac{\partial\varphi(x)}{\partial A}\right\}, \tag{D.15}$$

and the replacement, in Eq. (D.15), of the derivatives with respect to A by derivatives with respect to b gives the corresponding expression for $\partial P_H(x)/\partial b$. The formulas for $\partial\gamma/\partial A$ and $\partial\gamma/\partial b$ are given in Appendix D.3; $\partial P_p/\partial A$ and $\partial P_p/\partial b$ are given by Eqs. (D.10) and (D.11). Using Eq. (D.14), we obtain from Eq. (D.13)

$$\frac{\partial\varphi(x)}{\partial A} = \frac{\rho_0}{x}\left[\frac{\partial E_p(1)}{\partial A} - \frac{\partial E_p(x)}{\partial A}\right]. \tag{D.16}$$

A similar expression is found for $\partial\varphi(x)/\partial b$. The derivatives of $E_p(x)$ with respect to A and b can be found from Eqs. (D.11) and (D.2):

$$\frac{\partial E_p(x)}{\partial A} = \frac{3}{\rho_0}\left(\frac{1}{b}\exp[b(1-x^{1/3})] - x^{-1/3}\right), \tag{D.17}$$

$$\frac{\partial E_p(x)}{\partial b} = \frac{3A}{b\rho_0}\left(1 - x^{1/3} - \frac{1}{b}\right)\exp[b(1-x^{1/3})]. \tag{D.18}$$

The quantities $\partial E_p(1)/\partial A$ and $\partial E_p(1)/\partial b$ are calculated from Eqs. (D.17) and (D.18) by substituting x = 1.

D.3. Grüneisen Parameter

We shall use the generalized formula (2.45) for γ:

$$\gamma = -\frac{x}{2}\frac{\partial^2(P_p x^{2m/3})/\partial x^2}{\partial(P_p x^{2m/3})/\partial x} + \frac{1}{3}(m-2) + \delta. \tag{D.19}$$

Calculations are more convenient when the first two terms of Eq. (D.19) are combined into one. Replacing P_p in Eq. (D.19) by the expression given by Eq. (D.7), we obtain

$$\gamma = \frac{1}{6}\frac{r(x)}{\zeta(x)} + \delta, \tag{D.20}$$

where the normalization constant δ is found from the condition*

$$\delta = \gamma_0 - \frac{1}{6}\frac{r(1)}{\zeta(1)}, \tag{D.21}$$

where γ_0 is the thermodynamic value of the Grüneisen parameter for $x = 1$. The following notation is used in Eq. (D.20):

$$r(x) = A\left[b^2 x^{1/3} + 2b(1-m) + 2(1-m)x^{-1/3}\right]\exp\left[b(1-x^{1/3})\right] - 6K(2-m)x^{-1}, \tag{D.22}$$

$$\zeta(x) = A\left[b + 2(1-m)x^{-1/3}\right]\exp\left[b(1-x^{1/3})\right] - 2K(2-m)x^{-1}. \tag{D.23}$$

The quantities $r(1)$ and $\zeta(1)$ are found from Eqs. (D.22) and (D.23) by substituting $x = 1$. Thus, δ depends on A and b. The derivative of γ with respect to A is of the form:

$$\frac{\partial\gamma}{\partial A} = \frac{1}{\zeta(x)}\left[\frac{1}{6}\frac{\partial r(x)}{\partial A} - (\gamma-\delta)\frac{\partial\zeta(x)}{\partial A}\right] - \frac{1}{\zeta(1)}\left[\frac{1}{6}\frac{\partial r(1)}{\partial A} - (\gamma_0-\delta)\frac{\partial\zeta(1)}{\partial A}\right]. \tag{D.24}$$

Replacing the derivatives with respect to A by derivatives with respect to b, we obtain the corresponding expression for $\partial\gamma/\partial b$. Differentiating Eqs. (D.22) and (D.23) with respect to A, b, and using Eq. (D.2), we obtain

$$\frac{\partial r(x)}{\partial A} = \left[b^2 x^{1/3} + 2b(1-m) + 2(1-m)x^{-1/3}\right]\exp\left[b(1-x^{1/3})\right] - 6(2-m)x^{-1}, \tag{D.25}$$

$$\frac{\partial r(x)}{\partial b} = A\left[b^2 x^{1/3}(1-x^{1/3}) + 2b(1-m+mx^{1/3}) + 2(1-m)x^{-1/3}\right]\exp\left[b(1-x^{1/3})\right], \tag{D.26}$$

$$\frac{\partial\zeta(x)}{\partial A} = \left[b + 2(1-m)x^{-1/3}\right]\exp\left[b(1-x^{1/3})\right] - 2(2-m)x^{-1}, \tag{D.27}$$

$$\frac{\partial\zeta(x)}{\partial b} = A\left[b(1-x^{1/3}) + 2m - 1 + 2(1-m)x^{-1/3}\right]\exp\left[b(1-x^{1/3})\right]. \tag{D.28}$$

The derivatives of $r(1)$ and $\zeta(1)$ with respect to A and b are found from Eqs. (D.25)-(D.28) by substituting $x = 1$.

If the Grüneisen parameter is calculated without normalization ($\delta = 0$), we must then substitute $\delta = 0$ and drop the second terms, containing $\zeta(1)$, both in Eq. (D.24) and in the corresponding expression for $\partial\gamma/\partial b$.

---

*In solving the system (D.5) the quantity δ is calculated again at each iteration step but this does not affect the convergence of the method.